U.S. Pacific Islanders and the Sea

A History of the Western Pacific Regional Fishery Management Council (1976–2020)

by

Michael Markrich & Sylvia Spalding

Contributors: Paul Dalzell, Joshua DeMello, Asuka Ishizaki, Charles Ka'ai'ai,
Eric Kingma, Mark Mitsuyasu, Marlowe Sabater and Rebecca Walker

Western Pacific Regional Fishery Management Council
Honolulu

Western Pacific Regional Fishery Management Council (WPRFMC)
1164 Bishop Street Suite 1400
Honolulu, Hawai'i 96813 USA

Prepared for and published by the WPRFMC under
NOAA Award # NA20NMF4410013

Library of Congress Control Number: 2021908724

U.S. Pacific Islanders and the Sea
Copyright © 2022 by Western Pacific Regional Fishery Management Council

Front cover art: Hawai'i longline vessel. Caleb McMahon photo. Island ecosystem. WPRFMC illustration/Oliver Kinney. Carolinian sailing canoe off Saipan, Northern Mariana Islands. WPRFMC photo/Jack Ogumoro.

ISBN
978-1-944827-80-9 (Hardcover)
978-1-944827-81-6 (Paperback)
978-1-944827-82-3 (eBook)

'A'ohe pau ka 'ike i ka halau ho'okahi.

(All knowledge is not learned in one house.)

Hawaiian saying

Ua o gatasi le futia ma le umele.

(We must be of one mind in the understanding.)

While the fisherman swings the rod, the others must assist by paddling hard.

Samoan saying

Ayek i hinanåomu mo'na lao mungnga maleffa nu i tiningo'mu ginen i man mo'fona.

(Choose your path forward, but do not forget the lessons from those before you.)

Chamorro saying

Aramasal Metawal Wool, nge aramas iye eew peschel e lo wóól falúw, nge eew peschel e lo llól sáát. Paghúw eew peschel, nge aramasa la essóbw mmwelil úú. Paghúw ló i me ruwoow peschel nge aramasala e mérúghúló.

(A Northern Mariana islander or descendant is a person with one foot on land and the other in the ocean. Take away one foot, and the person cannot stand; take away both feet, and that person does not exist.)

Carolinian saying

Table of Contents

PART TWO:
THE FISHERY MANAGEMENT PLANS (1983–2009)

PART THREE:
EVOLVING BEYOND SPECIES-BASED MANAGEMENT BY RETURNING TO AN ECOSYSTEM APPROACH (2010–2020)

List of Illustrations

List of Tables

List of Abbreviations

ACL	annual catch limit
BiOp	biological opinion
BRFA	Bottomfish Restricted Fishing Area (State of Hawai'i)
CDPP	Community Demonstration Project Program
CMSP	coastal and marine spatial planning
CNMI	Commonwealth of the Northern Mariana Islands
COVID-19	coronavirus
CRE	Coral Reef Ecosystem
CRER	Coral Reef Ecosystem Reserve
DMWR	Department of Marine and Wildlife Resources (American Samoa)
EEZ	exclusive economic zone
EFH	essential fish habitat
EPAP	Ecosystem Principles Advisory Committee
EPO	Eastern Pacific Ocean
ESA	Endangered Species Act
FAD	fish aggregating device
FCMA	Fishery Conservation and Management Act
FCZ	fishery conservation zone
FEP	fishery ecosystem plan
FMP	fishery management plan
GFCA	Guam Fisherman's Cooperative Association
GPS	global positioning system
HDAR	Hawai'i Division of Aquatic Resources
HLA	Hawaii Longline Association
LVPA	Large Vessel Prohibited Area

MCP	marine conservation plan
MFCMA	Magnuson Fishery Conservation and Management Act
MHLC	Multilateral High Level Conference
MNM	Marine National Monument
MPA	Marine Protected Area
MPCC	Marine Planning and Climate Change
MSA	Magnuson-Stevens Fishery Conservation and Management Act
mt	metric ton
NGO	nongovernmental organization
NIOSH	National Institute of Occupational Safety and Health
nm	nautical mile
NMFS	National Marine Fisheries Service
NMSA	National Marine Sanctuaries Act
NMSP	National Marine Sanctuaries Program
NOAA	National Oceanic and Atmospheric Administration
NWHI	Northwestern Hawaiian Islands
PBDC	Pacific Basin Development Corporation
PMP	preliminary fishery management plan
REAC	Regional Ecosystem Advisory Committee
SAFE	Stock Assessment and Fishery Evaluation
SFA	Sustainable Fisheries Act
SFF	Sustainable Fisheries Fund
SSC	Scientific and Statistical Committee
TAC	total allowable catch
UNCLOS	United Nations Convention on the Law of the Sea
USFWS	U.S. Fish and Wildlife Service
VMS	Vessel Monitoring System
WCPFC	Western and Central Pacific Fisheries Commission
WCPO	Western and Central Pacific Ocean
WPRFMC	Western Pacific Regional Fishery Management Council

Foreword

Based on my congressional background in the Washington, D.C., office of U.S. Senator Hiram Fong (R- Hawai'i), and a subsequent 37 years of experience as executive director of the Western Pacific Regional Fishery Management Council, I believe that the Fishery Conservation and Management Act (commonly known as the Magnuson-Stevens Act) exemplifies the American principle of a government "of the people, by the people, for the people." The Act encourages local participation, responsibility and authority over fisheries through the establishment of the regional fishery management council system.

The Western Pacific Council draws heavily from the Pacific Island way of doing things, blending the local ocean values of our region with contemporary science to forge our own distinctive path. Our Council has been the principal broker for balancing traditional fishing rights and practices of Pacific Island communities with the policies of fishery managers and lawmakers in Washington, D.C. As a Native Hawaiian with family roots on both sides from Maui and the Big Island, I see the parallels in the consultative processes of the traditional 'Aha Moku system and the regional fishery management council system.

Since 1976, the Council has based its policies on science and its ability to hold fishermen accountable. To protect fish stocks and the ecosystems on which they depend, the Council has sought to continually improve the collection and availability of data necessary for sound decision-making. The Council also incorporates traditional knowledge handed down through generations of indigenous Pacific Islanders.

These observations have given Western scientists and managers valuable insights on the region's fisheries—information that is applied by the Council in its deliberations.

Increasing demand for fish over the years has led to successive waves of fishing vessels entering the international waters of the Western and Central Pacific Ocean from Russia, Asia, Europe and the Atlantic and Gulf Coasts of the United States. Each rise in fishing effort added to the pressure placed on fish stocks and escalated campaigns to discontinue or reduce commercial fishing in the interest of environmental protection. For this reason, it is critical to establish and build on binding international agreements to promote environmentally friendly management, to ensure fair allocation of resources (including to traditional small-scale fisheries) and to recognize those fisheries meeting international management objectives. Against this background, the Council has always recognized the importance of the needs and rights of the current and future generations of island people to harvest, to share and to consume the fish from throughout their island waters. The Council recognizes these activities as the birthrights of Pacific Islanders.

The work of the Council is ongoing. Balancing environmental protection with competing cultural, recreational and commercial uses, the Council employs an ecosystem approach to management. This place-based management of natural resources incorporates the findings of both intergenerational observations and modern scientific research in order to sustain the health, resilience and diversity of marine ecosystems.

In its first four and a half decades, the Council has consistently insisted that the concerns of affected fishermen are addressed in the management process. The Council, as it moves forward, will not abandon its responsibility to faithfully protect natural resources and the rights of future generations of Pacific Islanders to benefit from them.

Much of the Council's success is owed to the dedicated efforts of the Council's effective leadership team, beginning with the first Council chair, the late Hawai'i State Senator Wadsworth Yee, who guided the organization during the crucial early years of its existence. Additionally, the late U.S. Senator Ted Stevens (R-Alaska), U.S. Senator Daniel K. Inouye (D-Hawai'i) and Congresswoman Patricia Saiki (R-Hawai'i) garnered the support of other members of Congress on key legislative authorizations to conserve and manage our region's fisheries for the benefit of all Pacific Islanders. To them and the many others who believed in the Council's promise, we owe a debt of gratitude. *Mahalo, fa'afetai, si yu'os ma'ase, ghilisow* and thank you!

Kitty M. Simonds

Kitty M. Simonds
Executive Director
Western Pacific Regional Fishery Management Council

Preface

The Western Pacific Regional Fishery Management Council (the Council) originated 45 years ago as part of a great experiment in democracy. In 1976, the Congress of the United States passed the Fishery Conservation and Management Act (FCMA), which is based on the belief that responsible representatives from both the government and fishing industry, as well as ordinary citizens, could manage the offshore waters of the United States in the best interests of both the nation and communities reliant on fisheries.

That time period had very different fisheries priorities. Instead of concerns about climate change and the collapse of marine ecosystems, people felt the ocean possessed unlimited bounty. The dominant belief system was that the best interests of the United States were served by exploiting the untapped wealth of the seas for the greater good of its people. The mandate of the Council was to do this in the most equitable way possible. Unfortunately, at that time almost nothing was known about the lifecycle of the affected fish and marine ecosystems and the word "sustainable" was seldom used in the fishery context.

It soon became apparent that the FCMA legislation creating the Council was imperfect; that there were widely conflicting government mandates; that big-game fishermen, commercial fishermen, biologists, indigenous peoples and environmental advocates often had contrary interests; and that the ocean resources were finite. Consequently, Council members found that reaching an equitable consensus in the best interests of all communities was a difficult and sometimes almost impossible task.

This book is an account of how the men and women of the Council persevered in their efforts to establish a scientific process to manage the fisheries of the U.S. Western Pacific Region. As with any new organization, the members learned as they went along. They grappled with difficult problems of allocation, such as how to satisfy a public eager to eat more fish while balancing conservation and user needs. In this struggle, the Council pioneered many of the conservation and management tools widely used today by international fishery organizations.

However, as the Council tried to walk the middle ground, it increasingly became the target of professional environmental advocates and big-game sportfishermen who demanded ever higher standards of ocean management. The work of NOAA fishery biologists and commercial fishermen were progressively at odds with a new generation of environmental advocates who saw the extraction of marine resources from the open ocean as morally flawed and were not interested in compromise. As these political issues intensified, basic fishery management decisions that the Council was authorized to make under the FCMA were replaced by legal mandates for large-scale closures of commercial fishing grounds in U.S. waters, which were already tightly controlled by federal fishery regulations.

Today, many of the strongest criticisms of the Council come from young people who were not yet born when the Council began its work. They are rightfully concerned that they and their children may inherit a world of diminishing resources. The following history does not strive to justify everything that the Council did or did not do during its first four and half decades, but it asks the reader to understand the chain of events and the process that brought us to where we are today.

There is a frustrating irony in the history of the Council. At many stages in its earlier years, the Council tried to introduce precautionary measures because of the uncertain knowledge about the fish stocks in rapidly growing new fisheries. However, at that time, neither National Marine Fisheries Service managers nor National Oceanic and Atmospheric Administration attorneys would agree to such measures, absent information demonstrating an existing problem. In their eyes, information indicating that a problem warranting action existed had to be solid and "uncertainty" was not a basis for taking action. In later years, the Council was frequently criticized by some environmental organizations for *not* being sufficiently precautionary in the face of uncertain information. In short, the Council had to face conflicting signals: one day, precautionary controls could not be implemented; the next day, they should be implemented. In fact, as the Council and the National Marine Fisheries Service try to balance the imperfectly aligned beliefs and perceptions of fishery participants, environmentalists, lawyers and the public, this continues to be a challenge.

Nonetheless, the Council's actions during its first four decades have had a huge impact on fisheries management in the region and the nation.

1) The policy on drift gillnets helped to change international perception and use of this non-specific fishing technique.

2) Ecological modeling of pelagic and other species were undertaken to a degree unsurpassed by other fishery managers.

3) The Council's persistence led to the inclusion of tuna in the Magnuson-Stevens Act.

4) The creation of the 50-nautical-mile Protected Species Zone in the Northwestern Hawaiian Islands and the development of a coral reef ecosystem fishery management plan became the impetus for the Northwestern Hawaiian Islands Coral Reef Ecosystem Reserve and marine national monuments.

5) International regional fishery management organizations have adopted the Council's pioneering use of a satellite-based vessel monitoring system and its protected species bycatch mitigation methods.

6) Indigenous and local fishing communities have been empowered as decision makers to participate in the management and harvesting of fish in their regions.

7) The Council's professional education and outreach program has reached regional, national and international audiences.

For thousands of years, Pacific Islanders have voyaged across the open ocean to create regional networks that share cultural traditions, including those of fishery management. As the world faces an era of scarcity and climate change, this nexus of cultural knowledge, fisheries management and island community life grows in importance. Over its more than 40-year history, the Council has created a legacy as an indispensable hub for the exchange of information on fisheries management, economic growth and the rights of the indigenous peoples in the U.S. Western Pacific Region.

Michael L Markrich

Michael Markrich
Kailua, Hawai'i
March 2022

Acknowledgments

Acknowledgments and words of gratitude are due to the many people who agreed to be interviewed for this book and/or who reviewed the manuscript. These people include Paul Bartram, Paul Callaghan, John Calvo, Kelvin Char, Jim Cook, John Craven, Gary Dill, Manny Duenas, Frank Farm, Mark Fitchett, Svein Fougner, Eric Gilman, Stuart Glauberman, John Gourley, Richard Grigg, Judith Guthertz, Marcia Hamilton, Martin Hochman, Robert Humphreys, Nate Ilaoa, David Itano, Robert Iversen, Tim Johns, Gary Karr, Kurt Kawamoto, Jarad Makaiau, Roy Morioka, Jay Nelson, Domingo Ochavillo, Jack Ogumoro, Arnold Palacios, Frank Parrish, William "Bill" Paty Jr., Sarah Pautzke, Jeffrey Polovina, Samuel Pooley, Craig Severance, Kitty M. Simonds, Robert Skillman, Alo Paul Stevenson, Gerald A. Sumida, Timm Timoney, Ufagafa Ray Tulafono and Wadsworth Yee. This book would not have been possible without their assistance.

Chapter 1

Prologue—Before There Was a Council

1.1 Western Pacific Indigenous Fishing Communities

In the Pacific Ocean, the United States oversees the State of Hawai'i, Territory of American Samoa, Territory of Guam and the Commonwealth of the Northern Mariana Islands (CNMI). Each of these island entities shares a similar history of settlement by the first people who depended on the marine environment for their survival for millennia and then faced eventual colonization by the Western world.

In Hawai'i, the first people settled the islands about AD 300–600. The indigenous Hawaiians traditionally held their fisheries privately under a land tenure system. The government administered some fisheries, and the reigning Hawaiian monarch held certain fisheries in a private reserve. Following Western contact, King Kamehameha III in 1839 divided the nearshore fishing grounds into three, giving one portion to the common people, one portion to the landlords and retaining one portion for himself. This decree is recorded in the Laws of 1840, chapter III, Section 8, and stipulates that "[t]he fishing grounds from the coral reef to the sea beach are for the landlords, and for their tenants of their several lands, but not for others."

The Kingdom of Hawaiʻi was overthrown in 1893, upon which time a provisional government was instituted. This was replaced in 1894 upon the creation of the Republic of Hawaiʻi, until four years later (1898) when Hawaiʻi was annexed by the United States as the Territory of Hawaiʻi through the Newlands Resolution. In 1901, the U.S. Commission of Fish and Fisheries, which provided annual reports to Congress, sent John N. Cobb to inventory the fisheries of the new territory. After the inventory, legal instruments were set in place to undermine the traditional land tenure, to which fishery rights were tied, with a system that made access to the fisheries a public right and not subject to the traditional, prior tenant rights. This action impoverished the native community and moved Hawaiian society from a resource-based economy toward a cash-based economy.

A large group of people participate in a *hukilau,* pulling a long net onshore at Hamoa, Hana, Maui (circa 1936). Communities in Hawaiʻi continue to practice these and other traditional fishing methods today. *©Bishop Museum photo/Harold Stearns.*

In the Mariana Archipelago, Guam and the CNMI were settled by the first people (Chamorro) around 2000 B.C. In the late 1500s A.D., the islands fell under Spanish control and became a colony of Spain for more than three centuries. Cultural and ethnic genocide efforts by the Spanish were pronounced on some islands, e.g., Guam and Saipan, though not so much so on the smaller outer islands, such as Rota. Following Spain's defeat in the Spanish-American War of 1898, Guam was ceded to the United States, and the Northern Mariana Islands were sold to Germany. Then, during World War II, the Japanese military occupied the islands for nearly three years.

In 1944, the islands were freed from Japan by the United States, and they came under American control. In 1950, Guam was established as an unincorporated organized territory of the United States, and, in 1976, the Northern Mariana Islands became a U.S. commonwealth. The official languages of both are Chamorro and English. In the CNMI, the local government also officially recognizes the language of the Refaluwasch, who came from the Caroline Islands to a largely depopulated Saipan (the largest island of the Northern Mariana Islands) in the early 19th century.

Sails from 14 seafaring canoes from Yap, Chuuk, Palau, the Northern Mariana Islands and Guam fill the horizon at Paseo de Susana in Hagatna during the traditional canoe welcome ceremony to open the 12th Festival of Pacific Arts in Guam on May 22, 2016. Once banned from use in Guam and the Northern Mariana Islands during the colonial era, these traditional sailing outrigger boats, better known in Western culture as flying proa, are experiencing a renaissance among the Chamorro people today. *©Manny Crisostomo photo.*

Fish are distributed after a communal *chenchulu* (traditional surround net) harvest at Tanapag Village, Saipan, CNMI, in 2007. Sharing of fish with family and community members continues to be practiced throughout the Pacific Islands today. *WPRFMC photo/Jack Ogumoro.*

The Samoa Archipelago, settled by the first people around 1000–2000 B.C., was divided in the late 19th century with Germany taking Western Samoa and the United States taking Eastern Samoa, valued for its excellent harbor at Pago Pago on the island of Tutuila. In 1899, the chiefs of Eastern Samoa deeded their islands to the United States in two deeds of cession executed in Tutuila and in the Manuʻa island group, and Eastern Samoa became the U.S Territory of American Samoa. The American Samoan people, however, retained their rights to *faʻa Samoa*, the Samoan way of life, which is rooted in communal land ownership and the *matai* title (nobility) system. For this reason, American Samoa arguably retains more of its traditions than the Hawaiʻi and Mariana archipelagos.

Following traditional practices, members of Fagasa village, American Samoa, in 2018 use *lau* (braided coconut fronds) to trap a school of fish near the shore where they are collected in nets and shared among the residents. *National Park Service–American Samoa photo.*

The traditional islander ways of managing and utilizing resources are empirical, time-tested methods, and their success can be measured by the survival of the cultures that developed them. Ineffective methods and practices did not survive. In modern times, the return of economic benefits from natural resources to the native communities that once owned and were stewards of them has been slow, as has acknowledgement of their vast traditional knowledge about managing and utilizing their unique island resources.

Today, the U.S. regional fishery management process of public participatory decision-making—instituted under the Fishery Conservation and Management Act of 1976 (now known, after several reauthorizations, as the Magnuson-Stevens Fishery Conservation and Management Act,[1] or MSA)—can level the playing field for conflict

---

[1]   The FCMA was renamed to honor its co-drafters, U.S. Senators Warren Magnuson (Washington) and Ted Stevens (R-Alaska).

resolution related to marine resource use and management that arose due to colonization activities by the United States in the Pacific Islands. This process, implemented by the Western Pacific Regional Fishery Management Council (WPRFMC), also provides an avenue for indigenous rights and knowledge to be used to deliver benefits to native communities and to improve fisheries in the U.S. Pacific Islands.

1.2. Geopolitics and Fisheries

Passage of the FCMA by Congress was the result of a sweeping international demand for change in ocean resource management that began in the second half of the 20th century. When World War II ended in 1945, many areas of the ocean had been largely unfished for several years and were teeming with healthy fish stocks. Post-war competition among war-ravaged nations hungry for protein soon led to friction about access to these stocks. Nations with distant-water fishing fleets wanted their fishermen to be able to catch fish in whatever waters they chose, as they had done prior to the war, while nations without such fleets—but with now abundant fish stocks off their coasts—wanted priority access to those stocks.

The clash intensified as the economic value of fisheries became apparent. Between 1946 and 1950, Peru, Chile and Ecuador asserted jurisdiction over all fishery resources off their coasts out to 200 nautical miles (nm) from shore. These nations felt that the 3-nm limit of sovereignty, which had been in effect since the 17th century, benefited only wealthy countries with distant-water fleets.

Later, in the 1970s, a series of confrontations between Britain and Iceland regarding fishing rights in the North Atlantic (known as the "Cod Wars") occurred. In 1972, Iceland unilaterally declared a fishery zone extending beyond its territorial waters and policed its water with coast guard vessels. After a series of net-cutting incidents with British trawlers, Royal Naval warships and tugboats were employed to act as a deterrent against any future harassment of British fishing crews by Icelandic crafts. In retaliation, Iceland threatened to close a major North Atlantic Treaty Organization base. In 1975, a compromise between the two nations was reached that allowed a limited number of British trawlers access to the disputed 200-nm limit (Spalding and Dalzell 2009).

At the time, the U.S. position was somewhat divided on this issue. Arguing against the 200-nm limit were two forces: the Department of Defense and the U.S. tuna industry. The latter was dominated by the purse-seine fishing and canned tuna industries. The Defense Department argued that the extension of jurisdiction beyond 3 nm could threaten the ability of U.S. military forces to transit through the waters claimed by foreign countries. This view was strongly influenced by experiences in World War II, the Korean War and the Vietnam War. The tuna industry argued that, if the United States extended its jurisdiction to 200 nm, it would be recognizing the legitimacy of other nations to assert similar rights, to the detriment of U.S. fishermen. For the tuna industry, this would mean it would either have to abandon coastal waters off foreign countries (where most of the tuna were caught) or pay fees for access. This was a serious national concern, as the tuna industry was the largest U.S. fishing industry at that time.

In 1954, Congress passed the Fishermen's Protective Act, which authorized the U.S. government to pay the fines of U.S. fishing vessels (mainly tuna fishing vessels) seized or arrested in waters considered by the United States to be high seas but regarded by the countries concerned as territorial waters or exclusive fishing zones. The Act also allowed fishing vessel owners to file claims for reimbursement of losses incurred by such seizures. This was a tacit acknowledgement by the

United States that foreign governments had acted lawfully with respect to fisheries jurisdiction, but it was not an explicit acceptance of the legitimacy of those laws.

However, other forces at play in the 1960s and early 1970s led to further calls for a change in U.S. ocean governance. Some coastal states saw large increases in fishing pressure by foreign fleets off their coasts—especially off the U.S. coasts in the Northeast, the Northwest and Alaska. Vessels from the Soviet Union, Poland and Japan were systematically harvesting stocks in these waters, unconstrained by catch limits or other rules, and they did not have to pay access fees. Furthermore, they had no interest in long-term stock sustainability; they fished hard and moved on when catches dropped, leaving behind depleted stocks.

Besides the complaints of commercial fishermen—who were hard hit by large-scale foreign-flag factory ships operating nearshore U.S. coastal waters—were the protestations of politically influential big-game sportfishermen who feared that the foreign fleets were ruining the multi-million dollar U.S. saltwater recreational fishing industry. The big-game fishermen sought a political alliance with non-tuna commercial fishermen and environmental advocates to change the status quo.

Concern was also growing in Hawai'i. In the 1960s, longline tuna fishing fleets operating out of Japan, South Korea and Taiwan were expanding throughout the Pacific and harvesting yellowfin (*Thunnus albacares*), bigeye (*T. obesus*), skipjack (*Katsuwonus pelamis*), bluefin (*T. orientalis*) and albacore (*T. alalunga*) tuna with minimal controls on catch and fishing effort. Hawai'i Governor John A. Burns (1962–1974) was interested in documenting the fishery resources in the waters surrounding the state because he realized Hawai'i could also soon experience significant foreign fishing pressure. Hawai'i charter, recreational and commercial fishermen were alarmed that the large marlin they might otherwise catch were being intercepted by Japanese vessels to be sold in Japan for sashimi. In addition, there were mounting concerns about the impact of Japanese, Russian and Taiwanese trawl vessels that fished with impunity in the Northwestern Hawaiian Islands (NWHI).

These arguments for change were being made in the United States during a period when the nations of the world were developing a new international convention that would alter the landscape for ocean resources management. The Third United Nations Conference on the Law of the Sea took place between 1973 and 1982 and culminated in the United Nations Convention on the Law of the Sea (UNCLOS). Among several significant elements, UNCLOS affirmed and cemented the practice of extending national jurisdiction over marine resources out to 200 nm. UNCLOS also allowed for the extension of the territorial sea of each nation to 12 nm from shore, rather than the former 3 nm.

Although the United States was initially opposed to UNCLOS for geopolitical reasons (and has still not ratified it), the coalition of recreational and commercial fishermen and associated industries argued that change was inevitable. It was, therefore, in the best interests of the United States to preemptively declare its own 200-nm fishery conservation zone (FCZ), later renamed and now known as the exclusive economic zone (EEZ), before the rest of the world adopted UNCLOS.[2] This led to Congressional efforts that culminated in the passage of the FCMA in 1976.

---

[2]   The FCZ was recognized as the U.S. EEZ through Proclamation 5030 by President Ronald Reagan on March 10, 1983; the United States claimed sole ownership over all living and non-living resources within its boundaries.

However, in order for the FCMA to receive Congressional approval, a compromise with the politically powerful U.S. Tuna Foundation first had to be reached. The Foundation agreed to support the FCMA's passage only if tuna was excluded from management on the basis that it was a "highly migratory species."

1.3 The Fishery Conservation and Management Act of 1976

The FCMA was signed by President Gerald Ford on April 13, 1976 (16 U.S.C. sections 1811-82). It established a foundation for federal fishery management with the following components.

First, eight regional fishery management councils were created with clear authority to develop regulations for fishing (except tuna) in federal waters. Previously, to the extent there was any management in offshore waters, such management was carried out under state or territorial authority. Now, while states and territories retained jurisdiction in state waters,[3] the U.S. government held the responsibility for implementing fishery regulations seaward of state waters, based on recommendations of the councils. The eight councils are the New England, Mid-Atlantic, South Atlantic, Caribbean, Gulf of Mexico, Pacific, North Pacific and Western Pacific.

The inclusion of the Western Pacific Council in the FCMA was largely due to the insistence of Governor Burns. His administration firmly believed that the economic development of Hawaiʻi's ocean resources was important to the state's future. This policy goal was carried forward by his successor, Governor George R. Ariyoshi (1974–1986),

---

[3]   State waters extend generally 3 nm from shore, but some state waters extend 6, 9 or 12 nm from shore. However, CNMI did not have claim to any state waters until January 16, 2014, with Presidential Proclamation 9077.

who also supported the FCMA legislation. The Act was further supported in Congress by U.S. Senator Daniel K. Inouye (D-Hawai'i), U.S. Senator Hiram L. Fong (R-Hawai'i) and U.S. Representative Spark M. Matsunaga (D-Hawai'i), who together made a concerted effort to push for a Western Pacific Council headquartered in Hawai'i.[4],[5] Their political ambitions prevailed, and, as a result, the FCMA signed by President Ford included the WPRFMC.

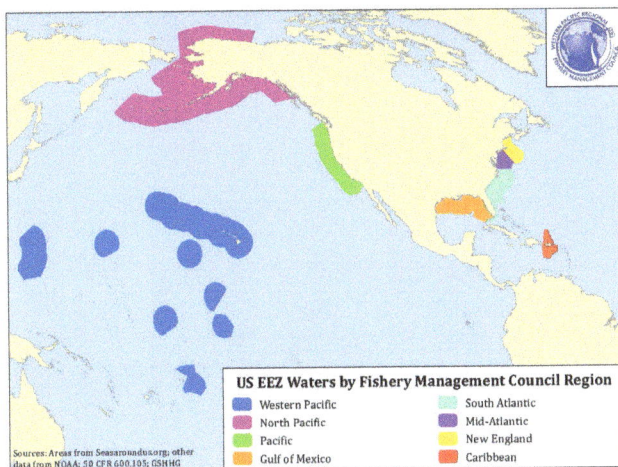

The Western Pacific Region contains approximately half of the U.S. waters. This is not obvious in the map showing the jurisdiction of the nation's eight regional fishery management councils because Mercator projection inflates the size of objects away from the equator. *WPRFMC illustration.*

---

[4]   John Craven, marine affairs coordinator for Governor Burns, in discussion with Michael Markrich, February 14, 2009.

[5]   U.S. Senator Hiram Fong previously played a key role in the Central, Western and South Pacific Fisheries Development Act (Public Law 92-444), which established the Pacific Tuna Development Foundation in 1974. This federally funded program fostered U.S. tuna fishery development in the Western and Central Pacific at a time when the U.S. purse-seine tuna industry was being affected by the tuna-dolphin issue in its Eastern Pacific fishing grounds off Central America and was looking for new fishing grounds in the Pacific.

Gerald R. Ford, the 38th president of the United States (1974–1977), signed the Fishery Conservation and Management Act into law on April 13, 1976. *Ford Presidential Library & Museum photo/David Hume Kennerly.*

Governor John A. Burns, who served as Hawai'i's governor from 1962 to 1974, advocated for inclusion of the Western Pacific Region in the FCMA of 1976. *Hawai'i State Archives photo.*

Hawai'i Governor George R. Ariyoshi (1974–1986) supported inclusion of the Western Pacific Region in the FCMA. He believed that the economic development of Hawai'i's ocean resources was important to the state's future. *Hawai'i County photo.*

Hiram L. Fong (R-Hawai'i), the son of Cantonese immigrants, became the first Asian-American U.S. senator, serving from 1959, the year Hawai'i became a state, to 1977. *U.S. Senate Historical Office photo.*

Daniel K. Inouye (D-Hawai'i) served in the U.S. House of Representatives as the first Japanese-American in Congress and became Hawaii's first full member in 1959. He was the first U.S. senator of Japanese heritage when sworn into office in 1963, and he served as president pro tempore of the U.S. Senate (third in the presidential line of succession) from 2010 until his death in 2012. *Office of U.S. Senator Daniel K. Inouye photo.*

U.S. Representative Spark M. Matsunaga (D-Hawai'i) succeeded Daniel Inouye as the state's sole member of the House of Representatives in 1963. He would go on to serve in the U.S. Senate from 1977 until his death in 1990. *gpo.gov photo.*

Patsy Takemoto Mink, the first woman of color elected to Congress, represented the State of Hawai'i from 1965 to 1977 and from 1989 until her death in 2002. *U.S. Library of Congress photo.*

Second, the National Marine Fisheries Service (NMFS) research laboratories were picked to be the principal source of scientific advice and information about fisheries, and the NMFS regional office was tasked with providing the administrative and staff support necessary to establish enforceable regulations.

Third, rather than a "top-down" system in which the U.S. government determined fishing rules, the Council system—which includes advisory panels consisting of fishery participants and interested citizens—provides the opportunity for active ocean users and the general public to have a say in the regulation of offshore fisheries and use of local marine waters.

Fourth, under the new law, U.S. fishermen were given fishing preference in U.S. waters from 3 to 200 nm from shore. If additional fish could be taken without overfishing the fish stocks, then foreign fishermen would be given an opportunity to fish the waters, subject to regulations and fees (Nickerson 1976).

Fifth, two forms of formal management plans were developed. Initially, preliminary fishery management plans (PMPs)—meant to cover only foreign fishing—were prepared and implemented by the Secretary of Commerce. PMPs were later supplanted by fishery management plans (FMPs) developed by the councils to regulate both domestic and foreign fisheries in FCZ waters.

Sixth, a set of national fisheries standards mandates that management plans adhere to the following:

a) Prevent overfishing while achieving, on a continuing basis, the optimum yield from each fishery.

b) Are based on the best scientific information available.

c) Manage individual stocks as a unit throughout their range.

d) Take into account and allow for variations and contingencies in fisheries, fishery resources and catches.

e) Not discriminate between residents of different states.

f) Minimize costs and avoid duplication.

g) Promote efficiency in the utilization of fishery resources but not have economic allocation as their sole purpose.[6]

Seventh, each regional fishery management council is required to have a scientific and statistical committee (SSC) and advisory panel(s) made up of people knowledgeable about each region's fisheries.

These FCMA requirements indicated a clear commitment to obtaining and using the best scientific information and data available to inform, guide and support the development of the PMPs and FMPs.

---

[6] Three additional standards would be added in the 1996 reauthorization of the FCMA, known as the Sustainable Fisheries Act.

TABLE 1: SUMMARY OF KEY POLICIES AND CONCEPTS THAT BECAME OPERATIONAL WITH IMPLEMENTATION OF THE FISHERY CONSERVATION AND MANAGEMENT ACT

- Federal jurisdiction over fishery resources (except tuna) in FCZ waters seaward of state waters.

- Eight regional fishery management councils to formulate management strategies and measures for the domestic and foreign fisheries operating in their respective regions.

- NMFS to be the principal scientific advisor to the councils.

- Council decision-making process to allow for public input.

- U.S. fishermen to have first priority to whatever portion of the optimum yield they can harvest, with the remainder made available to foreign fishermen.

- PMPs to be developed by NMFS for foreign fisheries, to be supplanted by FMPs developed by the councils for domestic and foreign fisheries.

- FMP adherence to national standards.

NATIONAL STANDARDS FOR FISHERY CONSERVATION AND MANAGEMENT

Any fishery management plan prepared and any regulation promulgated to implement any such plan pursuant to this title shall be consistent with the following national standards for fishery conservation and management.

1 CONSERVATION AND MANAGEMENT measures shall prevent overfishing while achieving, on a continuing basis, the optimum yield from each fishery for the United States fishing industry.

2 CONSERVATION AND MANAGEMENT measures shall be based upon the best scientific information available.

3 TO THE EXTENT PRACTICABLE an individual stock of fish shall be managed as a unit throughout its range, and interrelated stocks of fish shall be managed as a unit or in close coordination.

4 CONSERVATION AND MANAGEMENT measures shall not discriminate between residents of different States. If it becomes necessary to allocate or assign fishing privileges among various United States fishermen, such allocation shall be (A) fair and equitable to all such fishermen; (B) reasonably calculated to promote conservation; and (C) carried out in such manner that no particular individual, corporation, or other entity acquires an excessive share of such privileges.

5 CONSERVATION AND MANAGEMENT measures shall, where practicable, consider efficiency in the utilization of fishery resources except that no such measure shall have economic allocation as its sole purpose.

6 CONSERVATION AND MANAGEMENT measures shall take into account and allow for variations among, and contingencies in, fisheries, fishery resources, and catches.

7 CONSERVATION AND MANAGEMENT measures shall, where practicable, minimize costs and avoid unnecessary duplication.

8 CONSERVATION AND MANAGEMENT measures shall, consistent with the conservation requirements of this Act (including the prevention of overfishing and rebuilding of overfished stocks), take into account the importance of fishery resources to fishing communities in order to (A) provide for the sustained participation of such communities and (B) to the extent practicable, minimize adverse economic impacts on such communities.

9 CONSERVATION AND MANAGEMENT measures shall, to the extent practicable, (A) minimize bycatch and (B) to the extent bycatch cannot be avoided, minimize the mortality of such bycatch.

10 CONSERVATION AND MANAGEMENT measures shall, to the extent practicable, promote the safety of human life at sea.

Magnuson-Stevens Fishery Conservation and Management Act as amended through October 11 1996

The FCMA (also known as the Magnuson-Stevens Act) includes national standards to which each regional fishery management council must adhere when developing its region's fishery management plans. The last three standards were added during the Act's 1996 reauthorization. *WPRFMC illustration.*

1.4. The Local Picture in the Western Pacific

When the FCMA was passed by Congress in 1976, commercial fishing in Hawaiʻi was in an economic trough. Operators of the local aku (skipjack tuna) pole-and-line sampans were affected by the combination of an economic recession, falling fish prices, declining catch rates, increasing fuel and insurance costs, and restructuring of the U.S. tuna processing industry. While the aku boats had been the most important component of the Hawai'i fishing fleet from the 1930s through the early 1970s, the fleet and the local cannery it supported seemed destined for closure.

Fifty years earlier, the aku fishery had been thriving in Hawai'i. Founded by skilled immigrant fishermen from Wakayama, Japan, it combined Native Hawaiian fishing techniques, such as using live nehu (Hawaiian anchovy, *Stolephorus purpureus*) as bait for aku, with techniques garnered from their more advanced industrial fishery. The result was an export-driven canned tuna industry, which was ethnically diverse. Although dominated by Japanese nationals, it included Native Hawaiians, other Pacific Islanders, Okinawans, Koreans, Chinese and Caucasians. It was recognized as one of the most modern tuna fisheries in the world, and, during the 1920s and 1930s, it supported the local Hawaiian Tuna Packers cannery, which employed up to 500 workers and produced nearly 10 million cans of tuna per year (Young 2020).

However, the fishery was devastated after the Japanese attack on Pearl Harbor in 1941, when fishing boats belonging to Japanese-Americans and Japanese nationals were sold or nationalized. When the war ended, the Japanese-American fishing community attempted to restore the fleet to its former prestige, but it lacked both the investment capital and the labor to do so, especially as young people from the community pursued better paying employment in the post-war era. These circumstances were exacerbated when catches declined after 1974, and costs increased due to requirements that local fishing sampans carry insurance when entering naval port facilities, such as Pearl Harbor, to catch baitfish. This combination of higher costs and lower revenues caused the fishery to decline.

Conditions were scarcely better for the rest of the Hawai'i commercial fishing industry. At the time, it included a small-scale longline fishery (known locally as the flag-line fishery), which used "basket gear" (sections of tarred rope, stored in baskets, that were connected end to end to constitute a set). The small 40- to 60-foot (12- to 18-meter) vessels were not robust or diverse enough to expand into a high seas longline fishery. While the market and trade for high-end sushi and sashimi was growing, the capacity to service that market was limited. The number of flagline vessels based in Honolulu and Hilo rose to 42 after World War II but declined to 14 by the late 1970s (Hawaii Seafood Council n.d.).

Further, some of the leading operators of commercial bottomfish boats complained that their traditional fishing grounds for deepwater snappers and other bottomfish off the main Hawaiian Islands were being fished out by recreational fishermen.

Fishermen use bait and spray to attract aku (skipjack tuna) to the bow and stern of a historical pole-and-line sampan vessel (circa 1970). When Congress passed the FCMA in 1976, the once prominent Hawai'i aku fleet was in a state of decline. *NOAA Pacific Islands Fisheries Science Center photo.*

Employees at Hawaiian Tuna Packers in Honolulu can locally caught skipjack tuna. Like the local aku fleet, the cannery, which began operations in the 1920s and grew to nearly 500 employees, was in a downturn when the FCMA came into effect in 1976. *Hawai'i State Archives photo/Maude Jones Collection.*

Tarred segments of rope and the baskets used to store them lie in front of wooden flagline sampans in Hawai'i (circa 1970). The vessels were introduced by Japanese fishermen at the turn of the century and had a limited seafaring range compared to the modern longline vessels that would replace them. *NOAA Pacific Islands Fisheries Science Center photo.*

The economic prospects for modern fisheries in other U.S.-flag territories in the Western Pacific Region (American Samoa, Guam and the as-yet unincorporated Northern Mariana Islands) were also limited.[7] In those areas, there was little infrastructure in the way of fishing ports, dock space or ice houses to support a fishing industry. In American Samoa, two large canneries were established for the packing of tuna landed in Pago Pago by U.S. purse seiners and Asian longliners. However, there was no local commercial fishing fleet. The homeport for the few U.S. purse seiners was San Diego in the mid-1970s.

Tuna cannery workers complete their shift at StarKist Samoa in Pago Pago. During the time of the FCMA passage, foreign and U.S. mainland vessels provided fish for the cannery as there was no local commercial fleet. *Tom Coffman photo.*

---

[7]   After World War II, the United States administered the Northern Mariana Islands as part of the United Nations Trust Territory of the Pacific Islands. In 1970, the people of the Northern Mariana Islands sought closer union with the United States rather than independence. A covenant to establish a commonwealth in political union with the United States was approved in a 1975 referendum. The new government and constitution came into effect in 1978.

A U.S. purse seiner traverses Pago Pago harbor (circa 2016). In the mid-1970s, the homeport for the U.S. purse seiners that offloaded in American Samoa was San Diego. *WPRFMC photo.*

Asian longliners dock near the cannery in Pago Pago harbor (circa 2015). *WPRFMC photo/Sylvia Spalding.*

Offshore fishing in the U.S. Pacific Islands was unregulated except for minimal regulations in the territorial seas of some of the islands. For example, Hawaiʻi had size limits for some fish and invertebrate species.

The people of Guam and the Northern Mariana Islands were aware that their waters were frequented by substantial schools of tuna, which offered potential for jobs and revenue for their growing populations. However, they had no means of accessing these resources, lacking as they did the capital for vessels and infrastructure. All they could do was watch foreign vessels fish on the horizon.

Meanwhile, Japanese and Taiwanese fishing vessels in the NWHI used traditional techniques for coral harvesting—dragging tangle nets and heavy concrete buckets with inset iron bars along the seafloor to break and dislodge precious deepwater coral. The 230 tons of broken bits of pink coral they harvested in 1969 were worth $4 million.

In the Hawaiian–Emperor seamount chain, Soviet and Japanese vessels targeted groundfish—a fishery and stock the United States didn't even know existed—taking more than 877,000 metric tons (mt) of fish in huge trawl nets. Japanese pole-and-line tuna boats fished freely in the NWHI, as well as in many U.S. territories and possessions to within 3 nm of land. There were even longliners from Japan fishing for tuna and associated species in waters off Diamond Head in the main Hawaiian Islands. Foreign boats were taking 20 million to 30 million pounds of fish a year within the 200-nm zone around Hawai'i (WPRFMC 1984a).

At the time, there was no discussion of how this foreign fishing was affecting fish stocks and the catches of local fishermen or how many birds, sea turtles, dolphins and Hawaiian monk seals (*Monachus schauinslandi*) were killed each year. "Freedom of the seas" was the dominant principle that allowed foreign fishing vessels to ply their trade as they saw fit. Meanwhile, Hawai'i had become a major tourist destination and needed high-quality fresh fish to feed its visitors and a growing local population. But still there was little movement to change the status quo regarding foreign fishing.

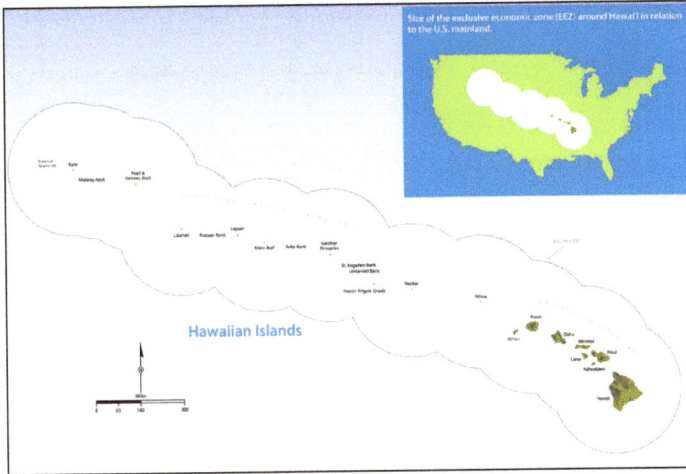

The U.S. waters surrounding the Hawai'i archipelago is a 400-nm swath spanning 1,300 nm. *WPRFMC illustration.*

In addition, there was less sensitivity at this time about threats to coral reefs or the sustainability of stocks in island economies that were small and largely undeveloped. People were too concerned about feeding themselves to think long-term. Some fishermen in American Samoa, Guam and the Northern Mariana Islands were reportedly using unsustainable and damaging fishing methods on coral reefs, such as chlorine, explosives and derris root used as fish poison. Hawai'i saw "high grading" (selectively landing only the best quality fish and discarding the others) of bottomfish for the lucrative tourist trade.

Many viewed the ocean as a vast dumping ground. Fishermen heard stories that Japan had plans to dump nuclear waste in the Western Pacific Region, but they felt powerless to make their voices heard in protest. The large canneries in Pago Pago dumped their waste directly into the harbor.

Further, throughout what would come to be known as the Western Pacific Region, indigenous peoples played a small part in the management of fishery resources that historically had been theirs for millennia. Virtually all employees of NMFS were white or Asian.

These were the circumstances prevalent in the Western Pacific Region when the FCMA was passed by Congress in 1976 and the 200-nm FCZ was established.

PART ONE

THE EARLY YEARS OF THE FISHERY CONSERVATION AND MANAGEMENT ACT (1976–1980)

Chapter 2

Organizing and Addressing Fisheries

2.1 Overview

The FCMA marked the end of uncontrolled fishing in the FCZ off the coasts and islands of the United States. Congress accepted the principle that U.S. jurisdiction over coastal waters needed to be extended to conserve living coastal marine resources for the benefit of the people of the United States. The United States would join the large majority of nations that had already taken action to claim jurisdiction over waters out to 200 nm adjacent to their shores.

The Western Pacific Council was one of eight regional fishery management councils established by the FCMA to manage fisheries seaward of state waters. The Council's jurisdiction was originally the FCZ around Hawaiʻi and the U.S. Pacific flag territories of American Samoa and Guam, but, in 2006, the CNMI and the waters around the eight Pacific Remote Island Areas were added.[8]

The Council's area of concern comprised islands thousands of miles apart, each with its own indigenous peoples who had their own unique legal relationship with the United States. The Western Pacific Region,

---

[8]   The Pacific Remote Island Areas include Baker, Howland, Jarvis, Johnston, Kingman Reef, Midway, Palmyra and Wake.

encompassing more than 2.2 million square miles (5.8 million square kilometers), is made up not only of islands and atolls but also numerous isolated seamounts, nearly all of which contain some form of marine resource—such as precious corals or deep-sea fish, e.g., armorhead (*Pseudopentaceros wheeleri*)—that required conservation from foreign fisheries.

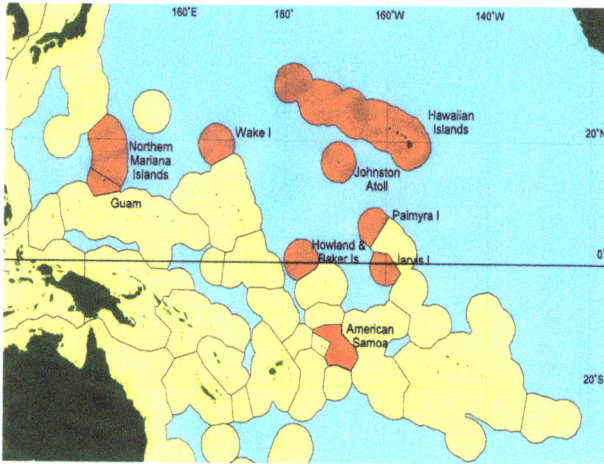

The U.S. waters surrounding the U.S. Pacific Islands is known as the Western Pacific Region (in red). It spans both sides of the equator and both sides of the dateline and is surround by the waters of other countries (in yellow) and the high seas (in blue), making the Western Pacific Council the nation's most international regional fishery management council. *WPRFMC illustration.*

The exclusion of tuna from the MSA, however, had an impact on the Western Pacific Council because tuna (collectively) was the most valuable stock in the region. Tuna is found in all waters under the Council's jurisdiction, and, in 1976, it represented 90% of the commercial value of fisheries of the Western Pacific Region. However, under the new law, fishing for tuna was excluded from Council control. This had the initial effect of making the Western Pacific Council the smallest of the eight councils in terms of managed fisheries but the largest in terms of the area of ocean under its jurisdiction. More disturbing to local fisheries, it meant that foreign fishing for tuna could continue with minimal federal regulation. It was not until the 1990 amendment to the FCMA that tuna species became subject to the Council's management process (see Chapter 8).

2.2 The Council—A Unique Entity

The eight regional councils created under the FCMA were tasked with specific new fishery conservation and management responsibilities. To fairly administer these responsibilities, the councils included a combination of local ocean users, scientific experts and government managers as part of an open political process funded by Congress through the National Oceanic and Atmospheric Administration (NOAA).

Under the FCMA, the Western Pacific Council initially was composed of 11 voting members (now 13 after the inclusion of the CNMI) and three non-voting members. Seven (now eight) of the voting members are private-sector individuals, nominated by the governors of their island entities and appointed to three-year terms by the Secretary of Commerce. They represent commercial and recreational fisheries as well as indigenous, environmental or other community interests in their island areas. The remaining four voting members (now five) are the designated state and territory fishery management agency directors and the NMFS regional administrator. Representatives of the U.S. Fish and Wildlife Service (USFWS), U.S. Coast Guard and U.S. Department of State are designated non-voting federal members. Legal advice to the Council is provided by the NOAA Office of General Counsel, initially from the Southwest Section in Long Beach, California, and currently from the new Pacific Islands Section in Honolulu.

From October 19 to October 21, 1976, the Council held its first meeting, and former Hawai‘i State Senator Wadsworth Yee was elected chair. Appointed members from Hawai'i included Native Hawaiian commercial fisherman Louis "Buzzy" Agard, sportfisherman Peter Fithian and commercial fishing representative Frank Goto. Guam appointees included Guam Department of Agriculture representative Francisco Aguon and commercial boat and fishing supply business owner Paul J. Bordallo. Businessman Lealaifuaneva Peter Reid Jr. represented American Samoa, and Joaquin Villagomez held observer status as a non-voting member from the Northern Mariana Islands (one year after the government of this former U.S. Trust Territory began negotiating for commonwealth status). Also participating were designated members Richard Wass of American Samoa, Isaac Ikehara of the Guam Department of Agriculture, Michio Takata of the Hawai'i Department of Fish and Game, Gerald Howard from the NMFS Southwest Regional Office, Lorry Nakatsu from the U.S. Department of State, Eugene Kridler of the USFWS and Rear Admiral J. W. Moreau of the U.S. Coast Guard.

Inaugural members of the eight regional fishery management councils met in Virginia in September 1976. The group photo of the Western Pacific Regional Fishery Management Council includes (front row, from left) Isaac Ikehara, Frank Goto, Michio Takata, Wadsworth Yee, Frank Aguon, Richard Waas, Paul Bordallo; (back row, from left) Peter Fithian, John Caffrey, Floyd Anders, Lealaifuaneva Peter Reid Jr. and Louis Agard. *WPRFMC photo.*

The Council members were drawn from varied recreational, conservation and commercial fishing backgrounds. Yee's experience was typical. An attorney and insurance executive, as well as a Republican senator in the Hawai'i Legislature, Yee was an avid big-game sportfisherman.

Wadsworth Yee served as the Council's first chair from 1976 to 1987.

Appointees from Guam and American Samoa also came from backgrounds that mixed fishing, conservation, local politics and business. However, compared to the other seven councils, the composition of the Western Pacific Council was unique. Due to its relatively small commercial fishing sector, it became one of the few U.S. regional fishery management councils where the representation and influence of noncommercial and indigenous fishermen was equal to that of commercial fishermen.

Under the interim direction of Robert Iversen—the Honolulu regional representative of the Western Pacific Program Office under the California-based NMFS Southwest Regional Office—the Council rented room 1506 at 1164 Bishop Street in downtown Honolulu. Iversen bought furniture and supplies (these were the days of typewriters, chalk boards and overhead projectors, long before computers and PowerPoint presentations), began holding meetings and worked with the personnel committee to coordinate the hiring of Wilvan Van Campen as the Council's first executive director.[9] At the second Council meeting, held December 15–16, 1976, approval was granted to hire Edwin Lee as administrative officer and Kitty M. Simonds as secretary. The following year, Ellen Reformina joined the staff as an administrative assistant.

After Van Campen was hired, Iverson returned to his former NMFS position. NMFS continued to provide "interim support staff" until the Council was fully up and running. Under the MSA, administrative and scientific support to the Council is provided by the NMFS Grants Office and fisheries science centers, respectively.

Robert Iversen of the NMFS Southwest Region served as the interim director of the Council and coordinated the hiring of its first executive director. *WPRFMC photo.*

---

[9]   Van Campen would be succeeded as executive director by Jack Marr (1979), Svein Fougner (1980–1982) and Kitty M. Simonds (1982–present).

Wilvan Van Campen (2nd from left) served as the Council's first executive director from 1976 to 1978 and was assisted by Council staff members (left to right) Kitty M. Simonds, Ellen Reformina and Edwin Lee. *WPRFMC photo.*

John "Jack" C. Marr served as the Council's second executive director in 1979. *International Center for Living Aquatic Resources Management photo.*

Svein Fougner served as the Council's third executive director from 1980 to 1982. *WPRFMC photo.*

Kitty M. Simonds, the Council's third staff person to be hired, would serve as the Council's fourth executive director in the coming years. *Paul Callaghan photo.*

During the first few meetings held in 1976 and 1977, administrative details governing the Council structure were established. Under the organizational plan, Council meetings would be managed by the chair, an elected position. Elections were to be held annually. The day-to-day work of the Council would be run by the executive director. Fishermen and other interested parties would be represented on advisory panels. Technicians, scientists and local managers would be members of plan teams, engaged in the work of turning ideas into management policy alternatives through the development of FMPs.

Because the members of the Council were not from contiguous areas, the Council made the decision to hold meetings in locations throughout the region, rather just in Honolulu, as this would allow the Council to hear directly from residents and fishery participants before developing or approving management measures in each island area. It was proposed that the Council meet six times per year and that four of those meetings convene in Honolulu. The remaining two meetings would be held in Guam and American Samoa. The meetings in Honolulu would take two days, and the meetings in Guam and American Samoa were scheduled to take at least three days because of the extended travel time involved.

Yee expressed the rationale for this by asking, "Why should I, from Hawai'i, tell the people who live on other islands how to fish, how to preserve their stock and what to do?"[10] In taking this position, Yee demonstrated how the new FCMA embodied America's democratic roots of public participation. Under the new law Council members were to see themselves as equals, operating in a public process and sharing a common management interest in fisheries and ocean resources. Reflecting this, the WPRFMC was one of the few U.S. councils that, from the beginning, recognized the knowledge and needs of indigenous fishermen and placed them in positions of influence alongside scientists,

---

[10]   Wadsworth Yee, former Council chair, in discussion with Michael Markrich, December 15, 2007.

federal and state administrators, commercial fishermen and big-game sportfishermen.

Travel by Council members to various locations in the Hawaiian Islands, Samoa, Guam and the Northern Mariana Islands added to the Council's operating expenses and eventually limited meetings to only four times per year (whereas councils on the U.S. mainland met as many as nine times per year), but the extensive travel ensured that Council members could learn firsthand about the areas affected by their decisions. For example, the third meeting of the Council (on from February 1 to 4, 1977) was held in Guam so Council members could see firsthand the fishery management challenges of Guam and the Northern Mariana Islands. Although such travel is now a normal method of conducting business, in the 1970s Council efforts to meet in the Pacific Islands drew harsh criticism. The Council persevered and continued to follow its meeting schedule.

As things got underway, questions arose regarding how the Council would be involved in NMFS's development of PMPs (which would contain initial regulations governing foreign fishing until the Council completed the FMPs that would regulate foreign and domestic fishing. Much had to be done to gain the information needed to prepare the PMPs and FMPs for the Western Pacific Region, and it was widely believed that fishery statistics for the region were either inaccurate or unavailable. This was a critical problem for the Council, which was charged under the national standards of the FCMA to manage based on "the best available scientific information." If information was poor or nonexistent, then management decisions could easily be wrong. This highlighted the importance of establishing the Council's SSC. The minutes of the Council's December 1976 meeting refer to the daunting job ahead for the government and others to "catalogue the resources and fisheries between the limits of State jurisdiction and 200 nautical miles."

The Council decided at its first meeting in October 1976 that the SSC would have representation from Hawai'i, American Samoa and Guam and, to remove any possible political bias, would not include Council members. One of the criticisms of the Bureau of Commercial Fisheries, which preceded NMFS, had been that the scientific work performed was sometimes not peer reviewed or free of the economic or political influence of special interest groups. The decision-making approach adopted by the Council was to strictly adhere to the new guidelines of the FCMA national standards.

Current Executive Director Kitty M. Simonds recalled the very high qualifications of the early SSC members. She described Richard Shomura, John Craven, Alexander Spoehr and others as "dynamos, brilliant thinkers who had good ideas and were able to work on many levels."[11] Shomura directed the NMFS Honolulu Laboratory (a part of NMFS Southwest Fisheries Science Center), which Congress established in 1948 to revive Hawai'i's commercial tuna fishery. Craven was the marine affairs coordinator for the State of Hawai'i and also the dean of marine programs at the University of Hawai'i. Spoehr was a noted anthropologist who specialized in Oceania and who had served as director of the Bishop Museum, the first chancellor of the East–West Center and president of the American Anthropological Association.

The first SSC meeting was held in Honolulu from February 24 to 25, 1977. Under the Council's new organizational plan, the SSC was to meet six times per year in the same locations as the Council meetings. SSC members were to be appointed for one-year terms (this was eventually changed to terms of indefinite length and then,

---

[11]   Kitty M. Simonds in discussion with Michael Markrich, December 15, 2008.

more recently, to three-year terms). The number of meetings has now been reduced to three or four per year unless circumstances require a special meeting. There was some initial debate as to the SSC composition. Eventually, it was decided that the SSC would not be limited to academics alone but would also include those with specialized knowledge from government, law and industry. There was particular interest in placing a bank economist on the SSC who had not only academic credentials but also expert knowledge of the financing needed for fisheries development. The initial task of the SSC was to identify the species within the fisheries under the Council's jurisdiction, assess them and set management and research priorities for them.

Richard Shomura, director of the NMFS Honolulu Lab, served as the Council's Scientific and Statistical Committee vice chair 1977–1978 and as its acting chair 1979–1980. *WPRFMC photo.*

John Craven, an inaugural member of the Council's Scientific and Statistical Committee in 1977, was the marine affairs coordinator for the State of Hawai'i under Governor John A. Burns. At the University of Hawai'i, he also served as the dean of marine programs and director of the Law of the Sea Institute. *University of Hawai'i Foundation photo.*

Alexander Spoehr, an early SSC member (1981–1985), was a renowned anthropologist specializing in Oceania, a former chancellor of the East-West Center and a former director of the Bishop Museum. *Spoehr family photo.*

Paul Callaghan, a University of Guam economics professor, served one term as a Council member (1979–1981) and then three decades as its SSC chair (1981–2011). *WPRFMC photo.*

Among the new management priorities of the Council was improving the economic situation Hawai'i fisheries. Towards this end, the Council requested that the University of Hawai'i Sea Grant program research the quantity of precious corals in the NWHI. It also requested that the U.S. Navy allow Midway, in the NWHI, to be utilized as a temporary berthing and resupply center for the Hawai'i fishing fleet. According to the minutes of the Council's first meeting in October 1976, the NWHI resources were seen as untapped and a potential source for growth for Hawai'i fisheries.

The new Council also took steps to legally delineate and monitor its area of jurisdiction. A formal request was made to the U.S. Department of Transportation to provide aerial surveillance of Hawai'i federal waters by the U.S. Coast Guard, while another request was made to Congress to include the U.S. possessions in the Western and Central Pacific Ocean (WCPO) within the Council's management area. The U.S. State

Department was asked to provide statistical information on foreign fisheries as a prerequisite to their being considered for permits to fish within the U.S. FCZ.

The delegate to the Council from the Mariana Islands, Joaquin Villagomez, held only a non-voting observer status. However, he soon made his voice heard, asking that a definite boundary be drawn between the U.S. FCZ surrounding Guam and the FCZ waters of the Northern Mariana Islands. In 1976, the future status of the Northern Mariana Islands as a U.S. commonwealth was not a certainty. Villagomez was worried that, because of the uncertain status, the people of Saipan might be classified as non-resident U.S. aliens when fishing in U.S. waters surrounding Guam (WPRFMC 1976b). He also let the Council members know that Japanese fishing vessels frequently fished in waters around the Northern Mariana Islands without paying any licensing fees. Villagomez persuaded NOAA attorney Martin Hochman to write a letter certifying full MSA benefits to the Northern Mariana Islands as soon as commonwealth status was achieved.

Similarly, Guam expressed concern about observed foreign fishing. However, some of the ships spotted by fishermen just a few miles outside Apra Harbor, Guam's primary port, were suspected of being Soviet spy ships, disguised as fishing trawlers. In addition, Guam had concerns that the U.S. purse-seine fleet—which was relocating to the western Pacific and benefiting Guam economically through the purchase of fuel, repairs and provisions—might contribute to local tuna stock depletion and adversely impact ex-vessel prices received by local small-scale fishermen.

The Council decided during its eighth meeting in 1977 to take the steps necessary to make the Northern Mariana Islands a full voting member. The ninth meeting of the Council took place on Saipan, the capital of CNMI, on January 10 to 12, 1978, so that members of the Council might participate in the inaugural ceremonies for the new CNMI government. The new government and constitution had come partially in effect in on January 9, 1978. In December 1978, Council Chair Yee wrote a letter to U.S. Senator Daniel K. Inouye asking him to re-introduce legislation amending the MSA to include full voting representation for CNMI.

While arrangements for the Council location and logistical details were being finalized, work began on the immediate task of assisting NMFS to complete the PMPs by March 1, 1977, when the 200-nm FCZ would become effective. While the Secretary of Commerce (through NMFS) had the legal responsibility to complete the PMPs, this could not be accomplished without the input and advice of the Council.

2.3 National Marine Fisheries Service

In concert with the formation of the Council is the history of NMFS. The agency's regional administrator serves on the Council as the only federal voting member, and the regional Fisheries Science Center serves as the Council's primary source of science and analyses.

Initially, Hawai'i and the U.S. Pacific Islands were a part of the NMFS Southwest Region, with the regional office based in Long Beach, California, and the Science Center in La Jolla, California. The Southwest Fisheries Science Center, previously known as the Fishery-Oceanography Center of the Bureau of Commercial Fisheries, had been operational since October 31, 1964, and included the California Current Resources Laboratory and the Tuna Resources Laboratory, plus several tenant agencies, including the Inter-American Tropical Tuna Commission. Also incorporated into the NMFS Southwest Fisheries Center was the Honolulu-based Pacific Oceanic Fishery Investigations, previously a part of the Department of the Interior's USFWS. It was created for the purpose of carrying out Public Law 329, enacted by the 80th Congress on August 4, 1947, "… to provide for the exploration, investigation, development, and maintenance of the fishing resources and development of the high seas fishing industry of the Territories and island possessions of the United States in the tropical and subtropical Pacific Ocean and intervening seas, and for other purposes" (Sette et al. 1954).

The NMFS Honolulu Laboratory, as the Pacific Oceanic Fishery Investigations came to be called, housed its main facility adjacent to the University of Hawai'i campus at Manoa. A smaller satellite research facility—with seawater capabilities for conducting research on large, live pelagic fish, monk seals and sea turtles—operated at Kewalo Basin in Honolulu. NMFS also maintained the Pacific Islands Area Office, an arm of the NMFS Southwest Regional Office, at the Manoa location.

For the first 27 years of the Council's history, NMFS's responsibility for managing marine resources in federal waters surrounding the U.S. Pacific Islands remained with the Southwest Regional Office and Fisheries Science Center and their satellite offices in Honolulu.

For nearly three decades, the Council's federal partner and science provider would be the NMFS Honolulu Laboratory and Pacific Islands Area Office, satellite offices of the California-based NMFS Southwest Fisheries Science Center and Regional Office. After Congress established the Pacific Islands as its own region in 2003, the new Pacific Islands Fisheries Science Center and Regional Office would remain in the same building near the University of Hawai'i at Manoa campus in Honolulu until 2014. *NOAA photo.*

Chapter 3

Developing and Implementing Preliminary Fishery Management Plans

3.1 Issues Related to Managing Foreign Fisheries in U.S. Waters

The fish species (fish stocks) chosen for the first PMPs were precious coral, seamount groundfish and billfish because each was affected one way or another by the actions of foreign fisheries. Precious coral was chosen because the Japanese and Taiwanese were fishing for precious corals within 200 nm of Midway, near the northwestern end of the Hawaiian archipelago. Groundfish were included due to harvests of armorhead and alfonsin (*Beryx splendens*) by Russian and Japanese trawlers on seamounts near Midway. Billfish were selected because foreign longline vessels targeting tuna (exempt under the FCMA) were also taking billfish in the FCZ. It was felt that that the harvests of billfish by Japanese longliners, to be sold for sashimi, were affecting catches by Hawai'i recreational fishermen (Nickerson 1976) and that control of foreign billfish catches would limit the damage inflected by foreign longline fishing on domestic recreational and commercial fisheries targeting billfish and tuna.

As the Council worked with NMFS to complete the PMPs, it had to develop an effective relationship not only with its new members but also with the local NMFS Honolulu Laboratory staff. Representatives sent to Hawai'i from the NMFS Southwest Regional Office in California further added to this complex group dynamic, and the differences of Council members posed a challenge that had to be dealt with, all while working on the business at hand.

Among the many issues that had to be addressed by NMFS and the Council was giving operational meaning to the determination of optimum yield for the various fisheries. No one had a firm understanding of what optimum yield was or how to estimate it for the PMP species, especially given the meager data available. It became the task of the new SSC and the Honolulu Laboratory to develop information and models that would aid management under the FCMA standards. But there was a perception that cultural nuances and practicalities of developing information and policy in the Western Pacific would be different than on the mainland because of such Hawaiian cultural traditions as not divulging where and when one goes fishing.

"We applied the standards in our own way," recalled Simonds. "The standards gave the United States a basis for what we should be doing, but it was very difficult to separate recreational fishing from commercial fishing on Pacific Islands where many of the [indigenous] people fished on a subsistence basis. It wasn't the same thing [as it was on the U.S. mainland]."[12]

In addition to the establishment of a policy direction, funds were prioritized and budgeted for U.S. Coast Guard aerial surveillance of the FCZ around Guam and American Samoa and for the establishment of NMFS offices in these areas. Funds were also set aside for fishery

---

[12]   Simonds ibid.

development projects. These projects were seen as an important way to generate jobs and revenue in American Samoa and Guam.

The combination of vague guidelines, lack of solid data, island conditions that differed markedly from the mainland and deeply held Native Hawaiian, Samoan and Chamorro cultural beliefs created both a dynamism and a struggle for organizational identity that would set the Council on its own unique course. The new members had differing goals, but all had a common interest in improving the lives and protecting the interests of their fellow Pacific Islanders. But, first, NMFS had to complete the PMPs. The first drafts were written by the NMFS in California and sent to Hawaiʻi for review, advice and editing, and the Council took an active role in this process.

3.2 Precious Corals Foreign Fisheries

Shallow-water reef building corals are today widely recognized as important indicators of ocean health, and they play an important role in the Council's coral reef ecosystem policy in the Western Pacific Region. Precious corals differ from reef-building corals in that they do not require light and are normally found hundreds of feet below the surface. In the U.S. FCZ waters, precious corals are mostly found living on isolated, submerged seamounts. Gold (*Kulamanamana haumeaae*), red or pink (Coralliidae) and bamboo (Isididae) coral are found at depths of about 1,000 to 5,000 feet (300 to 1,500 meters), while black coral (Antipathidae) is found at depths of about 100 to 330 feet (30 to 100 meters).

An underwater diver brings a harvest of black coral
to the surface. *Richard Grigg photo.*

Deepwater precious corals have been treasured by mankind for beauty and personal adornment for more than 25,000 years and were initially sourced primarily from the Mediterranean Sea. Valuable precious coral beds were discovered in the Pacific Ocean in 1803 in waters off Japan; however, it was not until 1868, during the economic reform of the Meiji era, that Japan developed its first commercial coral fishery. By the end of the 19th century there would be a hundred vessels in the Japanese precious coral fishery, and this would soon have an impact on Hawai'i (Grigg 2010).

Precious coral has been used for millennia to fashion into jewelry and engravings. Pink coral (*Corallium secundun*) is the most abundant deepwater precious coral in the Hawai'i Archipelago. *D. Doubilet photo.*

The Japanese method of harvesting deepwater precious coral in the 19th century involved first dropping exploratory "tangle nets" into the ocean at random sites and great depths until coral bits were brought to the surface. Once a precious coral bank was discovered, heavy stones attached to ropes and surrounded by tangle nets were dropped into the deep water. The stones would smash the coral into small pieces, which would float in the water column and then be captured in the nets dragged by the vessels. Because the precious corals are slow growing, the heavy stones would smash decades of coral growth. After one area's precious coral was exhausted by this method and it was no longer economically feasible to harvest there, the coral fishermen would sail off to hunt for a new site.

Over time the Japanese coral draggers had refined their gear from heavy stones to "coral nets" (basically heavy concrete buckets to which iron rings had been bolted and heavy tangle nets attached). A single Japanese or Taiwanese vessel might methodically crisscross the seafloor dragging 16 such concrete dredges and trawling tangle nets behind them, harvesting tons of coral and causing great damage. When five vessels or more worked an area, the damage was greater.

As the coral draggers increased their efficiency, the process of wiping out coral beds, which had once taken many years of individual voyages, could now be done in only a few visits. These destructive practices were being driven by rapidly growing economies in Japan and Taiwan. During the late 1970s, foreign vessels could potentially make millions of dollars per trip.

For 80 years Japanese and Taiwanese fishing vessels moved eastwards across the Pacific searching for new coral fishing grounds. In 1963 and 1965, Japanese coral draggers made two important precious coral bank discoveries in the Western and Central Pacific, one on the Oza Banks, about 86 nm south of Okinawa, and the second on the Milwaukee Bank in the Emperor Seamount Chain, about 430 nm northwest of Midway. Precious coral banks were also found on several other seamounts in the Emperor Chain. Buoyed by a thriving Japanese economy and strong customer demand, by 1969 almost 230 tons of red and pink coral worth more than $4 million were being harvested annually (ibid.).

The success of the Japanese and Taiwanese fisheries in finding new beds off the Emperor Seamounts prompted two researchers from the University of Hawai'i, professors Ted Chamberlain and Vernon Brock, to explore for coral in the main Hawaiian Islands. In 1969, they discovered large gold coral beds off the eastern coast of O'ahu. Chamberlain and Brock were not the first, however, to discover precious corals in Hawai'i. Prior to their efforts, the Challenger Expedition (1872–1876) and the Albatross Expedition (1902) had found black coral beds in Hawai'i.

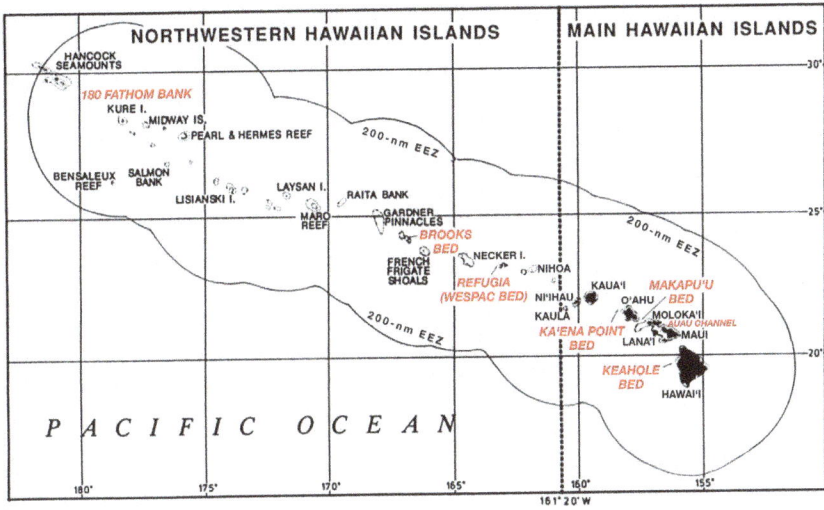

Precious coral beds are located in the NWHI and main Hawaiian Islands. The known beds are highlighted in red. *WPRFMC illustration.*

The Problem

How to develop a PMP for a fishery with zero documentation and little scientific information? That was the problem NMFS and the Council faced in addressing the precious coral fishery by foreign fleets, which was one of the first three candidates for PMPs. Given a long history of precious coral harvesting by foreign vessels in the Pacific and the recent discovery of precious corals around Hawai'i, the Council expected that more could be located—and possibly harvested profitably in the future by domestic fisheries—if there were some incentive to reward exploration while also controlling excessive development. Foreign fishermen might submit permit applications to fish in the FCZ if a PMP were in place and the terms were right.

The problem of managing the deepwater precious corals in the FCZ coincided with an ongoing issue of managing black corals within the state waters of Hawai'i. In 1958, Lahaina divers Jack Ackerman and Larry Windley found a large bed of black coral at depths of between 19 and 55 fathoms off Lahaina, Maui. The Lahaina Bed consisted of two species of black coral, *Antipathes dichotoma* and *A. grandis*. The discovery gave birth to a new fishery that soon grew to more than 10 participants (ibid.).

Although the harvest rates remained relatively low, the growth of the fishery and associated jewelry industry led to calls for the management of the black coral beds by the State of Hawai'i Department of Fish and Game. Because almost nothing was known about the ecology and biology of black coral in Hawai'i, there was a need for studies to be conducted on the life history of the black coral to guide their management (ibid.).

The first study was conducted by Richard Grigg and was the subject of his 1964 master's thesis for the Department of Zoology at the University of Hawai'i. In time, Grigg would earn his PhD at the University of California at San Diego's Scripps Institute of Oceanography and return to the University of Hawai'i to become an internationally recognized expert on the ecology of precious corals. Later, as a result of Grigg's pioneering work and mentoring, Garth Murphy and Vernon Brock would develop extensive expertise in precious coral life-history. The work by these researchers for the University Hawai'i over the next 30 years provided the main foundation for the Precious Coral PMP and subsequent FMP. Grigg's contributions to the field were so great that the shallow-water black coral, formerly *A. dichotoma,* was renamed to *A. griggi* (Opresko 2009).

However, prior to the PMP's implementation in 1978, there was no comprehensive set of regulations for precious corals, and there was a rising sense of urgency that something needed to be done

(Tenbruggencate 1988). It was estimated that approximately 10 mt (10,000 kilograms) of pink coral were harvested illegally in the NWHI each year by Taiwanese fishing vessels alone (WPRFMC 1978b). U.S. Coast Guard resources to patrol the area were limited, and violations in American waters by the coral draggers were flagrant.

Robert Iversen, a former NMFS staff member, recalled an incident in May 1978 when he flew as an observer on a routine U.S. Coast Guard patrol plane and discovered a group of illegal coral draggers in U.S. waters. "I flew up to Midway and then over Hancock Seamount, which [was] just inside the U.S. [FCZ]. The moment we got there I saw five tuna longliners all in a bunch on top of the seamount. Then I realized they were coral draggers. As soon as they saw us they pulled in their gear and headed straight for Japan. The pilot flew low so I took a bunch of photos, but the coral draggers escaped. Then I wrote up the report and forgot about it." Five years later in 1983, while Iversen was working as a regional U.S. fishery attaché in Japan, he received a telegram from the U.S. State Department with a diplomatic note that he was to take to the Japanese foreign ministry demanding $25,000 for each Japanese vessel photographed illegally fishing in U.S. waters. When the Japanese diplomats requested evidence, Iversen pulled copies of the photographs from his pocket. They asked him how he knew the photographs were genuine. "Because I took them," Iversen replied. The Japanese government paid the fine (Iversen 2020).

These large rocks from a Japanese coral dragger weigh roughly 75 to 100 pounds each. They were used to tow old salmon and trawl nets across the bottom of seamounts to snag precious corals and were located at a private residence in Hawai'i in 2008. *Robert Iversen photo.*

The Japanese 349-ton stern trawler *Koshin Maru No. 21* waits outside Midway Harbor before being seized by the U.S. Coast Guard in May 1978 for illegal fishing inside the U.S. FCZ in the NWHI. *NMFS photo/Special Agent Bill Streeter.*

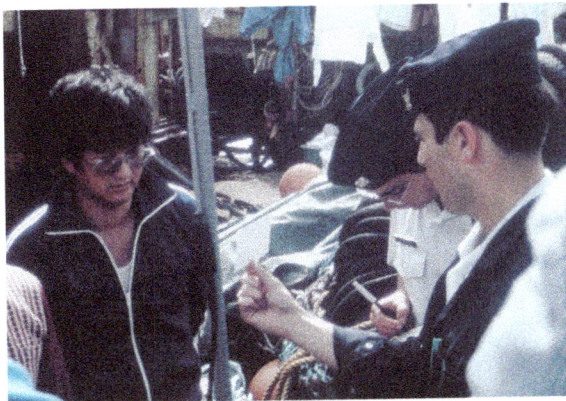

Bruce Stubbs of the U.S. Coast Guard (right) holds a card informing the captain of the Japanese stern trawler *Koshin Maru No. 21* that the ship is being seized for illegal fishing inside the U.S. FCZ on May 4, 1978, near Kure Atoll, NWHI. *NMFS photo/Special Agent Bill Streeter.*

The Solution

Members of the Council knew that development of a strong PMP could control the incursions until the FMP could be developed. Prior to the formation of the Council in 1976, limits on coral harvesting in U.S. waters were set by the Department of the Interior's Bureau of Land Management. After the FCMA was passed by Congress, NOAA and the Bureau agreed to a memorandum of understanding that the Bureau would not exercise its authority over corals that were under the jurisdiction of a PMP or FMP. After this was established, NMFS and the Council began collaboration on the PMP.

At first there was little consensus within the Council as to what should be done. As the Precious Coral PMP was being developed, some believed the Council should take strict protective actions, while others, particularly Pacific Islanders, found this to be too restrictive. They wanted to encourage exploration in their areas so that their islands could potentially benefit from newly discovered resources. The PMP would follow a middle path allowing regulated foreign and domestic precious coral harvesting but prohibiting damaging practices within the U.S. FCZ.

It became Grigg's task to use what was known of precious coral ecology and adapt that knowledge to the newly promulgated FCMA standards for maximum sustainable yield and optimum yield. He would place emphasis on the management of gold and pink precious corals because they were the most heavily impacted by coral harvesters. In addition, unlike black coral, they were found exclusively in the federal waters under Council jurisdiction and seaward of the 3-nm limit of state waters.

The key was to develop a fishing plan for precious coral that was sustainable. As a basis, Grigg used the Beverton and Holt Model, a method for estimating population dynamics that had been used in developing fisheries management plans since 1957. Using elements of the model and the information he had about the lifecycle of precious coral, he was able to create calculations for maxiumum sustainable yield and optimum yield. As he developed the mathematical model based on the lifecycle of the coral, its growth patterns and its estimated reproductive capacity, he began making several immediate recommendations.

Perhaps his most significant recommendation was to prohibit coral dredging because of the enormous damage it was doing to the sustainability of the precious coral stocks. However, he urged that harvesting not be completely stopped as controlled harvesting could be done selectively by undersea submersibles (Grigg 2010).

The PMP was completed in 1977. Under it, four management area categories were designated: 1) Established Beds (known areas in which optimum yield had been determined; 2) Conditional Beds (known to contain precious corals and for which the optimum yield is calculated based on the area of the conditional beds relative to the area of the Makapuʻu Bed); 3) Refugia (no-harvest areas); and 4) Exploratory Areas (open for exploration).

The intent of the new PMP was not to end fishing for precious corals but to encourage it in an ecologically responsible manner. The PMP set a two-year maxiumum sustainable yield of 2 mt for harvest from all Exploratory Areas combined. Half the allotment was designated for foreign fishermen, and the other half was reserved for domestic fishermen, of which there were none in Hawaiʻi.

The Council saw this approach as an opportunity to gain a greater understanding of the precious coral resources by encouraging regulated explorations throughout the region. It was the intent of NMFS and the Council to encourage U.S. vessels to explore and harvest corals in areas that would otherwise be completely dominated by foreign vessels, and, if possible, to use foreign experience and information to aid in the development of a domestic fishery. Ultimately, there were no applications for foreign fishing.

3.3 Deepwater Seamount Foreign Fisheries— Saving Armorhead and Alfonsin

The Problem

How to manage FCZ fishing, including foreign fishing, for undocumented fish stocks on seamounts in remote areas with little information about the productivity of the stocks and the impacts of fishing on those stocks? Once again, this was the problem that NMFS and the Council faced in responding to foreign fishing for armorhead and alfonsin, primarily at Hancock Seamounts at the far northern end of the NWHI.

Large aggregations of armorhead were discovered over the nearby summits of the southern Emperor Seamount Chain by a Soviet trawler in 1967. According to an interim report produced by NMFS scientists, Soviet catches from this area totaled 728,500 mt between 1968 and 1975, and the Japanese harvested 147,635 mt between 1969 and 1975 (WPRFMC 1984c). The Soviet and Japanese fishing effort is believed to have removed more than 90% of the armorhead biomass from the Emperor Seamount Chain, including in the area of the Hancock Seamount, which lies at the northern most boundary of the Hawaiian Ridge Seamount Chain.

Alfonsin, a bycatch, was well documented by the Japanese fleet but not the Soviet fleet. There has never been a U.S. fishery for these species, and, because less than 5% of the armorhead habitat lies within U.S. jurisdiction, efforts to rebuild the stock rely on international cooperation and management.

Armorhead fish are common in the northernmost part of the Hawaiian Archipelago. Soviet and Japanese fisheries are believed to have removed more than 90% of the armorhead biomass from the Emperor Seamount Chain, including in the area of the Hancock Seamount, which lies within the U.S. waters. *NOAA photo/Ocean Explorer 2003 NWHI expedition.*

A deep-sea alfonsin is seen swimming in the Gulf of Mexico during a NOAA 2012 expedition. Also known as alfonsino, the species is found circumglobally and was a bycatch of the foreign armorhead fishery at Hancock Seamount prior to establishment of the U.S. FCZ. *NOAA photo/Okeanos Explorer Program.*

In 1977, NMFS considered the situation at Hancock so severe that it made the PMP for this fishery a priority. Prior to the PMP, each vessel was allowed to harvest 2,000 mt, and the Soviet applications were required to note only that they fished in an ambiguous "Hawai'i region."[13] The situation was further complicated because very little was known about the productivity, maxiumum sustainable yield or optimum yield of armorhead generally and within the FCZ specifically. It was assumed that these species were (like most bottomfish species) long lived, slow growing and late to mature, meaning that they could be overfished fairly quickly and then take a long time to recover. Ultimately, Hawai'i found help with species identification and information via the U.S. embassy in Japan.

NMFS, at the Council's request, sent the NOAA research vessel *Townsend Cromwell* to the NWHI to survey armorhead and alfonsin

---

[13]   Yee op. cit.

in the area from Necker Island to the Hancock Seamount. This would be the first of many *Townsend Cromwell* surveys on the Hancock Seamount.

The Solution

Based on information from NMFS and foreign fishing records, the Council recommended that the PMP specify exactly the location and amount that foreign fleets could harvest. It proposed that 2,000 mt be allowed per country per year for armorhead and alfonsin combined. In addition, the Council recommended that all foreign trawling be limited to FCZ waters around Hancock Seamounts west of 180 degrees longitude and north of 28 degrees north latitude. It was the Council's view that this level of fishing would be sufficient to entice foreign interest, which would generate both foreign fishing fees and data through logbooks and observers, while at the same time be conservative enough that overfishing would be unlikely.

However, before the PMP was finalized, the Council learned that in Japan alfonsin was selling for $1,000 per mt while the U.S. government was granting permits for only $139 per mt. Moreover, the Council was made aware that Taiwan would soon be entering the seamount fishery, putting even greater pressure on the stocks (WPRFMC 1978c).

Richard Shomura, the head of the NMFS Honolulu Laboratory, noted at the Council's May 22 and 23, 1978, meeting that one of the ways the SSC members obtained information on the seamount fishery was through translating Japanese reports and sending Japanese speaking fishery specialists Wilvan Van Campen and Tanio Otsu to Japan to learn about the fishery. It was by undertaking this kind of personal hands-on information gathering that the Honolulu Laboratory learned of the plans by the Taiwanese to enter the fishery with bottom longlines (ibid.). Van Campen also translated Soviet reports on the fishery and the initial studies of armorhead biology that the Soviets began writing in the 1970s.[14]

For these reasons, the Council ultimately recommended that foreign fishing around Hancock Seamounts be limited to 1,000 mt per country annually and that foreign permits be withdrawn completely if catch rates of armorhead did not increase.

The PMP for the seamount-groundfish fishery resources within the U.S. FCZ around the Hawaiian Archipelago was implemented in early 1977 (NMFS 1977). The regulations temporarily restricted the foreign harvest of pelagic armorhead and alfonsin to a 2,000-ton catch quota by trawling or bottom longlining and to 60 vessel-days of effort. In addition, the regulations established a licensing procedure for foreign vessels and required the submission of detailed catch and effort data, as well as the placement of U.S. observers on any foreign fishing vessel.

In 1978, an application submitted for a trawling survey at Hancock Seamounts was approved, after which two commercial trawlers received approval to fish at Hancock. On each trip, including the survey trip conducted in 1978, a U.S. observer onboard measured and weighed representative samples of the catch and collected detailed data on catch by species and fishing effort. As time permitted, the observer recorded food, feeding habits and gonad conditions of the fish (Shomura and Tagami 1984).

Today it is believed that the unregulated fishing that took place through the mid-1970s, combined with the episodic nature of armorhead recruitment, has significantly delayed Pacific stock recovery.

---

[14] Robert Humphreys Jr., NMFS Pacific Islands Fisheries Science Center, in discussion with Michael Markrich, September 7, 2016.

3.4 Billfish Foreign Fisheries

The Problems

One of the most significant and difficult problems that NMFS and the Council faced was considering how to control catches of billfish incidentally caught in fleets targeting tuna. It was difficult to establish meaningful controls for billfish when the FCMA did not authorize management of tuna under the PMP and FMP. Nonetheless, that is what had to be done.

Tuna are generally considered to be highly migratory species, their range extending across ocean basins. However, in the FCMA, Congress expressly forbade the management of "highly migratory species," a common euphemism for tuna. Many Council members felt this exemption needed correction. They worried that unchecked and unmonitored catches by large U.S. and foreign vessels threatened the long-term health of tuna stocks in the Western Pacific Region. They argued that, even if the stocks were not depleted throughout their migratory range, they could be depleted on a local scale such that domestic, island-based fishermen would be adversely impacted.

A second difficulty in developing the Billfish PMP was the problem NMFS and the Council faced in obtaining good fishery-independent scientific data on stocks and reliable catch-and-effort data in order to estimate maxiumum sustainable yield and optimum yield. There could be no useful policy direction if there were no reliable datasets on which to base it. Data from active domestic tuna fishermen, mostly local flagliners who caught billfish as bycatch, needed to be obtained and analyzed. Otherwise, working only from Japanese fishing data meant that any analysis would be incomplete. Unfortunately, the American Tunaboat Association refused to share data on billfish caught as bycatch in the domestic tuna purse-seine fishery.[15] In addition, reporting requirements of commercial catches were frequently ignored by small-scale fishermen who sold a portion of their catches to cover their "recreational" expenses.

Besides incomplete data, NMFS and the Council faced a third difficulty—deciding what species to include under the Billfish PMP. Billfish included a number of species, such as blue marlin (*Makaira mazara* and *M. nigricans*), striped marlin (*Kajikia audax* and *Tetrapturus audax*), black marlin (*M. indica*), broadbill swordfish (*Xiphia gladius*), shortbill spearfish (*T. augustirostris*) and sailfish *(Istiophorus platypterus)*. Other migratory fish important to local fishermen were regularly taken by longline fishing, including mahimahi (dolphinfish, *Coryphaena hippurus*), wahoo (*Acanthocybium solandri*), moonfish (*Lampris guttatus*) and oceanic sharks.

The Solutions

The Council response to the first problem was to emphasize the development of rules to manage foreign vessels known to catch billfish in the region's FCZ. Controlling billfish catches by foreign vessels through quotas and non-retention zones would likely reduce tuna harvest and foreign competition in the FCZ. At its fifth meeting, the Council voted to call for the FCMA to include tuna (WPRFMC 1977b). The vote was non-binding but reflected what was becoming a dominant view among Council members—that exempting the management of tuna was not feasible for the long-term sustainability of tuna stocks in the Western Pacific Region.

The impetus for this Council action was NMFS's move to encourage the different Atlantic councils to defer development of their Billfish PMPs

---

[15]   Paul Callaghan, former Council SSC chair, in discussion with Michael Markrich, September 12, 2016.

and FMPs to the newly formed NMFS Highly Migratory Species Division. This new division was specially created at NMFS national headquarters level to resolve the problem of the migratory species not covered under the FCMA and to address the overlap between domestic management under the FCMA and international management under the International Commission for the Conservation of Atlantic Tunas. Development of a billfish plan in the Atlantic was further complicated by the overlapping interests of the five regional councils on the East and Gulf coasts (New England, Mid-Atlantic, South Atlantic, Gulf of Mexico and Caribbean).

To overcome the problem of data gaps, strategies were developed to use the available data and statistical information and to undertake new research to estimate the maxiumum sustainable yield and optimum yield for billfish and other pelagic species. It became the task of the Council, working with the NMFS Honolulu Laboratory, to analyze available longline catch data to determine the extent of the billfish bycatch in the FCZ.

One of the first billfish modeling studies commissioned by the Council involved a computer analysis of Japanese data on the effects of foreign longlining in the FCZ around Hawaiʻi (Lovejoy 1977, 1981). The data were used to estimate the catch of blue and striped marlin and to assess the potential impacts on local fishermen of closing various areas to fishing by foreign vessels. Other related research undertaken at the same time by the NMFS Honolulu Laboratory included albacore stock assessments, efforts to remedy the shortage of live-bait for the domestic tuna pole-and-line fishery in the islands and experiments with anchored fish aggregating devices (FADs). For many years, Honolulu Laboratory researchers hand recorded billfish and tuna sale receipts and observations from the Honolulu fish auction. These data were used to construct an extensive database on the weight and size of the fish sold and led to the length-weight relationships still used by fishery scientists today. This demonstrated the value of working cooperatively with the fishing industry as the auction and fishermen were not obligated to share this information.

The difficulty of deriving workable estimates of maxiumum sustainable yield and optimum yield from these unconventional data sources slowed the development of both the PMP and FMP. A fishery data workshop held in December 1977 provided statisticians at the Honolulu Laboratory access to a significant amount of historical Japanese longline data. Based on available information, the maxiumum sustainable yield for blue marlin was estimated at 22,000 tons; for striped marlin 22,000 tons; and for swordfish 20,000 tons (Shomura 1978).

To solve the third problem—what species to manage under the billfish plan—the NMFS Southwest Regional Office began an official determination as to which species would be considered "highly migratory," settling on marlin, oceanic sharks, sailfish and swordfish (and, eventually, tuna).

The PMP for Billfish and Oceanic Sharks in the Pacific Ocean was published on July 21, 1978. However, public concerns caused it to be withdrawn on September 14, 1978. Among the concerns were insufficient recognition of the socioeconomic differences among the human populations of Hawai'i, American Samoa and Guam and the lack of management measures for wahoo, dolphinfish and little dolphinfish (little mahimahi, *C. equisetis*) (NOAA 1978).

Meanwhile, in Hawai'i, Japanese longliners continued to take tuna and billfish within the U.S. FCZ until 1980 and reported catches of between 1,300 and 5,000 tons per year (Bienfang n.d.).

On April 1, 1980, the PMP was finalized and implemented. The intent of the PMP was essentially to regulate foreign fishing through a permitting system. The optimum yield was set at the estimated catch expected under the restrictions and limits imposed by the PMP until such time as the Council could complete the FMP and NMFS could determine a more science-based optimum yield (Skillman et al. 1993).

The PMP allowed and managed otherwise prohibited foreign longline fishing for pelagic species within the U.S. FCZ of the Pacific, excluding Alaska, under the following measures:

a) Allowed an unlimited amount of foreign longlining in the newly declared FCZ;

b) Specified quotas on the amounts of billfish and oceanic sharks;

c) Established non-retention zones from shore out to 12 to 100 nm (depending on the area) in which all billfish had to be released;

d) Provided a system of reserves if domestic vessels did not meet expected levels;

e) Required data reporting by foreign vessels; and

f) Subjected foreign vessels to U.S. observers for onboard inspection and sampling.

Because the non-retention areas were not as extensive as had been recommended, the Council was concerned that substantial foreign fishing might adversely affect local fishermen by causing localized stock depletion. The consequence, however, was a dramatic shift in fishing patterns. No foreign vessels ever fished in the FCZ under the PMP. The administrative requirements were simply too burdensome.

SOCIOECONOMIC CASE STUDY: AMERICAN SAMOA

The socioeconomic situation as regards foreign fishing was particularly unique in American Samoa. With a population of about 65,000 people, it is comprised of

77 square miles (199 square kilometers) of land and has 126 miles (203 kilometers) of coastline. American Samoa has been a U.S. Territory since 1900. When the Council was created in 1976, there was little formal regulation of the fishing industry, and what existed centered on two large tuna canneries and a small-scale artisanal fishing fleet located in rural, outlying areas. Pollution, runoff and waste came from the canneries and clouded nearby waters. Data collection efforts were rudimentary.

The American Samoans lived a largely subsistence existence. They had little financial wealth, and, because they did not have the economic means to gather large numbers of fish in other ways, they sometimes used dynamite and Clorox to harvest fish on coral reefs. Most of the local fishermen used canoes at that time, and few had modern vessels with engines and fishing equipment. In 1976, the management of ocean resources in American Samoa was an amalgam of emerging modern methods and the traditional village system.

From 1953, when the first canneries opened, the economy of American Samoa became 80% dependent on that industry. The tuna companies had situated their facilities there in order to take advantage of the inexpensive tuna that could be bought from fishing vessels—mainly foreign vessels from Japan, Taiwan and Korea—that operated in the waters of the western Pacific (including waters that became the U.S. FCZ around American Samoa).

It was, and still is, legal for foreign vessels to land their catches in American Samoa. Also, a special law enacted by Congress in the early 1950s allows tuna processed in American Samoa to enter the U.S. duty free if the foreign contribution to the final product is less than 50%. This made American Samoa canned tuna competitive with tuna canned in Hawai'i and California—even with the higher shipping costs involved. The canneries were soon the largest employer in the territory. But by 1976, cannery owners worried that changes to the status quo of tuna management resulting from the newly formed Council management system might disrupt their business and force them to leave American Samoa. Ed Stockwell, speaking for StarKist Samoa, pointed out at a Council meeting, "It was feared that the effect of the [Billfish] PMP restriction on the economics of longliner operation might be severe enough to cause the vessel operators to shift to another base" (WPRFMC 1978c).

These issues were uppermost on the minds of American Samoa fishermen and business owners when the Council held its first meeting in American Samoa (the Council's fourth), April 19 to 22, 1977. It was co-chaired by American Samoa businessman and Council member Lealaifuaneva Peter Reid Jr. When the proposed Billfish PMP was presented by NMFS at the meeting, it immediately raised issues for the canning companies. While Hawai'i and Guam Council representatives were asking for a billfish non-retention zone of 150 nm surrounding their islands, the people of American Samoa were informed by the managers of the tuna canneries that only a 12-nm area would be acceptable. The managers thought the larger area would have too great an impact on the longliners and purse seiners that brought fish into American Samoa.

American Samoa Governor Peter Tali Coleman (with bow tie) inspects the Van Camp cannery at Pago Pago (circa 1950s). In the early 1950s, Congress allowed tuna processed in American Samoa to enter the U.S. duty free if the foreign contribution to the final product was less than 50%, which made American Samoa canned tuna competitive with tuna canned in Hawai'i and California. *Coleman family photo.*

Lealaifuaneva Peter Reid Jr. at the 41st Council meeting in the Manu'a Islands, American Samoa, July 27, 1983. Reid served as the Council vice chair from American Samoa from 1976 to 1985, during the time when American Samoa called for a 12-nm rather than a 150-nm billfish non-retention zone so as to minimize impacts to the longline and purse-seine vessels that brought fish to the territory's canneries. *WPRFMC photo.*

A related concern was the proposed inclusion of wahoo (*Acanthocybium solandri*) and mahimahi (*Coryphaena hippurus*) as proposed managed species. Wahoo was canned on a fairly large scale in American Samoa and had significant economic importance as it was canned for domestic sale and consumption. The cannery managers worried that measures restricting foreign operations could reduce fish deliveries or raise their costs.

Other significant issues that affected the people of American Samoa involved the fisheries and controls of independent Pacific Island nations with bordering waters, such as Western Samoa and Tonga. As American Samoa Governor Uifa'atali Peter Tali Coleman (1956–1961, 1978–1985, 1989–1993) said at the 11th Council meeting, "Foreign fishing vessels can fish freely for tuna in the waters of American Samoa, but American Samoa-based vessels may be restricted from fishing within 200 miles of other islands. … There is concern over who will enforce the fishery jurisdiction of such countries as Western Samoa and Tonga" (WPRFMC 1978c).

Governor Coleman wanted parity and protection for American Samoa vessels that historically fished in the waters of its neighbors. He was particularly concerned about the U.S. Department of State's position on the Line and Phoenix Islands of Kiribati as well as the FCZ boundary near Tokelau, between the Line Islands and north of Swains Island. The concern was one of the first international issues that was effectively mediated by the Council.

PART TWO

THE FISHERY MANAGEMENT PLANS (1983–2009)

Chapter 4

Overview of the Species-Based Plans

In its first decade of existence, the Council worked with NMFS in the development of the PMPs and also began development of the more permanent FMPs to manage both foreign fishing within the FCZ as well as regional domestic fisheries seaward of state and territorial waters. The Council started with FMPs for those fisheries thought most in immediate need of conservation and management. Development of the FMPs ran on parallel tracks.

The first FMP was for spiny lobster fisheries of the Western Pacific Region (later renamed the Crustacean FMP). This FMP came into effect on March 9, 1983, implementing a suite of measures for the NWHI, including permit, data reporting and observer requirements within the FCZ waters around the main Hawaiian Islands, American Samoa and Guam. On the following day, President Ronald Reagan established the U.S. EEZ, which is now the term for the 200-nm zone previously known as the FCZ.

On the heels of the Crustacean FMP, the Precious Coral FMP came into effect on September 29, 1983, establishing the plan's management unit species, management area and the classification of several known beds.

The next FMP came into effect on August 27, 1986, for bottomfish and seamount groundfish, prohibiting the use of explosives, poisons, trawl nets, bottom-set gillnets and other protentially destructive fishing techniques; establishing a moratorium on the commercial harvest of seamount groundfish stocks at the Hancock Seamount; and implementing a permit system for fishing for bottomfish in the waters of the EEZ around the NWHI.

The following year, on March 23, 1987, the Pelagic Fisheries FMP was implemented to manage open-ocean species, such as billfish, wahoo, mahimahi and oceanic sharks. Measures prohibited drift gillnet fishing throughout the U.S. EEZ waters in the region and foreign longline fishing within certain areas.

The Council's early work culminated in the FMP for Coral Reef Ecosystem Fisheries—the nation's first ecosystem-based FMP—which was finalized in 2001 and implemented in 2004.

Implementation of the FMPs was only a start, however. For a quarter of a century, the Council would amend each FMP to take into account new scientific information, fishery developments and regulatory requirements—including the mandates of applicable laws, such as the Endangered Species Act (ESA), Marine Mammal Protection Act and National Environmental Policy Act and their amendments; presidential executive orders; and the growth of and fluctuations in the fisheries.

Additionally, the FMPs had to be amended to meet new requirements each time the FCMA was reauthorized. In 1990, the Act (now known as the Magnuson FCMA, or MFCMA) included tuna as a federally managed species.[16] In 1996, it became known as the Sustainable

---

[16]   The Act was officially retitled "The Magnuson Fishery and Conservation and Management Act" in 1980 by Public Law 95-561, to honor the late U.S. Senator Warren Magnuson, who was instrumental in developing the original legislation.

Fisheries Act (SFA) and included requirements to designate essential fish habitat (EFH) and define bycatch and fishing communities.

The SFA also had enhanced recognition of and support for indigenous fishing communities in the Western Pacific Region, through establishment of the following:

a) Western Pacific Community Development Program;

b) Western Pacific Community Demonstration Projects Program (CDPP);

c) Pacific Insular Area Fishing Agreements;

d) Marine Conservation Plans (MCPs); and

e) Western Pacific Sustainable Fisheries Fund (SFF).

These latter three items addressed concerns by the U.S. Territories that, unlike independent Pacific Island countries, they could not license foreign vessels to fish in the waters surrounding their islands.

PACIFIC INSULAR AREA FISHING AGREEMENT CASE STUDY: THE NORTHERN MARIANA ISLANDS

The United States had made no fishing agreements with foreign fishermen to operate in the U.S. Pacific islands, and, as a result, the U.S. Territories were not benefitting from the potential revenue from the fisheries. This was of particular concern for the fledgling CNMI government, which had few funds but could not engage in joint-venture licensing with foreign fishing fleets because, during its treaty negotiation with the U.S. government, it lost jurisdiction over both its 3-nm state waters and its 200-nm EEZ. To make matters worse, it would not have voting rights on the Council for many of the early years.

On March 14, 1979, the first governor of CNMI, Carlos S. Camacho, pleaded for economic relief during his opening remarks to the Council at its 16th meeting convened in Saipan, saying, "In this ocean of presumed abundance .... Fish, fish everywhere, but no fishing vessels, no capital, no processing plants, no suitable reefers, no commercially trained fishermen ... and maybe not very much commitment nor expectation." He then clearly laid out what the new CNMI administration hoped to accomplish:

1) Full participation and voting rights in the Council;

2) Fishery surveillance and enforcement by the U.S. government;

3) A marine resources assessment;

4) A solution to shark problems that were plaguing the industry;

5) The ability to use foreign hull-fishing vessels in the Northern Marianas (i.e., an exemption from U.S. fishing vessel documentation standards);

6) Capital obtained through the NMFS financial programs; and

7) Access to the federal Intergovernmental Personnel Act through the Cooperative Program for Fishery Manpower

Development Program, so the CNMI could get funds to bring in experts to train local youth to become fishermen (WPRFMC 1979, Attachment A).

In response to requests from the CNMI government and in an effort to develop fisheries in Guam and CNMI, in 1980 the Council proposed a Guam-CNMI integrated management plan. Although never implemented, this was nonetheless the first plan of its kind to develop commercial fisheries in these areas.

From 1980 to 1985, the NMFS Honolulu Laboratory sponsored the Resources Assessment Investigation of the Marianas Archipelago. The assessment focused on the potential for fisheries for tuna using a night handline technique and on species research into bigeye scad (*Selar crumenophthalmus*, known locally as atule), bottomfish and deepwater shrimp (*Heterocarpus* spp.) in Guam and the CNMI (Polovina 1981).

When CNMI's efforts to obtain the legal right to its EEZ were defeated in the federal district court for the Northern Mariana Islands by Chief Judge Alex Munson on August 7, 2003, the CNMI government began to look into the possibility of using a Pacific Insular Area Fishing Agreement as an alternative form of compensation. At the request of CNMI Governor Juan N. Babauta (2002–2006), the Council helped facilitate the completion of a fishing agreement in the CNMI in 2004. It was considered a means to obtain revenues from foreign fishing that could then be used to close the Puerto Rico Dump, an open dump site created by the U.S. military in Tanapag Lagoon. The dump site was under an U.S. Environmental Protection Agency administrative order, and it was believed that toxic waste was leaching into Saipan Lagoon. The fishing agreement was never signed.

In order to provide benefits not realized through any Pacific Insular Area Fishing Agreement to the U.S. Pacific Territories, the SFA required each territory to develop a three-year MCP that would identify how funds deposited into the SFF would be utilized. Such funds could come from a fishery agreement and/or from other specific sources (e.g., fines/penalties for illegal foreign fishing in the U.S. EEZ around the Pacific Remote Island Areas and contributions received in support of conservation and management objectives). The Council would work with the governors to ensure that the MCPs were compatible with the FMPs. Monies from the funds would be made available to the Council to implement approved MCPs for the U.S. Pacific Territories.

In 2006, the FCMA celebrated its 30th anniversary and was reauthorized as the MSA. Among the new responsibilities for the regional fishery management councils was the requirement to manage all federal fisheries (with a few exceptions) through annual catch limits (ACLs) by 2012. This posed particular challenges in the Western Pacific Region, which has thousands of coral reef fish species about which little is known. More research was needed about their life histories and stock status.

The 2006 reauthorized MSA also established, through Section 306j, the Western Pacific Marine Education and Training Program, which put formal emphasis on education and outreach activities that the Council had been undertaking since its inception.

Past and current Council chairs and early Council members pose during a gathering on March 12, 2007, in Honolulu to celebrate the 30th anniversary of the Council, the reauthorization of the Magnuson-Stevens Act and the 200th anniversary of NOAA. Pictured (from left) are Edwin Ebisui Jr., Frank Farm Jr., Jim Cook, Paul Stevenson, Lealaifuaneva Peter Reid Jr., Wadsworth Yee, Frank Goto, William Paty Jr., Kitty M. Simonds (executive director), Frank McCoy, Judith Guthertz, Peter Fithian, Sean Martin and Roy Morioka. *WPRFMC photo.*

Council actions in the coming decades would also have to take into account the increasing involvement of the United States in international regional fishery management organizations. In 2004, the Western and Central Pacific Fisheries Convention entered into force with the United States as a member and with American Samoa, Guam and the Northern Mariana Islands as Participating Territories. In 2005, the U.S. Senate ratified the Antigua Convention, a 2003 treaty that strengthened the Inter-American Tropical Tuna Commission and set forth legal obligations and cooperative mechanisms for the long-term conservation and sustainable use of highly migratory species (such as tuna and swordfish) in the Eastern Pacific Ocean (EPO).

Between the late 1980s and 2009, the five FMPs would incorporate nearly five dozen amendments, regulatory amendments and framework measures to address developments in the fisheries, knowledge about them and the mandates of the MSA, executive orders and applicable law. Additionally, the Council would be actively implementing the new indigenous, fishery development and education opportunities provided for in the 1996 and 2006 MSA reauthorizations.

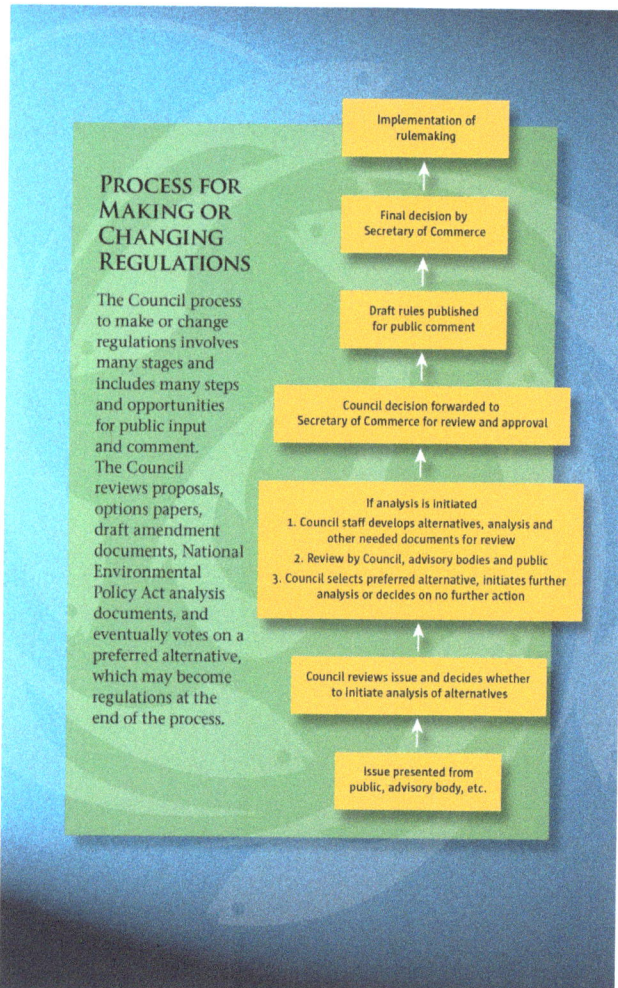

PROCESS FOR MAKING OR CHANGING REGULATIONS

The Council process to make or change regulations involves many stages and includes many steps and opportunities for public input and comment. The Council reviews proposals, options papers, draft amendment documents, National Environmental Policy Act analysis documents, and eventually votes on a preferred alternative, which may become regulations at the end of the process.

Implementation of rulemaking

↑

Final decision by Secretary of Commerce

↑

Draft rules published for public comment

↑

Council decision forwarded to Secretary of Commerce for review and approval

↑

If analysis is initiated
1. Council staff develops alternatives, analysis and other needed documents for review
2. Review by Council, advisory bodies and public
3. Council selects preferred alternative, initiates further analysis or decides on no further action

↑

Council reviews issue and decides whether to initiate analysis of alternatives

↑

Issue presented from public, advisory body, etc.

The Council procedure for making or changing federal fishery regulations in the Western Pacific Region is a bottom-up process that provides fishermen and the public ample opportunities to participate. *WPRFMC illustration.*

During the last six years of the FMPs, the Council would also have to work with a new NMFS regional office and science center. In April 2003, NMFS transferred the responsibility for managing marine resources in federal waters surrounding the U.S. Pacific Islands from the Southwest Regional Office and Fisheries Science Center to the newly established Honolulu-based Pacific Islands Regional Office and Pacific Islands Fisheries Science Center. The newly created Pacific Islands Region was temporarily headquartered in downtown Honolulu before moving in March 2104 to the newly built Daniel K. Inouye Regional Center on Ford Island, located in Pearl Harbor, Oʻahu.

Finally the policy-making responsibilities and offices for the region were now all located in Honolulu. The Pacific Islands Fisheries Science Center, Pacific Islands Regional Office, the Council and the Pacific Islands Division of the NOAA Office of Law Enforcement would collaboratively monitor, conserve and manage marine fisheries in the region. NOAA also established a General Counsel's office in Hawaiʻi to support the Pacific Islands Fisheries Science Center, Pacific Islands Regional Office and the Council.

The Northwestern Hawaiian Islands—
Jurisdictional Challenges and Discovery of Fishery Resources

The NWHI, also known as the Leeward Islands, were said to have been used as a kind of backyard larder for the communities of Kauaʻi and Niʻihau. According to island traditions, residents would frequent the NWHI to fish and gather food. An ancient fish hook found on Necker Island (Mokumanamana) during the 1920s, by an archaeological expedition of the Honolulu-based Bishop Museum, provides evidence of long-standing Native Hawaiian presence in the area, which may stretch back a thousand years (Kikiloi 2012).

This ancient Hawaiian *heiau* (sacred site) of upright stones on Mokumanamana (Necker Island) is evidence that Native Hawaiians have used the NWHI for millennia. A fish hook and other artifacts found on the island indicate that the island was used as a place to live, worship, fish and gather resources. *Bishop Museum photo.*

Through much of the 19th and 20th centuries, arguments over control of the NWHI took place between Washington, D.C., interests—namely the Department of the Interior and the U.S. Navy—and the Kingdom/Territory/State of Hawai'i. The Territory and, subsequently, the State of Hawai'i was the trustee for the Native Hawaiian people of the lands (including submerged lands) that the Territory of Hawai'i ceded to the United States. From the perspective of Hawai'i, the owners of the NWHI and its resources are the people of Hawai'i and not the federal administrators. So, when President Theodore Roosevelt declared a part of the area as the Hawaiian Islands National Wildlife Refuge in 1909 to protect birds there, whatever the merits of his intent, it was viewed by some as an unjust taking of lands from the people of Hawai'i without compensation.

John Craven, the marine affairs coordinator for Governor Burns, recalled the governor's position on the NWHI. "He was very clear. He said that these islands belonged to the Kingdom of Hawai'i and the people of the State of Hawai'i, and he was not going to give them up under any circumstances to the federal government."[17]

Under the administration of Burns' successor, Governor Ariyoshi, the growth of Hawai'i's fisheries was made a priority in the State's economic development plan. One goal was to revive a 1950s effort to establish a fishing base and re-supply point using the World War II airfields at Midway and Tern Island. The members of the newly formed Council supported the State's efforts to develop a fishery in the NWHI, thus putting it in conflict with the Interior Departmnent's USFWS, which claimed the lands and surrounding waters and allowed only scientific research there. However, the boundaries of the USFWS-managed Hawaiian Islands National Wildlife Refuge were unclear, and the USFWS's constant attempts to expand its holdings created conflict. Attempts to reach an agreement between the State of Hawai'i and the USFWS were not successful during the governorships of Burns or Ariyoshi, primarily because of the State of Hawai'i's insistence on joint management authority (WPRFMC 1978a).

The USFWS administrators believed that the original intent of the Hawaiian Islands National Wildlife Refuge was to protect the wildlife (mostly birds) contained therein. However, they also believed that management jurisdiction should not be restricted to land but should also include the ocean waters and submerged lands from which the birds sought food. To Craven, a member of the Council's SSC, the USFWS expansionist actions could be characterized as "jurisdictional creep," which he explained occurs when a federal agency with sovereignty over an ocean area "seek[s] to expand and control all it can."[18]

---

[17] Craven op. cit.

[18] Craven op. cit.

Over time, the USFWS authority over the Hawaiian Islands National Wildlife Refuge resources evolved from successive interpretations of the Migratory Bird Act, the National Wildlife Refuge Administration Act and the ESA. The Migratory Bird Treaty with Japan specified that the USFWS must protect the ecological balance of unique island environments. Hence, according to legal interpretation, any action by the USFWS that allowed the depletion of food supplies for birds or disturbances of nesting sites would be considered a treaty violation. Eventually the USFWS came to see the entire area of the NWHI, not just the Hawaiian Islands National Wildlife Refuge, as a kind of real-life science laboratory, where research on biological interactions could take place.

In 1977, the year after formation of the Council, the USFWS contracted a detailed coral reef modeling study in the NWHI that compared untouched reefs to those impacted by development. That same year, the USFWS also conducted a study on shark predation on monk seals, paid for by the Marine Mammal Commission. The USFWS worked to expand its authority over the entire NWHI. In 1978, the U.S. House of Representatives passed H.R. 1907, proposing the placement of 302,000 acres of the NWHI, including submerged land, under the administration of the USFWS as part of a new National Wilderness Preservation System. The Council vigorously objected. Chair Yee saw H.R. 1907 as a direct challenge to the future of State ownership of the NWHI and as a threat to continued fishing in the NWHI. He encouraged the Council to register "an immediate expression of deep concern to the Hawai'i Congressional Delegation" (WPRFMC 1978a).

At the time, the State's official position on the issue was as it had been under Governor Burns—the submerged lands in the NWHI were all subject to the State's jurisdiction and were not to be transferred to the USFWS or any other federal agency. Hawai'i State Deputy Marine

Affairs Coordinator Stanley Swerdloff explained at the Council's 10th meeting in 1978 that, as part of the effort for the Interior Department to steadily expand control over the Leeward Islands, the Marine Mammal Commission was seeking to establish a 3-nm critical habitat for the monk seal, which could result in further loss of fishing rights in this area. From his perspective, *what was at stake was not only the rights to submerged lands within the NWHI but also the future long-term rights to the State's territorial sea in the NWHI, which he warned could one day be taken from the State by the federal government without compensation.* He said that future efforts would be taken to determine the constitutionality of the actions taken by the federal government (WPRFMC 1978b).

At the same Council meeting, Hawai'i State Senator T. C. Yim spoke in defense of the State of Hawai'i's rights on the issue saying there was "no reason for the people of Hawai'i to be treated like second-class citizens. There is certainly no need for the Washington bureaucrats to send experts out here to tell us how to run things in our own islands." He appealed to state officials to make plans for a campaign to reestablish the State's rights and promised that the State would hire the best legal talent available to help in the fight. The senator rejected the concept that "the Great White Father knows what is best for the islands." Council minutes show that his statements were "loudly applauded." Roy Mendelssohn of NMFS Southwest Fisheries Center pointed out in reply that Hawai'i was not the only state that had lands taken over by the federal government for national parks and preserves (WPRFMC 1978b).

Unbeknownst to the early Council members, the federal government's incremental removal of fishing access to waters and submerged lands in the name of preservation would eventually encompass more than half of the entire Western Pacific Region, beginning with the NWHI.

The impetus for the MSA was the desire for economic growth for U.S. fishermen, fish processors, seafood salesmen and ancillary businesses. Part of the early work of the Council and NMFS was discovering what fishery resources existed in the region. During the late 1970s, approximately $30 million, which became known as "extended jurisdiction" money, was allocated annually by the federal government for such research. Under Honolulu Laboratory Director Richard Shomura, this money—combined with funds from the University of Hawai'i Sea Grant College Program—supported the first large-scale scientific study of the NWHI, using the latest methods available to assess and develop new fisheries. Known as the NWHI Fishery Investigations, or the Tripartite Study, it was conducted by three agencies: NMFS, the Hawai'i Division of Fish and Game and the USFWS. Although the Council wanted to participate in the study because the work would include stock assessments, Brent Giezentanner, director of the USFWS Hawaiian Islands National Wildlife Refuge, made it clear that USFWS did not want the Council involved, even though the State did (WPRFMC 1978b). The USFWS intended to create a master plan for the Leeward Islands.

Approximately $10 million was spent over eight years on the NWHI research program, and it resulted in the publication of 115 papers. In addition, two symposia were presented, the first on April 24 and 25, 1980 (Grigg and Pfund 1980) and the second from May 25 to 27, 1983. The research program was not open-ended. It was organized with the intent of collecting data from the NWHI that could be used for effective management of the area's fishery resources. Long-range scientific planning was done through a tripartite Council for Coordinating Research in the NWHI, which set research priorities and coordinated spending.

Among the research priorities were those that provided baseline information for recovery plans for the Hawaiian monk seal and the

Hawaiian green sea turtle; detailed information needed for the creation of the Bottomfish, Crustacean and Pelagic FMPs; and much original research on corals. There was, at the time, an enormous sense of optimism, a sense that after so many years during which the Hawai'i fishing industry had gone through economic retrenchment, new opportunities were at hand. As the cooperative studies increased during the 1980s, many on the Council and in the scientific community felt that commercial fisheries and natural resource management could co-exist in the NWHI. Conservationist nongovernment organizations (NGOs), however, opposed the idea of commercial fishing in the region.

The encouraging news for the local fishing industry was the discovery of a potentially large lobster resource in the nearby NWHI by a NMFS survey vessel in 1975. In addition, the State of Hawai'i implemented fishing vessel loan programs throughout the early 1970s to encourage investment in commercial fishing and, in 1979, published a fishery development plan focused on these and other infrastructure needs (Hawai'i State 1979).

Issues Regarding Indigenous Fishing Communities

Unlike many other Native American peoples, Native Hawaiians and the Chamorro of Guam do not have a formal treaty with the United States that specifies their rights under the law, nor are they recognized as a distinct people and culture. In contrast, however, under Kingdom, Territorial and State law, the Native Hawaiians are entitled access to certain fisheries, based on traditional and customary practice.

When Native Hawaiian issues, such as preferential rights to resources, were brought up at Council meetings by fisherman Alika Cooper and others in the late 1970s and early 1980s, federal officials were often reluctant to address them. Such matters raised too many uncomfortable questions for which mid-level NMFS and USFWS officials had no answers. The Council was often the only venue where indigenous peoples in the Western Pacific Region could discuss in a formal government setting their issues and traditions without being subject to ridicule, such as their taking and eating green sea turtles, a species that was placed on the ESA threatened species list on July 28, 1978, because of the decline in its population.

Hawaiian people cut a captured turtle, Hilo, Hawai'i (circa 1890–1905). The Council has often been the only venue where indigenous peoples in the Western Pacific Region could discuss their issues and traditions in a formal government setting without being subject to ridicule. *Bishop Museum photo.*

The Council has demonstrated support throughout its history for "home rule" of its territorial and commonwealth members. When the EEZ of the CNMI was federalized to the high water mark on land, the Council allowed and encouraged its CNMI members to make regulations for 0 to 3 nm in what otherwise would have been the Commonwealth's waters.

The Council has long recognized and supported the region's indigenous fishing. It provided support for an archaeological analysis of ancient fishing village sites in Guam. The results shed light on pre-contact fishing and fish consumption and pointed to interesting connections with prehistoric sites in Taiwan (Amesbury and Hunter-Anderson 2003, 2008). From November 28, 2009, to December 31, 2009, the Council partnered with the Guam Fishermen's Cooperative Association to put on an exhibition of Guam cultural cuisine at the Guam Community College. Along with cuisine, the exhibition displayed seafood, plants, animals and artifacts important to Chamorro culture.

Among the Council's education and outreach efforts would be many that supported indigenous knowledge, including their traditional methods of time keeping. Lunar calendars play an important role in the lives of people in traditional cultures as the calendars help connect the people with natural changes that occur in the environment. While Native Hawaiians had retained knowledge about their traditional lunar calendar, other U.S. Pacific Islanders had lost much, if not all, of such knowledge. The Chamorro in particular were impacted by more than 400 years of Spanish, Japanese, German and U.S. governance, during which their language was at times forbidden.

To ensure the accuracy of the traditional lunar calendars that it produced, in 2011 the Council held a Traditional Lunar Calendar Workshop in the CNMI with participants from throughout the Western Pacific Region and other U.S.-affiliated Pacific Islands (e.g., Palau and Yap). Subsequently, the Council contracted the American Samoa Community College's Samoan Studies Institute to conduct research in American and Western Samoa. It also contracted a professional group to conduct focus groups and interview surveys throughout the region on the effectiveness of the calendars, with funding assistance from the NOAA Coral Reef Conservation Program. The traditional lunar calendar endeavor was so successful that the Council has continued to produce lunar calendars annually for each island area in both English and the indigenous languages of the region.

The Council's support for communities and indigenous people has remained steadfast. The formula used includes the following:

a) Determine the local communities' aspiration for their resources;

b) Help guide the realization of that aspiration;

c) Support the recognition of indigenous rights; and

d) Seek to deliver those rights for the benefit of the entire fishing community under a science-based decision-making process.

FISHING COMMUNITIES CASE STUDY:
THE ʻAHA MOKU SYSTEM

In Hawaiʻi, in 2006 and 2007, along with the Office of Hawaiian Affairs, the Hawaiian Civic Clubs, the Hawaiʻi Tourism Authority and Kamehameha Schools, the Council cosponsored a conference series on traditional ocean management practices of Native Hawaiian people (Simonds et al. 2008). The Hoʻohanohano I Na Kupuna Puwalu (Honor Our Ancestors Conferences) eventually grew into the ʻAha Moku traditional management initiative. In 2007, the initiative culminated in Act 212 in the Hawaiʻi State legislature, which recognizes the work done to identify the traditional ʻAha Moku system of natural resource management in Hawaiʻi. The Act further established an ʻAha Kiʻole Advisory Committee to advise the legislature on best practices for traditional management of natural resources in the islands. The ancient system in Hawaiʻi divided the lands and waters into moku (traditional land divisions), whose natural resources were managed by a moku council. The 'Aha Moku Council initiative reaffirmed the rights of local tenants to manage the resources in their district and provided the Council with an opportunity to engage with the Native Hawaiian and fishing communities through an officially recognized organizational structure. In the 2009, Senate Bill 1108 extended the official life of the 'Aha Ki'ole Advisory Committee through Act 39 to June 30, 2011. On April 30, 2009, Hawai'i Governor Linda Lingle (2002–2010) allowed Act 39 to pass into law without her signature.

On November 19 and 20, 2010, the Council partnered with the 'Aha Ki'ole Advisory Committee and the Office of Hawaiian Affairs to sponsor the Ho'o Lei Ia Pae'Aina puwalu, a public forum of 200 Hawaiians, fishermen and community members. Six months later, on May 3, 2011, the Hawai'i State legislature unanimously passed a bill to establish the 'Aha Moku Advisory Committee to advise the Hawai'i Board of Land and Natural Resource on issues related to land and natural resources management through the 'Aha Moku system. Hawai'i Governor Neil Abercrombie (2010–2014) on July 9, 2012, signed Act 288 into law, which formally established the 'Aha Moku Advisory Committee, placed within the Hawai'i Department of Land and Natural Resources. Since then, numerous 'Aha Moku meetings have convened at the island and archipelago level, many supported by the Council, to further the 'Aha Moku system. Reestablishment of this traditional system could help the Council facilitate the consultation process with the indigenous communities, incorporate their place-based intergenerational knowledge in the fishery management decision-making process and make management more adaptive. With Council support, a history of the ʻAha Moku initiative, *Traditional-Based Natural Resource Management: Practice and Application in the Hawaiian Islands,* was published by Palgrave Macmillan (Glazier 2019).

Supporters of the ʻAha Moku system surround Hawaiʻi Governor Neil Abercrombie following his signing into law Act 288, relating to Native Hawaiians, on July 10, 2012, at the Hawaiʻi State Capitol. The law officially recognizes the traditional ʻAha Moku system of natural resource management and established the ʻAha Moku Advisory Council. *WPRFMC photo.*

Informing the Public and Policymakers about Federal Fisheries and Their Management

The Council went to considerable effort to educate fishermen, the general public and policymakers about the FCMA and the region's fisheries issues and to encourage them to engage in the federal decision-making process. The Council began publishing a newsletter as early as 1977 (Rizzuto 1977).

The *Pacific Islands Fishery News* evolved into its present format in 1984 and eventually became a quarterly publication of Council activities and issues of interest to fishing communities in the region. In 1989, the Council started hosting Fishers Forums as a part of the Council meetings, scheduling them in the evenings and at venues accommodating to fishermen.

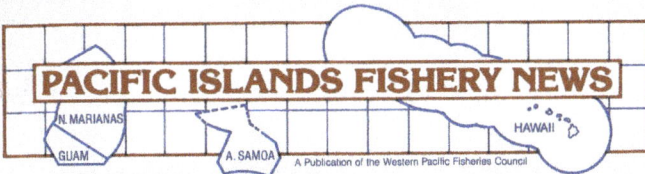

PACIFIC ISLANDS FISHERY NEWS

N. MARIANAS · GUAM · A. SAMOA · HAWAII

A Publication of the Western Pacific Fisheries Council

Council Has Two Jobs:
Fishery Development AND Conservation

The Western Pacific Fishery Council is the westernmost of the eight U.S. fishery management councils.

The main thrust of the Council is to prevent the over-fishing of heavily-fished species, and to limit foreign fishing within the 200-mile management zone.

Hawaii, up to 1976, was a prime example of the conditions which led to regulation.

At that time foreign boats were taking 20 to 30 million pounds of fish a year within Hawaii's 200-mile zone. Hawaii fishermen were taking only 10 to 15 million pounds.

Today the Council is increasingly aware of a second need. That is to encourage the fishing—by American fisherman—of underused species.

Hence the Council is involved both in fishery conservation and development. The Council's stance depends on what is appropriate for each species.

A 200-mile management zone in an area as huge as the Pacific may seem relatively insignificant.

But the total area is 1.5 million square miles of ocean. It includes the 200 miles offshore from all the islands, atolls and reefs under American jurisdiction in the Pacific.

The entire Hawaiian archipelago, American Samoa, Guam and the

Pacific Fishery Council Gets New U.S. Standard for Mercury

When the Western Pacific Fishery Management Council recently convinced the Federal regulators to give billfish fishermen a fair break on mercury, it set a new nationwide standard for all fish.

The new standard is 1 part per million methyl mercury (the most toxic form), as opposed to the old standard of 1 ppm total mercury.

As a result, the threat of a ban on billfish sales by the Hawaii State Department of Health has lifted. This ban was actually applied briefly in July of 1984 to all Pacific Blue Marlin and other billfish over 200 pounds.

But two commercial fishing companies, the United Fishing Agency (Honolulu) and Suisan, Inc. (Hilo) took the State to court and won a temporary injunction.

Taking advantage of this time, the Fishery Management Council set to work. It marshalled all the latest research on mercury accumulation, supporting the view that the more toxic methyl mercury could be measured separately from the less toxic inorganic mercury.

Then Wadsworth Yee, Council (continued inside)

... the total area is 1.5 million square miles of ocean ... all the islands, atolls and reefs under American jurisdiction.

Commonwealth of the Northern Marianas Islands are all part of it—plus such smaller islands as Wake, Jarvis, Baker, Palmyra, Kingman Reef and Johnston Atoll.

When you see it mapped (see fold-out, inside), the importance of the 200-mile zone around the American Pacific islands becomes more apparent.

The Council is made up of a cross-section of industry, government and scientific concerns.

On the Inside:
Limited Entry
Tuna
Promotion
Pacific Map

The Council's quarterly newsletter, *Pacific Islands Fishery News*, keeps fishermen and the public updated on recent Council actions and events of interest. The cover story of the 1984 issue highlighted the region's goals to prevent overfishing and to encourage the harvest of underutilized species by local fishermen. *WPRMFC photo.*

The Council formalized its outreach efforts in 1995 with the launch of its *Public Involvement and Outreach Plan*. The plan's overarching goal was to deliver the message of "sustainable fisheries" so as to ensure their existence for future generations. The plan's aim was threefold: 1) to promote awareness of biological, environmental, economic and social factors of or affecting the fisheries; 2) to establish an understanding of the relationships among the Council and other entities that share responsibility in managing the resources; and 3) to promote fishery diversification. It was to be implemented first in Hawai'i and then the most effective activities were to be tailored and applied to American Samoa, Guam and the CNMI. Fishing communities, the general public and regulatory/policy-setting agencies comprised the target audiences. The reason for targeting the general public was to promote community-based involvement in the management of the resources. School-aged children were seen as future fishermen, policy makers and resource managers and so were a part of the general public audience.

Using television and radio to reach fishermen and the public, the Council produced a series entitled *Fishing in Old Hawai'i*, which aired for several years on *Let's Go Fishing*, a popular television program viewed by fishermen. The Council also contributed regular columns and informational ads for publication in *Hawaii Fishing News, Lawai'a Hawaii Skin Diver* and *Marianas Fishing Magazine*.

Further, the Council has participated in the annual Hawaii Fishing and Seafood Festivals events organized by the Pacific Islands Fisheries Group in Honolulu since 2006. It was estimated that 11,500 people attended the first year, and the event grew in popularity each year, drawing 20,000 attendees in 2013. One of the most popular events at the 2007 Seafood Festival was that of food preparation and food safety.

Another community outreach effort, kick-started by the Council, was a fish drive to help feed Hawai'i's needy. The fish were donated to the Hawaii Foodbank as part of World Fisheries Day on November 20, 1999, and a proclamation recognizing the day was signed by Governor Benjamin J. Cayetano (1994–2002). Similar efforts to engage the fishing community to donate fish to feed the needy have been continued by the Hawai'i Fishermen's Alliance for Conservation and Tradition, spearheaded by former Council member Roy Morioka.

The Council's education and outreach efforts also targeted policymakers, including those in Washington, D.C., whose decisions have had a dramatic impact on the fisheries at the regional, national and international levels. The Council has regularly participated in the annual NOAA Fish Fry and Capitol Hill Ocean Week and has worked with the other seven regional fishery management councils and NMFS to orchestrate three Managing Our Nations Fisheries conferences in 2003, 2005 and 2013 (Witherell 2004, 2005; Gilden 2013). These conferences were particularly important during reauthorizations of the MSA and when new administrations came onboard with staff members who were unfamiliar with the nation's fisheries, their management and their issues.

The communication staffs of the eight councils worked together as a subcommittee of the Council Coordination Committee, which is comprised of the chairs, vice chairs and executive directors of the eight councils. The communications group maintains an all-Council website (fisherycouncils.org), which archives documents from the Council Coordination Committee's biannual meetings, as well as outreach and education products developed by the group. The proceedings of the Council Coordination Committee's national Scientific Coordination Subcommittee workshops are also on this website.

Council Executive Director Kitty M. Simonds, Secretary of Commerce Donald Evans, NOAA Administrator Conrad C. Lautenbacher and Honolulu Chef Russel Siu serve up dishes from the Western Pacific Region at the 2003 NOAA Fish Fry. The Council's education and outreach efforts were extended to policymakers in Washington, D.C., as their decisions can dramatically impact the region's fisheries. *WPRFMC photo/Sylvia Spalding.*

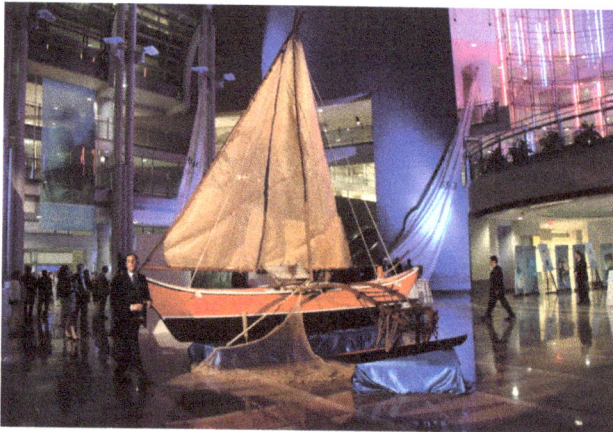

A Chamorro *proa* (traditional sailing canoe) on display at the 2012 Capitol Hill Ocean Week gala in Washington, D.C., presented an opportunity for politicians and environmentalists to learn about the indigenous cultures of the U.S. Pacific Islands. The Council organized the display with support from Matson Inc. and Tradition about Seafaring Islands. *WPRFMC photo/Sylvia Spalding.*

Responding to Demands for Diversity

One of the initial questions about educating islanders to become fishery scientist and managers was raised by a fisherman named Alika Cooper, who served at the time as a Council member. At the 54th Council meeting held from August 6 to 8, 1986, in Kailua-Kona, Hawai'i, he asked why no Native Hawaiians were working at the NMFS Honolulu Laboratory. Out of 75 full-time employees at the Honolulu Laboratory in 1977, not a single one was Native Hawaiian.

Izadore Barrett, the NMFS Southwest Fisheries Science Center representative, replied that Native Hawaiians were not hired because few were qualified (WPRFMC 1986b, p. 9). Although his answer, taken in context, was not meant to offend, it reinforced the perception that Native Hawaiians and other indigenous people in the region had been marginalized from much of the decision-making process of fisheries resource management, an area that affected them both culturally and economically. Barrett pointed out that few Hawaiians or other islanders were attending schools that trained students in marine biology and related fields. He claimed that NMFS had at one time searched for six months to find a qualified Native Hawaiian candidate for a position open in the Southwest Region.

Since its inception, the Council has actively pursued programs to support the formal education and training of Native Hawaiians, Chamorro, Carolinians and American Samoans so that they can play meaningful roles in Pacific fishery management. Council Executive Director Kitty M. Simonds said she could have hired anyone for these positions; however, over a number of years, she found young people from the indigenous communities whom she thought would be interested in and qualified for this kind of work. She trained them and put them in positions where, in the future, they would be able to make decisions on ocean issues affecting Native Hawaiian and Pacific Islander communities.[19]

Council efforts have included creation of fisheries curricula for classrooms. In 2002, the Council produced the *FishQuest* series in collaboration with Hawai'i Department of Education, Hawai'i Public Television and Pacific Resources for Education and Learning. Aired originally as a live interactive broadcast series, the programs on "Fishing for Facts," "Fishing for Food" and "Fishing for Solutions" were also made available in print and DVD. In 2009, some of the lessons were updated with National Science Education Standards and Ocean Literacy objectives and made available as part of a special issue of the National Marine Educators Association's *Current— Journal on Marine Education Current* (Spalding et al. 2009).

Students show cut-out copies of common fish species caught in the Western Pacific Region during production and live airing of *FishQuest* in 2002. The Council-sponsored television series featured a Hawai'i Department of Education teacher interviewing guest speakers, showing video clips and taking online calls from classrooms to help educate students about the region's fisheries and their management. *WPRFMC photo.*

---

[19]   Simonds op. cit.

Council efforts to address the full range of students from K–12 have included sponsorship of outreach programs, such as the Kewalo Keiki Fishing Conservancy (beginning in 2004) and art, poster and photo-essay contests on fisheries-related themes (beginning in 2005). The first contest began as a campaign to educate the fishing community about the Council's move toward ecosystem-based fisheries management by engaging students and, through them, reaching the parents. Teachers were provided with lesson plans to help students understand the concepts behind the contest's theme. Local businesses provided awards; local governments recognized the winners; and the local media publicized the results.

To extend the reach to the fishing community and larger public, the Council featured the winning art on traditional lunar calendars. The calendars led to the creation of Traditional Lunar Calendar festivals on Guam and CNMI, which for a decade became major attractions for residents, fishermen and visitors.

At the 2011 Chamorro Lunar Calendar Festival in Guam, freshman Ralph E. S. Patascsil of Father Duenas Memorial High School shows his winning artwork featured on the cover of the Council's 2011 *Fanha'aniyan Pulan CHamoru* (Chamorro lunar calendar). *WPRFMC photo/Sylvia Spalding.*

103

Since 2006, each summer the Council has provided annual marine resources and fishery management courses for high school students, offering one course in each of the island areas under its jurisdiction. Initially, these courses received funding support from the NOAA Coral Reef Conservation Program. In 2009, after the 2006 MSA reauthorization established the Western Pacific Marine Education and Training Program, the Council requested that the program provide support for these courses. The course in Hawai'i allows students to earn a credit in science from the State's Department of Education.

Students attending the Council's 2014 high school summer course on marine resources and fisheries management in the CNMI prepare for a boat excursion. Held annually throughout the region since 2006, the course combines hands-on and classroom learning to encourage students to pursue an education in fisheries science and management. *WPRFMC photo.*

The Council was also instrumental in the formation of the Traditional Knowledge Committee of the National Marine Educators Association, which offers an annual Traditional Knowledge Scholarship, and in the creation of the International Pacific Marine Educators Network (Simonds et al. 2008).

Anaseini Ban delivers a presentation on the Marine Environment Education Program via the web from Papua New Guinea during the International Pacific Marine Educators Conference held January 2007 in Honolulu. Organized by the Council, this gathering of more than 100 educators from 18 Pacific countries led to the creation of the International Pacific Marine Educators Network and similar regional networks in Asia, Europe, Latin American and Canada. *WPRFMC photo/Sylvia Spalding.*

Chapter 5

Crustacean Fisheries (1983–2009)

5.1 The Northwestern Hawaiian Island Lobster Fishery

Discovering the Fishery

In 1975, the discovery of large, virtually untouched stocks of spiny (*Panulirus marginatus*) and slipper (*Scyllaridies squammosus*) lobsters in the NWHI caused great excitement. In an otherwise sluggish local fishery, an opportunity arose for commercial fishermen to make substantial profits supplying slipper lobsters to Hawaiʻi's booming hotel and restaurant industry. A Gold Rush mentality spurred local and U.S. mainland fishermen to set traps in the newly discovered beds.

Some experienced local fishermen and state government officials warned of negative consequences if firm rules were not put in place immediately with regard to the lobster fishery. It was widely acknowledged that little was known about the lifecycles of Hawaiʻi's subtropical lobsters and that marine resources are vulnerable to rapid overfishing in Hawaiʻi's limited undersea habitat. It was a prescient concern that was to involve the Council for the next 12 years and set in motion a chain of events that would have a lasting effect on Hawaiʻi.

At the second Council meeting on December 16, 1976, Michio Takata, then director of the Hawai'i Department of Fish and Game, said that lobster fishing was taking place in the NWHI and that creating a management plan for it should be a priority of the new Council. He warned that the lobster stocks needed to be regulated or they would be quickly depleted (WPRFMC 1976a). Takata's concerns would be taken up by newly appointed Council member and long-time local commercial fisherman Buzzy Agard, who in 1977 warned of "disturbing rumors of waste of the catches through death and spoilage of the lobsters" and expressed concern that the resource "might be injured before adequate data have been acquired for its management" (WPRFMC 1977b). Approximately 77,000 lobsters were reported as having been caught that year.

Spiny lobster is one of two lobster species harvested commercially in Hawai'i. Discovery of lobster beds in the NWHI in the 1970s attracted local and U.S. mainland fishermen to the area as the newly established Council worked to develop a management plan for the fishery in short order with limited data. *Bruce Mundy photo.*

Slipper lobster is known in Hawai'i as ula papapa. It would become an important part of the NWHI lobster fishery. *Bruce Mundy photo.*

Unlike long established lobster fisheries on the U.S. mainland, where continental shelves can stretch for hundreds of miles and provide habitats for very large populations of lobster, NWHI lobsters grow in isolated clusters on small banks that surround low islands and pinnacles. Their lifecycle in the subtropics is very different from the continental lobsters and was much less studied. The Hawai'i spiny and slipper lobsters in the NWHI represented a large but virtually untapped resource. Nonetheless, the growing consensus among Council members was that, no matter how abundant the stocks, the technology available to modern commercial fishermen could potentially exhaust them in a relatively short period of time.

As former Council Chair Yee remembered, "We knew we had to do something because people were already going up there [to harvest lobsters]. But we were new. We were groping as to how to best work with the fishermen for the overall conservation and replenishment of the stock."[20]

---

[20]   Yee op. cit.

The issue for managers was how to manage a stock sustainably within the context of an economically profitable fishery. Knowledge of spiny and slipper lobster biology was so limited in 1976 that no one knew how they reproduced. The key question posed to the Council was "What number of lobster can be removed from the fishery without damaging the stock's ability to sustain its populations?" Unlike commercial fishery managers in New England, the Western Pacific Council did not have years of lobster research data on which to base their decision.

To deal with these concerns, the Council created a Crustacean Plan Team under the direction of Tim Smith from NMFS. The team worked with consultants Paul Struhsaker and then-graduate student Craig MacDonald to develop policy recommendations regarding spiny and slipper lobster harvesting.

Early on, the Crustacean Plan Team shared concerns that the lobster fishery could be easily overexploited. Intense discussions took place between the SSC and the plan team over how to set rules that would protect the lobsters from overexploitation while ensuring the optimum yield would be reached. Since little about the biology of the spiny lobster was understood, the maximum sustainable yield (and optimum yield) could not be quantified with much confidence. The Council and its committees worked to find data from which initial calculations could be made.

The Council members and the Crustacean Plan Team wanted the Hawai'i lobster fishermen to help them in their research by keeping accurate catch data logs. However, according to Takata, there was a regulatory impediment: the State had issued permits for vessels operating only within state waters in the NWHI, from French Frigate Shoals to Kure. Under this rule, those operating around Necker and Nihoa and elsewhere in the NWHI were not required to provide information to state fishery managers. As a result, information could be obtained from fishermen only on a voluntary basis, and for the most part there was no information. Four commercial lobster boats were known to be fishing in the NWHI in 1977 (WPRFMC 1977b). Three vessels worked out of U.S. mainland ports, meaning that they could legally harvest NWHI lobsters without providing any information on catch, effort or landings.

According to the minutes of the June 29, 1977, meeting, Chair Yee noted that only one of the three boats was willing to carry an observer for the purposes of data collection. He suggested that, because of the rumors of overfishing, an immediate moratorium be placed on lobster fishing in the NWHI, using the emergency powers of the Secretary of Commerce. He was opposed by NOAA attorney Martin Hochman, who said that the secretary "is very reluctant to take emergency actions unless it is to save a resource from imminent disaster" (ibid.).[21] Proactive action could not be justified based on the limited information available. Since Hochman's position was that restrictions could not be imposed until after an FMP was created, Yee told his fellow Council members that the best strategy was "to go ahead as fast as possible with development of a management plan" and in the meantime persuade the fishermen to keep accurate records and conserve the resource (ibid.).

Yee and the members of the Council were learning that implementing precautionary conservation rules under the FCMA was not easy, as it could involve a clash between federal policy and local interest. It was NOAA's policy that a need for regulations to constrain fishing activity had to be demonstrated by facts and analyses. In this case, a need for

---

[21] NOAA provides the attorney who advises the Council; that person also advises NMFS Regional staff about FCMA legal issues.

immediate action could not be demonstrated because management regulations requiring such records did not yet exist.

At the seventh meeting of the Council on September 29 and 30, 1977, the Crustacean Plan Team presented the first draft of the Crustacean FMP, which included the best available information on the biology of the lobsters, history of the Hawai'i lobster fishery, the economics of the NWHI lobster fishery at that time and estimates of the maximum sustainable yield.

At the same meeting, Council member Frank Goto asked whether a limited-entry plan should be imposed on the new NWHI lobster fishery.

Richard Shomura, director of the NMFS Honolulu Laboratory, replied that any limited-entry plan would be premature because the lobster fishery was in its first stage of development and not enough was known about the fishery to limit it for economic reasons. As for protecting the resource, he was confident it could be done by limiting the minimum size of the lobsters to be taken and prohibiting the harvest of egg-bearing females. Chair Yee then asked, given that there was so little information on the lobster resources, whether it would be prudent to limit vessels and traps. Michael Adams, the staff economist for the NMFS Honolulu Laboratory, responded that the economic potential of the lobster fishery appeared too small to require the organization of a limited-entry plan (WPRFMC 1977c).

Because data on the life history of Hawai'i lobsters were so limited, the draft Crustacean FMP depended heavily on the work of Craig MacDonald, who was then completing his doctorate on lobster lifecycles and had extensive data on the lifecycle of the lobsters from O'ahu and Midway Island (ibid.). MacDonald was hired to do a special study on the comparative lifecycle of the three species of lobsters found in Hawai'i.

It was anticipated at the time that the NWHI would be able to supply no more than 5% to 7% of the lobsters consumed in Hawai'i. The NMFS perspective was that fishing in the NWHI would be naturally limited to only a few months each year by the harsh winter wave conditions.

Hawaiian Monk Seals and the U.S. Fish and Wildlife Service

Another issue facing the Council in its development of the Crustacean FMP was the fishery's potential impact on the Hawaiian monk seal, a critically endangered species. The Hawaiian monk seal is a pinniped, the same order of marine mammals as seals, sea lions and walruses. It is considered the most primitive of all seal species and is related to the Caribbean monk seal (extinct since the 1950s) and the Mediterranean monk seal (now existing in only a very few areas). Hawaiian monk seals are one of two species (the other being the hoary bat) endemic to Hawai'i, meaning they are found nowhere else. The population size of monk seals in the Hawaiian Islands prior to and following human settlement in the archipelago is a subject of speculation. One synthesis of the available information on Hawaiian and other monk seals suggests that there may have been about 8,000 monk seals in the archipelago before human settlement, with two-thirds of that population in the NWHI (Watson et al. 2011).

The Hawaiian monk seal is a protected species that nests principally in the NWHI. The potential impact on the species by developing fisheries would be a center of controversy for decades. *NOAA photo.*

The large numbers of monk seals in the NWHI attracted hunters in the 19th century who killed the seals for fur and oil (ibid). Monk seals were also disturbed by other entrepreneurial efforts that took place on some of the NWHI, such as rabbit and oyster farming. The seal population was further disrupted during World War II when the U.S. government established airstrips and radar sites at Midway, Tern Island (part of French Frigate Shoals), Kure, and Pearl and Hermes. The Coast Guard brought dogs to the same islands, which interfered with the breeding habits of the monk seals. Between the 1950s and 1970s, the population declined by an estimated 50%, largely due to the constant disruption by military and civilian presence on the islands (Baker et al. 2012).

The first sporadic monk seal counts began in the late 1950s, and standardized beach counts, which tally all observed seals on land at any given time, began in 1983. The estimated number of monk seals over time for the entire Hawai'i archipelago has ranged from 1,000 to more than 1,400 individuals (Antonelis et al. 2006; Carretta et al. 2021). Population growth rate from 2013 to 2019 is estimated at 2% per year.

Prior to 1987, which saw the arrival of longline swordfish vessels in the NWHI (see Chapter 8), interactions between fishermen and monk seals were few. The first recorded death of a monk seal due to fishing was attributed to a gill net off Lehua Island in 1976. There were occasional interactions with bottomfish boats and vessels engaged in lobster fishing. However, there were no documented instances of harm or mortality to a monk seal due to these fishing activities, and there was no evidence to correlate commercial fishing activities in the NWHI with monk seal decline.

Conservation of monk seals is required under the 1972 Marine Mammals Protection Act. At the same time, since the monk seal is listed as endangered under the ESA, NMFS established a recovery team of government and nongovernment experts to provide advice on the status of the stock and to develop a "recovery plan" for protection and ultimate recovery of the population to an optimum sustainable level.[22]

A key part of the strategy for recovery was the designation of a "critical habitat" for the monk seal, which is defined under the ESA as "the specific areas within the geographical area occupied by the species at the time of listing on which are found those physical or biological features (i) that are essential to the conservation of the species and (ii) which may require special management considerations." Critical habitat

---

[22] The first recovery plan, completed in 1983, did not include any recovery criteria. A revised recovery plan, approved in August 2007, provides "delisting criteria" (conditions that must be met for the species to be downlisted from endangered to threatened) (NMFS 2007). The conditions were that the aggregate numbers of seals must exceed 2,900 individuals in the NWHI, with at least five of the six main subgroups in the NWHI being above 100 animals and the main Hawaiian Islands population being above 500 animals. Monk seals would not be completely removed from the ESA listing unless the downlisting criteria were met for 20 consecutive years without new crucial or serious threats being identified.

for the monk seals in the NWHI was initially proposed on December 9, 1976 (Gilmartin 1983).

From the start, critical habitat designations were controversial throughout the United States because their establishment could lead to federal restriction of economic activities (such as fishing) that might adversely affect the recovery of the ESA-listed species. For the Council, the proposed designation of critical habitat for monk seals posed a potential obstacle to the plans to develop the NWHI fisheries.

In 1977, Tim Smith, an NMFS employee and member of the Council's SSC, explained to the Council that the SSC was recommending a 10-fathom depth contour around the NWHI in order to provide a reserve of lobster spawners and a protective zone for monk seals (WPRFMC 1977c). There was considerable discussion regarding the relationship of monk seals and NWHI lobster stocks, as monk seals in the Mediterranean and Caribbean were known to eat lobsters. At several meetings during the Council's first year, members often expressed concern that the actions being taken by the NMFS and USFWS to create a critical habitat for the monk seal in the NWHI would restrict the development of commercial fishing there, particularly the lobster fishery.

On December 5, 1977, NMFS completed an environmental assessment for the proposed designation of a critical habitat for monk seals. On March 16, 1978, the Council discussed a 3-nm critical habitat for the monk seal proposed by USFWS (WPRFMC 1978b). Subsequently, the SSC was given a draft of the proposed critical habitat designation document for review.

The USFWS indicated that the agency would be willing to compromise with the Council over management of the NWHI and said it would potentially be willing to open some of the area to commercial fishing but insisted that the area first be studied to determine "whether fishing is feasible in the area and what the impact on other species might be" (WPRFMC 1978b).

It became apparent over time that the Council's ongoing work on the Crustacean FMP was proving to be incompatible with the monk seal critical habitat proposed by the USFWS and NMFS. The Council was concerned about whether fishing would be prohibited in areas designated as critical habitat. While user groups were not allowed input into recovery plans, designation of critical habitat had to go through the environmental impact statement process of public review required under the National Environmental Policy Act (ibid.). However, Council members and fishermen felt that this was too late in the process and that NOAA would declare the area a critical habitat, regardless of what fishermen wanted. It was also noted that the plan team for the NWHI, a culturally and historically important area for fishing to Native Hawaiians, did not include a single Native Hawaiian fisherman.[23]

The tension between some Council members on one side and NMFS and USFWS on the other continued to increase throughout 1978. There were accusations of collusion by the two agencies to marginalize Council involvement in the critical habitat designation. From the Council's

---

[23] In 1986, in a response to a statement by USFWS representative Allan Marmelstein that the USFWS master plan for the NWHI "was now final," Council Member Alika Cooper asked if the USFWS had permission from the State of Hawai'i to be at French Frigate Shoals. Marmelstein responded that he was not going to discuss that since it was not relevant to the Council. Council Chair Yee said the matter of whether the USFWS is legally in possession of Tern Island (at French Frigate Shoals) would ultimately be settled in court, and Marmelstein agreed (WPRFMC 1986b). The issue of the USFWS claim in the NWHI was never fully resolved and provided an undercurrent of resentment and defensiveness that would cloud future USFWS and Council relations.

viewpoint, USFWS intended to take the waters around the NWHI out of the Council's jurisdiction and away from the people of Hawai'i to be administered from the Department of the Interior in Washington, D.C., and through subordinates based in Hawai'i. This belief was reinforced when the Council heard that a meeting had been held on the critical habitat of the monk seal on April 28, 1980, in Washington, D.C. NGO groups attending included Greenpeace and the Sierra Club, among others.

Finalizing the Crustacean Plan

After much deliberation and advice from the Crustacean Plan Team and the SSC, the Council offered the following preferred alternatives for its developing Crustacean FMP:

1) Lobster fishing in all waters shallower than 10 fathoms would be prohibited in the NWHI except around Midway and Kure. This would provide a reserve of lobster spawners and a protective area for the monk seals.

2) Fishing for spiny lobsters would be prohibited within 20 nm of Laysan.

3) Spiny lobsters could be taken only by traps outside of 10 fathoms except at Midway and Kure, where they could also be taken by hand (to allow the military personnel living there to continue to take lobster).

4) Vessels would require permits, and captains would be required to complete official logbooks and carry observers when requested.

5) Vessels fishing in the NWHI would be required to submit their catches for inspection in Honolulu Harbor (in order to prevent U.S. mainland vessels from taking lobsters directly to the West Coast without reporting).

6) A minimum harvest carapace length of 8.5 centimeters (about 3.3 inches) for male lobsters and 9 centimeters (about 3.5 inches) for female lobsters.

7) Undersized and berried lobsters had to be returned to the sea alive.

8) There would be an annual review of the status of the lobster fishery and the effectiveness of the management measures for protecting the resource.

It was decided that after an environmental impact statement was completed, the FMP would be ready for implementation.

Although the Crustacean FMP assumed a single source of lobsters throughout the chain, it was understood that a large gap existed in the science and management of lobster resources (WPRFMC 1977c). Additionally, at one point NMFS proposed that the Council add a provision to the Crustacean FMP to establish a trigger mechanism (see 50 CFR 665.248 for full details) whereby any interaction resulting in the death of a monk seal that could possibly be attributed to the lobster fishery would be investigated and analyzed by the NMFS regional administrator. If he/she found that there was good reason to believe that the event was caused by the fishery, he/she would consult with the Council and the State of Hawai'i and would solicit public comment as to the potential for implementing measures that would prevent repetition of such an event. The regional administrator could then recommend that NMFS promulgate regulations to address this concern while the Council (under the FMP), or NMFS and USFWS (under the ESA), considered longer-term measures as appropriate. The Council accepted this approach with misgivings as it seemed to prejudge that the crustacean fishery had caused harm to monk seals without any evidence to that effect. Fortunately, this trigger mechanism never had to be invoked.

As the Council struggled to finish the Crustacean FMP, numerous disagreements broke out between lobster fishermen on the Advisory Panel who advocated a smaller size limit for lobster (as that would boost the landings and revenue of the lobster fishery) and members of the Crustacean Plan Team and SSC who favored limiting the take to larger lobsters. Those in favor of allowing smaller-sized lobsters to be taken were confident that the lobsters reproduced at a rate that was high enough to make up for their catch. Others, such as Yee, objected because the understanding of lobster population dynamics was still evolving. An uneasy compromise of a 4.9-centimeter (about 1.9-inch) tail width equivalent to a 7.7-centimeter (about 3-inch) carapace length was reached.

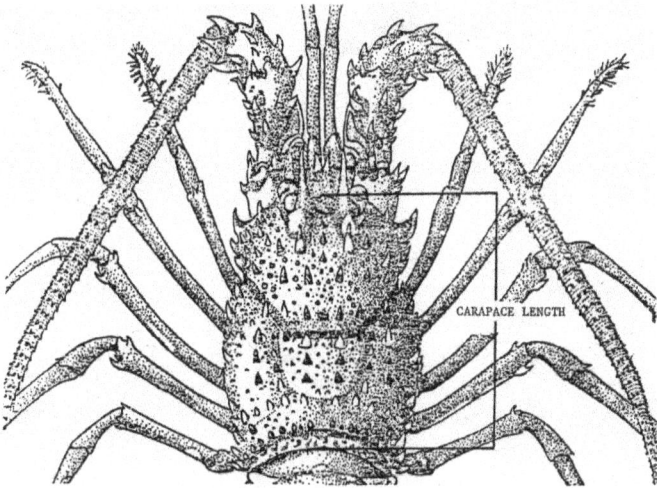

The carapace is the hard upper shell of the lobster. Early management measures for the NWHI lobster fishery included a minimum carapace length.

The Council also wrestled with the definition of optimum yield. It decided that regulations—such as minimum sizes, no retention of berried females, closed areas, mandatory permits, and inspection and reporting requirements—would allow participants to safely optimize their activities.

A draft of the Crustacean FMP was completed in May 1978. Unfortunately, the data and modeling to support a minimum size was limited. Solving this problem became the top priority of the Council (WPRFMC 1978d).

NMFS scientist Jeffrey Polovina, who joined the Crustacean Plan Team in 1979, recalled that that the process of developing the original Crustacean FMP was complicated. "What we had was data from 1978 and 1979 from Necker Island, and, based on the fishing from that one bank, we tried to extrapolate the fishing maximum sustainable yield for the whole archipelago. We assumed that [the lifecycle data] would be all the same for all the other banks at the same depth ranges. We were doing the best with what we had."[24]

By 1980, Polovina and Darryl Tagami combined catch-per-unit-effort and other data from nine NMFS lobster research trips to the NWHI conducted between 1976 and 1978 with commercial landings and information from MacDonald and others. Based on these data, Polovina and Tagami calculated a surplus production value of 10,000 to 21,000 legal-sized (Hawai'i State minimum 8.25-centimeter, or 3.25-inch, carapace length) spiny lobsters per year from Necker Island. This was then expanded to include all NWHI lobster grounds under several scenarios. The Council's FMP estimated the NWHI spiny lobster maximum sustainable yield to be in the range of 200,000 to 435,000 lobsters per year when managed under the State's minimum carapace length.

The Council approved the final draft of the FMP at its 29th meeting, held from March 31 to April 1, 1981. It included slipper lobster as well as

---

[24]   Jeffrey Polovina, NMFS Pacific Islands Fisheries Science Center scientist, in discussion with Michael Markrich, December 17, 2018.

spiny lobster as management unit species. Optimum yield was defined as between 168,000 and 420,000 lobsters annually. The 4.9-centimeter (1.9-inch) tail width (equivalent to a 7.7 centimeter, or 3-inch carapace length) was refined to include a tolerance level. Other additions included ghost-trap research and allowance for additional measures on the fishery under emergency conditions by the NMFS regional director, among others.

Approval of the Crustacean FMP and its regulations were published on February 7, 1983, and became effective March 9, 1983. That year, landings increased to 218,000 pounds.

Following the implementation of the FMP, the Council adopted the state minimum carapace length of 8.25 centimeters (3.25 inches) for EEZ waters around the main Hawaiian Islands to facilitate enforcement (Amendment 1). The Council also modified details regarding the construction of the NWHI lobster traps to prevent entrapment of monk seals due to the growing use of the plastic traps and changes in the configuration of wire traps over time due to stacking (Amendment 2).

Emerging Problems and Crustacean Plan Amendments

With Amendments 1 and 2 completed, MacDonald suggested that the Council organize field trials for the lobster traps' escape gaps so that undersized lobsters would not be taken. NMFS researchers Alan Everson, Robert Skillman and Jeffrey Polovina organized the trials, which were conducted on the commercial lobster vessel *Shaman*. They tested the difference between four 67-millimeter (2.6-inch) circular escape vents and two 49-millimeter (1.9-inch) by 285-millimeter (11.2-inch) rectangular escape vents. The researchers' finding was significant: the circular vents caught 83% fewer sublegal spiny lobsters and 93% fewer sublegal slipper lobsters than non-vented control traps (Everson et al. 1992).

Fishermen deploy plastic lobster traps in the NWHI. Because the traps are stackable, vessels can carry many more on deck compared to conventional wooden and wire traps, which led to a fourfold increase in production between 1983 and 1984. *NOAA photo.*

As the Council deliberated, word of the financially successful lobster fishery in Hawai'i continued to spread. In addition to vessels from California, crab boats from Alaska made their way to Hawai'i. The annual catch of spiny lobsters grew fourfold between 1983 and 1984 from 157,000 to 667,000 lobsters, and the catch of slipper lobsters grew exponentially from 26,000 to 285,000 lobsters. However, the joint 1984 annual report from the NMFS Southwest Region and Southwest Fisheries Science Center on the fishery was positive, stating that this catch was sustainable because scientific measures of recruitment and catch rates did not show significant decline.

The optimism of NMFS notwithstanding, some Council members were skeptical about the heavy fishing and outside interests, and they wanted a pause in fishing pressure. In May 1985, the Council members voted 10 to 1 (with the NMFS regional director voting "no") to develop an emergency moratorium on the spiny lobster fishery. The vote was taken because data collected by the NMFS Honolulu Laboratory indicated a sharp decline in catch-per-trap-night for spiny lobsters in the NWHI between the first quarter of 1984 and the first quarter of 1985. The Council also voted 10 to 1 (with the NMFS regional director voting "no") *not* to allow spiny lobster permit holders to fish more than the number of traps shown on their permit applications (WPRFMC 1986a). It should be pointed out that NMFS often voted "no" to avoid a unanimous vote and so preserve secretarial flexibility to disapprove Council action. Once again Council members were caught between the specific instructions of the NMFS to push for optimum yield and the natural caution they felt toward the conservation of the resource.

There had been early misgivings. A key argument in the development of the Crustacean FMP was the specification of regulations on carapace length. Council members, the Crustacean Plan Team and SSC wanted to set a larger size requirement as a precautionary measure to assure immature lobsters would not be taken. However, the carapace measurement would soon prove inconsequential as the lobster fishermen routinely removed and froze the tails and threw away the rest. The Crustacean Plan Team then had to change the measuring system to focus on lobster tails. Tail width was calculated relative to carapace length. Since the length of the carapace corresponded to a wide range of tail sizes, figuring out a tail size "was like trying to figure out somebody's height from their waist size," said Polovina, who was senior statistician and chair of the Crustacean Plan Team.[25]

Developing enforceable rules proved difficult. Federal enforcement officers had to check the size of the lobster tails with small rulers in freezers, which made their fingers numb. It was an enforcement nightmare because the original rules allowed a 15% tolerance from the minimum tail width of 4.9 centimeters (1.9 inches). Enforcement officers

[25]   Polovina ibid.

sometimes had to review 10,000 to 35,000 frozen lobster tails per trip to see if the catch was in compliance. Since they could not check the whole catch without defrosting the lobsters and ruining the value of the catch, they checked random samples, which introduced another large degree of error. Because of these factors, it was next to impossible to determine what percentage of the lobsters was undersized (WPRFMC 1984c).

As the Crustacean Plan Team tried to develop better, more enforceable regulations in the expanding fishery, catch rates steadily declined. In 1977, the catch per unit effort at Necker had been 4.92 lobsters per trap. By 1983, it had declined to 2.05 lobsters. In 1984 the number of areas fished increased from three to seven.

As catch rates slowed, fishermen began requesting that existing rules be eased to allow them to take smaller lobsters. By 1985, some members of the fishing industry were pressing for a further reduction in the tail width to 4.5 centimeters (1.77 inches) so they could continue to catch the same number of lobsters per trap. This drew objections from some scientists on the Crustacean Plan Team and SSC. Ultimately, a 5.0-centimeter (~2-inch) tail width was chosen by the Council and approved by NMFS in 1985 (Amendment 3).

Although it is now widely understood that errors were being made in the calculation of the maximum sustainable yield for the FMP, Polovina recalled the NMFS policy of that time was to err on the side of optimism: "Nowadays people tend to be precautionary, and the rule is, if we don't have the data, we assume the worst. [When I started] we had just had the Magnuson Act and the State was pushing to expand this valuable resource and the mindset was 'we don't know what's there, but let's not assume the worst case. Let's assume a reasonable case scenario.' That was true of the managers as well."[26]

---

[26] Polovina ibid.

It was still not widely recognized that the slipper and spiny lobster fishery in the NWHI was very different than lobster fisheries in other areas such as Maine. Hawai'i vessels equipped with freezers fished for weeks, while East Coast vessels were small and fished for only a day or two. So it was hard for NMFS headquarters to understand the scale of the Hawai'i problem.

Other issues with management measures and how they affected lobster populations soon surfaced. Honolulu Laboratory research using underwater film showed that, as undersized and berried lobsters drifted down through the NWHI water column, they were eaten by large ulua (*Caranx ignobilis,* or giant trevally) (Gooding 1985). The study concluded that mandatory escape gaps on lobster traps could reduce the number of undersized lobsters landed on the fishing vessels and thus reduce the potential for mortality from exposure and handling, as well as predation.

Ulua schools in the NWHI have been compared to buffalo herds in their size and abundance. One of the controversies with the developing lobster fishery was the soundness of a management measure requiring the release of undersized and berried lobsters as it was suspected that the ulua would consume the lobsters before they could reach the seafloor. *John Naughton photo.*

In addition, lobsters that could be thrown back became bruised, lost appendages and died when they were piled upon one another prior to being sorted (DiNardo and DeMartini 2002). Field trials of resettlement experiments in which undersized lobsters were stored in sea water until they could be lowered and released on the seafloor were unsuccessful as the process proved fatal to most test subjects.

Another problem had to do with the taking of berried females. Although the Crustacean Plan Team recommended a male-only fishery so that a greater number of spawners would survive, sexing the animals on the deck of the fishing boats was nearly impossible. It was not easy to tell whether lobsters had developed eggs, and sometimes they developed eggs later in the holds of the ships. In addition, the fishermen believed it was wasteful to throw any lobsters back because they could be eaten before they reached the seafloor.[27]

Managing Rapid Growth

When the lobster fishery began in the NWHI, relatively small lobster boats were deploying bulky wood and wire traps, which had to be laboriously set and then hauled up. The extent of a day's fishing was limited by the number of lobster traps that could fit on the boat's deck. Most of the Hawaiʻi boats had limited capacity to hold the catch. However, as the fishing boats switched to lighter, more efficient plastic traps that could be stacked within themselves and large vessels from the U.S. mainland entered the fishery, the catch soared. In 1983, each vessel carried an average of 471 traps. In 1986, that average increased to 708 traps. The increase could be immediately noted in the number of traps set. In 1983, a total of 6,600 traps were set; by 1986, the number was 36,090 for an average annual increase of 110% for three years.

---

[27]   Polovina ibid.

The increase in traps set in this three-year period was accompanied by an increase in the number of active vessels (vessels that set at least 100 traps during one trip), from three to 19 vessels.

Some Council members and fishermen supported more stringent rules. Under the FMP, spiny lobsters were not to be taken within 20 nm of Laysan Island, which was intended to be a lobster refuge zone. It was not long before fishermen began fishing in the area, claiming they were targeting slipper lobsters, which were not prohibited.

Laysan is the second largest island in the NWHI. In 1983, the Council established Laysan as a refugia with lobster fishing prohibited within 20-nm of the island. *USGS photo.*

In response to this obvious breach of at least the *spirit* of the regulation, Charlie Fullerton (then director of the NMFS Southwestern Regional Office) called for an emergency regulation to be recommended by the Council closing the loophole, taking out the word "slipper" and specifying that all lobster harvests at Laysan were prohibited. However, when the Council voted 10 to 1 for the measure at its 50th meeting in August 1986, Fullerton voted against it. When asked by Frank Farm Jr., vice chair of the Council's Advisory Panel, why he would vote against a measure that he himself introduced, Fullerton explained that he was under orders from Washington, D.C., to vote against emergency regulations. NMFS attorney Martin Hochman then explained that, under the complicated rules of the MSA, without a unanimous vote the Secretary of Commerce had the discretion to either go along with the emergency regulation or not. The mixed signals sent by NMFS exemplified the complexity of the MSA rules contributing to the difficulties of fishery management (WPRFMC 1986b).

By 1986, vessels were traveling 1,200 nm up the length of the entire NWHI chain looking for lobsters in and around remote islands and atolls like Kure. But even as fishing effort increased, the total reported spiny lobster catch (legal, sublegal and berried) continued to fall. In 1986, the catch per unit effort for legal spiny lobsters was 0.60 per trap, compared to the 1985 figure of 0.85. For slipper lobsters, the catch per unit effort fell to 1.02 per trap compared to 1.31 the year before.

Part of what drove the expansion was spiny lobster prices. In 1986, spiny lobster harvests were down in the fisheries that most often supplied the Hawai'i hotel and restaurant market—Australia, New Zealand and Florida. As a result, supplies were tight, and Hawai'i prices for spiny lobsters skyrocketed. More fishermen wanted to enter the market.

In 1986, the Council contracted with University of Hawai'i resource economics professor Karl Samples to review the catch and economic data of the lobster fishery. After reviewing the data and talking with fishermen, he concluded that too many spiny lobsters had been taken and the increases in reported catch were not due to any remaining abundance of spiny lobsters but were rather due to fishermen having switched to catching slipper lobsters. Samples warned that, when the slipper lobsters were depleted, nothing would replace them and the industry would find itself in a downward spiral of higher costs and lower catch rates. He explained that all the known fishing grounds had been fully exploited by 1986 and that "the mere fact that the physical expansion of fishing areas had come to an end raises questions about the continued economic viability of the industry" (Gates and Samples 1986). He argued that it made more sense to limit the fleet to the capacity of the fishery than it did to leave entry open so as to achieve some theoretical maximum sustainable yield or optimum yield set by government officials.

Some of the local fishery advisory groups agreed with him, and the Council helped organize a working group called Hui Ula, made up of lobster fishermen and members of the Crustacean Plan Team, to recommend lower catch limits and a limited-entry program to protect the fishery. This point of view was supported by most members of the Council. However, the NMFS Honolulu Laboratory had come to a different conclusion. At a Council meeting in 1987, Director Shomura replied, in response to a question from the Council about the sustainability of lobster stocks, "All the evidence indicates it is not overfished. It is a dynamic situation" (WPRFMC 1987b). Shomura mentioned that sardines had gone from 8,000 tons to a million tons and no one understood it, despite having studied their lifecycle for more than 50 years. He expressed confidence in the judgment of the NMFS scientists. "Scientists will blow the whistle when they have indications that there is overfishing ... you don't want to have overregulation of fishing" (ibid.).

Polovina, as head of the Crustacean Plan Team and the person in charge of its statistical data set, believed that what Samples and some others saw as a downturn was in fact only a temporary condition. Despite his concerns regarding overharvesting, he concluded that the trend and catch data available at the time supported Shomura and that the NWHI lobster fishery was sustainable.

Polovina had used a widely accepted statistical model to provide the basis for setting the NWHI spiny lobster maximum sustainable yield at 1,140,000 animals per year. Honolulu Laboratory scientists wrote in 1989, "the fishery regulations that are currently operational for the NWHI lobster fishery appear to be protecting lobster stocks at the current levels of fishing and natural mortality" (Polovina and Harman 1989).

The fishing industry asked the Honolulu Laboratory to recheck its findings. Council Executive Director Kitty M. Simonds and Council staff officer Bunny Lowman set up an open forum with 25 fishermen, 15 of whom were owners of fishing boats. The fishermen stated that effort levels at Necker and Gardener were not sustainable and that there had been a dramatic decline in catch rates for lobsters. The catch per unit effort had dropped from 3.1 lobsters per trap in 1982 to a little over 1 per trap in 1989 (Langraf and Pooley 1990). The fishermen were clearly worried. Although there were concerns that the Honolulu Laboratory was in error, the Council took the advice provided to it by the government scientists.

According to Samuel Pooley, an economist who had been hired by the Council to help complete the Crustacean FMP and who years later directed the NMFS Pacific Islands Fisheries Science Center (formerly the Honolulu Laboratory), "Some people were very angry at the Lab, but more so at the Council, which was taking our advice. But the Council efforts were based on the scientific information and models that many at the Lab were producing. The problem was that the science was not always correct. In retrospect, the stock assessments were fairly primitive and lacking knowledge about environmental conditions, but it was the best that we knew at the time. I wouldn't say our models were cutting edge, but, for the time, they were state-of-the-art, given the biological and environmental information we had at hand."[28]

In the meantime, the Plan Team persisted in searching for new data. A survey trip to the NWHI in the summer of 1990 discovered a sudden and dramatic decline in the number of small lobsters coming into the traps. This indicated the long-feared fall in recruitment (needed for the replenishment of the stock) had occurred. The Crustacean Plan Team called for an emergency reduction of fishing effort for 1991 and a system to adjust fishing pressure to manage the stocks (Crustacean Plan Team 1991). It would later be determined that growth of the NWHI commercial slipper lobster fishery had coincided with a regime shift in the Transition Zone Chlorophyll Front, a large oceanic area that separates subtropical areas from more productive North Pacific waters. When the front shifted, the availability of nutrients in the NWHI diminished, which in turn limited recruitment and reproductive activities of not only lobsters but also birds, fish, turtles and monk seals over a 10-year period.

Polovina described his reaction at the time. "We had no indication that anything like this was going to happen," he said. "Everything just collapsed so dramatically. Even places that had never been fished, like Laysan, showed a decline. There just weren't any more small lobsters entering the fishery."[29] This has since been known as the 1989 oceanographic regime shift (Polovina 2005). Polovina also thought that the shift of the Pacific Decadal Oscillation to a negative condition in the late 1980s could have impacted the lobster stock.

---

[28]   Samuel Pooley in discussion with Michael Markrich, December 18, 2019.

[29]   Polovina op. cit.

The Monk Seal Issue Resurfaces

The impact of the oceanographic regime shift and collapse of the NWHI lobster fishery refueled controversy on whether or not the Council, by allowing fishing under its Crustacean FMP, had caused the decline of the Hawaiian monk seal population. From the point of view of monk seal advocates, the fishery was depriving the seal of an important food source.

Several significant milestones in monk seal research occurred during this period (mid-eighties). In 1985, the Honolulu Laboratory discontinued its program of removing aberrant males (who mobbed females). During that year, it also was found that the highest natural mortality of monk seals took place at French Frigate Shoals, where sharks preyed on pups and where pupping beaches were shrinking. In the meantime, the number of monk seals remained within the same narrow population range (WPRFMC 1985b). It was estimated by NMFS that in 1986 that there was a population of 1,500 monk seals in the NWHI (WPRFMC 1986a).

On April 30, 1986, NMFS formally designated critical habitat for the Hawaiian monk seal in the NWHI to the 10-fathom depth line. In the meantime, largely as a result of actions taken by the USFWS and NMFS to protect newly born monk seal pups, circumstances seemed to be improving for the monk seal population. In August 1988, there was a report to the Council that overall a monk seal "head start" program had been a success. Begun in 1981, this program gave female monk seal pups a chance to survive by keeping them away from sharks and aggressive males. Largely as a result of this intervention, the number of births increased from six in 1987 to eight in 1988. Four of the females that gave birth had been protected as pups. For those involved in trying to stabilize the small monk seal population, this seemed a significant milestone.

Between 1983 and 1987, USFWS and NMFS personnel investigated the reasons for the low monk seal population levels. Six main causes were found:

1) The birth rate of seals is low because monk seals reach sexual maturity at six years of age and are capable of having only one birth every two years.

2) Many pups are eaten by sharks.

3) Males outnumber females and often bite the females, resulting in infection and death.

4) The stomachs of eels consumed by monk seals contain dinoflagellates that produce ciguatoxins that can be deadly to the seals.

5) Food supplies are inadequate.

6) Monk seals get entangled in marine debris, such as abandoned fishing gear or fishing and cargo nets that drift into the NWHI from the North Pacific, which leads to injury or mortality (Gilmartin 1983). In the period 1982–1998, seven monk seals were reported killed by marine debris—principally abandoned fishing gear not from Hawai'i (Henderson 2001).

By 1988, researchers were more optimistic that the population had stabilized and were confident that the monk seal population was improving. As the plight of the monk seal gained increasing public exposure, additional funds were made available for monk seal research programs.

However, in 1991, scientists became aware of a decline in both lobsters and monk seals. They were particularly alarmed by the decline in weight of monk seal pups, which were suddenly showing their lowest weights since 1983 when their monitoring began. Since the pups were unlikely to eat lobster as part of their diet, this indicated that some other event was affecting the ecosystem. The population, which had seemed to stabilize during the 1980s after substantial effort was put into breeding programs, now appeared suddenly to be even more threatened.

Some of coral reef biologists and other scientists thought it was not coincidental that the decline of the monk seal population after 1990 occurred simultaneously with the crash of the lobster population. However, NMFS scientists questioned the implied correlation for a number of reasons. The first was a question of methodology. The current means of estimating monk seal populations, based on beach counts, made it difficult to determine the total population because many seals were out swimming in the ocean; therefore, the monk seals ashore accounted for only an unknown fraction of the entire population at any point in time.

Second, while it was accepted that beach counts could be used to calculate long-term trends of monk seal populations, they were poor indicators of year to year changes in monk seal numbers (Eberhardt et al. 1999). For this reason, NMFS scientists found it difficult to conclude that a change in the lobster abundance led to a decline in monk seal numbers, particularly because the decline of the monk seal pup population took place even in areas where lobster fishing had never been permitted.

Third, when the NWHI lobster fishery was closed for 10 years and the slipper lobsters returned in abundance, the monk seal population still did not recover. For these reasons, a number of NMFS scientists repeatedly said that the health of the monk seals and the lobster fishery were not connected.

In 1998, a report on monk seal waste indicated that crustaceans made up just 5% of the contents of Hawaiian monk seal scat and spew. Tests showed the seals ate mostly benthic fish. However, this controversy was not put to rest until the NMFS commissioned a special test known as the Quantitative Fatty Acid Signature Analysis to determine what made up the monk seal diet. The report, published 13 years later, determined that lobsters contributed a minimal amount to the monk seal's diet based on analyses of monk seal spew, scat and quantitative fatty acid signatures (Iverson et al. 2011). Although it had been pointed out at an early Council meeting that lobsters were thought to be a part of the Mediterranean monk seal diet, there was never conclusive evidence to link Hawaiian monk seal health with lobster abundance. However, this argument resonated so deeply within the environmental and scientific community that it still periodically surfaces to justify measures to restrict commercial fishing.

Making a Correction

The impacts of the 1989 oceanographic regime shift and the collapse of the NWHI fishery would have a dramatic effect on the Crustacean FMP over the next 10 years. On January 13, 1991, the Council recommended a 90-day emergency closure. Subsequently, the NWHI lobster fishery was closed on May 18, 1991; reopened in 1992; and closed again in 1993. The Crustacean Plan Team recommended continued closure until the regime shift ended.

In 1992, Amendment 7 to the FMP implemented a limited-entry program with 15 transferrable permits and a vessel limit of 1,100 traps per vessel. Amendment 7 also established an adjustable fleet-wide NWHI harvest limit based on the catch and effort data of all lobsters in the entire archipelago to be announced by February 28 of each year. The quota for 1993 was set at zero. During this period, the Honolulu

Lab took several lobster research trips to the NWHI to gauge the size of the population and the catch per unit effort.

After it had been determined that the stock decline was associated with an environmental regime shift and not caused simply by overfishing, the Honolulu Lab revamped its models to take into account natural variations. The Council held several workshops with the specific aim of estimating stock abundance.

NMFS scientists (such as Andrew Rosenberg, a lobster specialist brought in from the northeast by the Council) noted that the population dynamics of lobsters differed on the various banks within the NWHI archipelago. Most islands in the archipelago west of Maro Reef had poor lobster recruitment, while other areas, such as Necker, had good recruitment. They came to believe that, in the event the lobster fishery reopened, they would need to determine the lobster populations over the entire archipelago and then set catch limits for specific areas.

Under Amendment 9, implemented in July 1996, the allowable harvest rate was set as a means of ensuring a successful spawning potential ratio. The harvest guideline was based on a 10% risk of overfishing, which was associated with a constant harvest rate of 13% per year. The population was estimated at 2.2 million, which provided a harvest guideline of 286,000.

The archipelago was divided into four zones. Region 1, from Nihoa to French Frigate Shoals, was believed to have very good spiny lobster natural recruitment. Region 2, from Brooks Banks to Maro Reef, had good recruitment of slipper lobsters and poor recruitment of spiny lobster. Region 3, Gardner Pinnacles, had very large spiny lobsters. Region 4, from Laysan to Kure, had very little data. NMFS lobster expert Gerard DiNardo testified to the Council that this method to determine harvest guidelines was scientifically sound (WPRFMC 1998).

The Council also instituted a "retain all" policy for the NWHI after researchers found levels of discard mortality approaching 100%. The "retain all" policy forced the fishermen to count every lobster brought on board as part of the annual quota and reduced the practice of "high-grading," discarding animals with low market value in order to maximize the quality and value of their catches.

In 1998, area-specific quotas were implemented for the first time. The lobster population was estimated at 2.2 million, which provided a harvest guideline of 286,000. The lobsters would come from individual banks in the following numbers: 70,000 lobsters from Necker Island; 80,000 from Maro Reef; 20,000 from Gardner Pinnacles; and 116,000 from all other banks. By that time, only six vessels still participated in the fishery that was experiencing steadily declining catches.

In 1999, the lobster fishery was closed from January to June. At the end of June, when the boats finally went fishing, the quota for Necker Island, the most productive area, would be reached and the fishery closed by July 17; Gardner Pinnacles would close on July 20; and Maro Reef by the end of August. The combination of tight regulations and low catch rates made the NWHI lobster fishery uneconomical by the fall of 1999 (WPRFMC 1999b).

Additionally, in January 2000, Greenpeace, the Center for Biological Diversity and the Turtle Island Restoration Network sued the Secretary of Commerce, alleging that commercial lobster and bottomfish fishing was depleting the monk seal's food resources and invading its critical habitat. The plaintiffs were successful in obtaining an injunction against the lobster fishery, and, in June 2000, NMFS closed the NWHI lobster fishery due to "shortcomings in understanding the dynamics of the NWHI lobster populations, the increasing uncertainty in model parameter estimates and the lack of appreciable rebuilding of the lobster population despite significant reductions in fishing effort" (NMFS 2000). NMFS set the annual harvest guideline for the NWHI lobster fishery at zero for the year. Due to the injunction and the NMFS closure, there was no active NWHI lobster fishery in 2000, which would doom the fishery's future.[30]

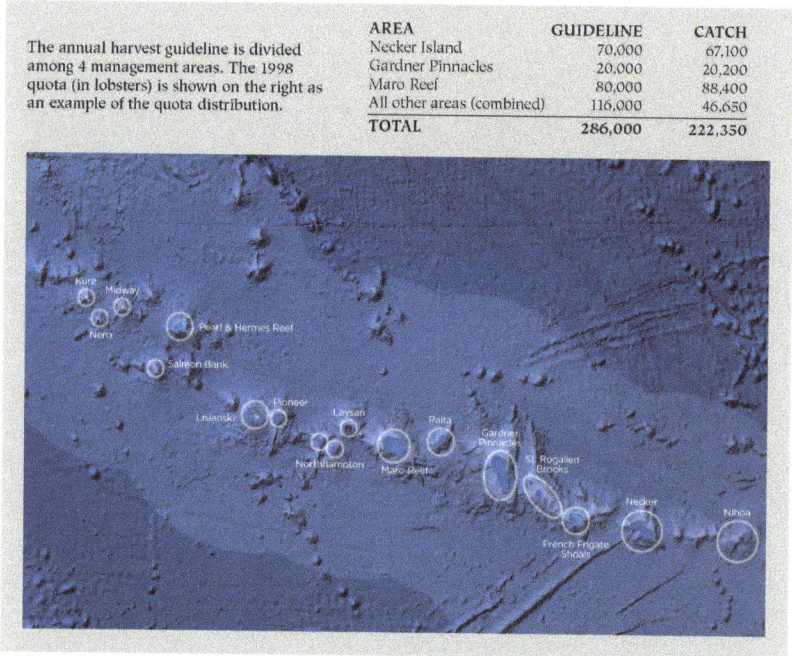

	AREA	GUIDELINE	CATCH
The annual harvest guideline is divided among 4 management areas. The 1998 quota (in lobsters) is shown on the right as an example of the quota distribution.	Necker Island	70,000	67,100
	Gardner Pinnacles	20,000	20,200
	Maro Reef	80,000	88,400
	All other areas (combined)	116,000	46,650
	TOTAL	286,000	222,350

Lobster recruitment is not the same throughout the NWHI chain. After fishery scientists and managers became aware of this, they divided the NWHI lobster quota among four management areas for the 1998 season. The harvest guideline was based on a 10% risk of overfishing, which was associated with a constant harvest rate of 13% of the estimated population per year. *WPRFMC illustration.*

---

[30] Although NMFS stated that it would continue fishery research to re-open the NWHI, this did not occur. To this day, NMFS, pursuant to Amendment 9 of the Crustacean FMP, announces the zero harvest guideline for the NWHI lobster fishery by February 28 each year.

TABLE 2: LANDINGS (NUMBERS) OF SPINY AND SLIPPER LOBSTER FROM THE NORTHWESTERN HAWAIIAN ISLANDS, 1983–1999. Source: Kawamoto and Pooley 2000.

Year	Spiny Lobster	Slipper Lobster
1983	218,000	25,600
1984	991,000	288,000
1985	1,490,000	1,249,000
1986	1,293,700	1,181,800
1987	727,300	488,900
1988	1,281,700	284,700
1989	1,481,000	340,100
1990	1,236,400	309,900
1991	352,600	49,900
1992	607,700	242,700
1993	fishery closed due to stock concerns	
1994	185,300	84,800
1995	89,700	11,500
1996	165,200	22,400
1997	175,800	134,400
1998	82,350	140,000
1999	85,858	150,257

Closing the Fishery

The long-standing interagency conflict simmered between USFWS biologists who felt they had an obligation to ban commercial fishing within the entire NWHI, an ocean area the size of California, and NMFS biologists who believed the area should remain open as a managed fishery. This controversy became increasingly politicized.

Environmental NGOs used sophisticated advertising campaigns to generate negative publicity about NWHI fishing interests and to leverage goals directed towards ending that fishery. With fishermen, environmentalists and agency scientists having such polarized positions, it became difficult for the Council to seek middle ground on the future of the NWHI fisheries. Further, in none of these discussions was any recognition given to the rights of Native Hawaiians to determine what should be done in their own waters—nor was reparation offered to the Native Hawaiian people should the area be closed. The emotional argument centered only on whether the area should be completely closed.

What had initially been a purely regional argument reached national recognition when William Brown, a former head of the Bishop Museum in Honolulu and chairman of the Ocean Conservancy, was hired by Secretary of the Interior Bruce Babbitt in 1997 to be a science advisor tasked with finding ways to increase and make more visible the Interior Department's role in protecting the ocean. For the first six months he was at a loss, before discovering that the Department of the Interior had jurisdiction over coral reefs in "the Island territories," which made him decide to promote coral reef protection within the department (Brown 2006). He was subsequently asked by the staff of the White House's Council on Environmental Quality to write an executive order for the president and an action plan for resource protection.

On June 11, 1998, President William "Bill" Clinton signed Executive Order 13089 on Coral Reef Protection, which created the interagency U.S. Coral Reef Task Force, made up primarily of government officials with NGOs serving on working groups. Clinton announced this order in his remarks to the National Ocean Conference, held June 11 and 12, 1998, in Monterey, California, as part of the U.N. International Year of the Ocean activities.

The question then became, what coral reefs to protect, and where? A number of sites were put forth from all over the United States. One of the areas Brown chose for protection was the NWHI. His idea was to use the Antiquities Act to declare the NWHI a marine monument, thereby ending commercial fishing by presidential fiat.[31] The final sites selected for coral reef protection consideration were Navassa (35 nm from Haiti), the NWHI, Palmyra, Kingman Reef and Wake.

In 1999, the long-standing concerns of USFWS and other coral reef biologists about the Council's development of a Coral Reef Ecosystem (CRE) FMP (see Chapter 9) coalesced into an appeal from the USFWS Hawai'i Regional Office directly to Babbitt. As one USFWS employee put it, "We ran it up the flag pole." The complaints from the Hawaiian Islands National Wildlife Refuge field staff about NWHI management, which had simmered on the local level for 25 years, were directed to the Secretary of the Interior and from him directly to Clinton at the very moment his administration was looking for a large reef area to formally protect with an executive order (Babbitt 2006).

Wake was dropped from the list because of its remoteness. Navassa, Palmyra and Kingman Reef were then designated as National Wildlife Refuges, a designation for certain protected areas of the United States managed by the USFWS. On September 2, 1999, the USFWS established the Navassa Island Wildlife Refuge. On January 18, 2001, Babbitt created the Kingman Reef and Palmyra Atoll National Wildlife Refuges during his final days in office with the Department of the Interior Secretary's Order 3223 and 3224.

Ultimately, the NWHI were not chosen for refuge protection for a specific reason. Brown recalls, "It emerged that (according to the Department of Justice) the secretary of the interior or the president

---

[31]   Tim Johns in discussion with Michael Markrich, December 9, 2009.

could not make a marine refuge in the NWHI. ... [However,] the president could establish a national monument within the U.S. territorial sea and also within the U.S. exclusive economic zone. We still had a problem though—we could not get a national monument out of the president without agreement from the Department of Commerce through NOAA" (Brown 2006).

Armed with Brown's ideas and a plan of action, Babbitt requested that Clinton issue an executive order to turn the entire NWHI into a marine national monument to be managed by the Interior Department. He would later say that he would "draw a circle around the Hawai'i territorial waters" and declare them a monument (Babbitt 2006).

The unusual choice of monument designation came about for a specific purpose. Under the normal government participatory process dealing with the acquisition of public lands by the federal government, numerous safeguards are in place to protect the public interest. Among these are required scoping meetings in response to Federal Register publication of a Notice of Intent. However, a monument can be designated through presidential proclamation under the authority of the Antiquities Act of 1906, thus bypassing the public process. This Act was signed into law by President Theodore Roosevelt to address concerns about protecting mostly prehistoric Native American ruins and artifacts—collectively termed "antiquities"—on federal lands in the West, such as at Chaco Canyon, New Mexico. The law allows the president to declare historic landmarks, historical and prehistoric structures and other objects of historic or scientific interest as national monument and "may reserve a part thereof lands, the limits of which in all cases should be confined to the smallest area compatible with proper care and management of the objects to be protected."

Babbitt went to Clinton with a list of monuments established by Roosevelt. Clinton responded positively. Brown wrote, "The president bought in. Legacy readily takes a grip when the end of power is near" (Brown 2006). Hawai'i was particularly susceptible to the idea of monument declaration. Other areas, such as Florida, had reefs, and, as Brown recalls, "Florida was more central politically, but Hawai'i's reefs were bigger, even if their inhabitants mostly voted with their beaks and fins" (ibid.).

The statement that Hawai'i reefs were bigger than Florida's was controversial, though often used by monument proponents who claimed that 70% of the coral reefs in the United States are located in the NWHI. That message was based on a study (Hunter 1995) commissioned by the WPRFMC that defined coral reef habitat in the Western Pacific Region as the substratum adjacent to coastlines (or on shoals) from depths of 0–300 feet (0–100 meters) that is primarily composed of hard bottom. In 1998, other jurisdictions around the United States began to assess the extent of coral reef habitat in their area. However, methodologies were not consistent and some jurisdictions reported the extent of coral reef habitat while others reported the extent of living coral.

Regional and national studies (Miller and Crosby 1998) began to compare these individual findings and concluded that Hawai'i contained 85% of the nation's coral reef and the NWHI alone contained 70%. NOAA scientists would later state that the coral reefs in the NWHI make up at most 10% of the U.S. total, while the south Florida shelf contains the greatest percentage, at 83% (Rohmann et al. 2005). The NWHI percentages may even be as low as 5% (Grigg 2006). Another study estimated benthic habitats considered as coral reefs in the NWHI at 29% of the nation's coral reefs based on satellite imagery data (depths of 0 to 500 feet, or 0 to 150 meters, including unmapped areas) (Monaco et al. 2012). Florida contains the most coral reefs estimated at 35% using the same set of parameters.

As part of the process to get the monument designation in place, it was necessary to change the management structure of the ocean in the Hawai'i region. Brown wrote that at first it was not easy to get NOAA's blessing. NOAA was "split at the root" between NMFS and the National Ocean Service. Although the NOAA administrator had called for "One NOAA," these two agencies within NOAA had different and often conflicting goals (Brown op. cit.). NMFS, which Brown has described as "user friendly to industry," is characterized by critics as being ethically compromised—too close to the commercial fishing industry to do its job well; yet, it describes itself as regulatory in nature and scientifically oriented.

Likewise, the National Ocean Service, which manages national marine sanctuaries and undertakes public information campaigns under the National Marine Sanctuaries Program (NMSP), is described by critics as being ethically compromised due to its close ties to the commercial ocean tourism industry. The valuation of reef areas as marine parks can displace fishing effort to foreign countries, which may lack the means to manage their resources, so as to supply the U.S. market.

On May 26, 2000, Clinton, in response to his advisers, directed the Secretaries of Commerce and the Interior, in cooperation with the State of Hawai'i and in consultation with the WPRFMC, to develop recommendations for a new, coordinated management regime to permanently protect the coral reefs of the NWHI and provide for sustainable use of their resources. The president's directive included a public outreach process, with hearings to gather public comments to shape the final recommendations.

On July 10 and 11, 2000, the Council convened its 105th meeting at Midway in the NWHI. The meeting provided Council members with an opportunity to experience the Leeward Islands firsthand—their remoteness, wildlife, tourism attractions and land and harbor facilities. At the time, Aloha Airlines offered scheduled air service to Midway, and three companies sold packages to the atoll. The company Midway Phoenix, under contract with USFWS, ran all tourism aspects of the island, which included diving and snorkeling, fishing, wildlife viewing, history tours, three restaurants, tennis, bowling, a gymnasium and a move theater. Accommodations ranged from a three-room suite with private bathroom to a room with a shared bathroom (Oakley 2000).

Council members enjoy sportfishing at Midway on July 9, 2000, before the start of the 105th Council meeting. *WPRFMC photo.*

At the Council meeting on Midway on July 10 and 11, 2000, Council Executive Director Kitty M. Simonds took the opportunity to discuss the future of the NWHI with Robert Schallenberger, USFWS Midway Atoll National Wildlife Refuge manager (left), and Robert P. Smith, USFWS Pacific Islands Ecoregion manager. *WPRMFC photo.*

In accordance with Clinton's executive order, seven public "visioning sessions" were held between July 21 and August 1, 2000, and a 21-day comment period was open July 12 to August 2, 2000, for public input on the future management and use of the NWHI. Council staff worked with the State of Hawai'i, NOAA and USFWS to prepare visuals for the sessions. Exhibit panels developed by the Council demonstrated that the NWHI, far from being "pristine" as claimed by some environmentalists, had been exploited and developed by humans for centuries. It was now used for ecotourism, education and managed fisheries—and the potential for bioprospecting for compounds and/or pharmaceutical agents was possible, using new technologies such as deep-water, manned submersibles with highly selective harvesting capabilities.

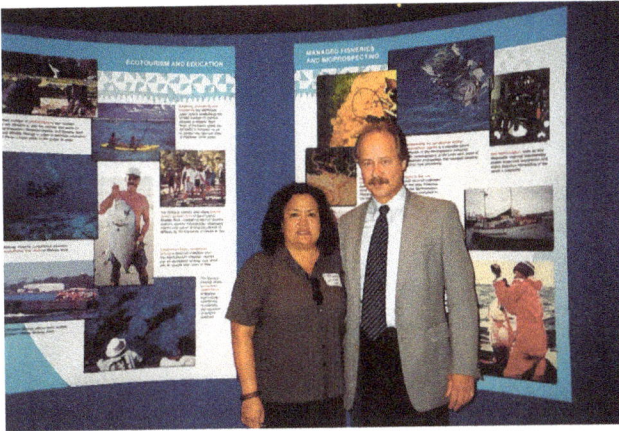

The visioning session display developed by the Council shows ecotourism, education and managed fishery activities, as well as potential bioprospecting opportunities in the NWHI. Council Director Kitty M. Simonds (left) accompanied the display to the first of seven sessions, held July 21, 2000, in Washington, D.C. *WPRFMC photo.*

Council staff attempted to share information on the Coral Reef Ecosystem FMP that it was in the process of finalizing. The FMP was applicable to the NWHI and other areas in the Western Pacific Region. However, at the first visioning session, held in Washington, D.C., the associate director of the Council on Environmental Quality (for oceans, coasts and environmental policy) prevented the Council from presenting information on the FMP, which had been in the works since the early 1990s. The Council was told that the FMP hadn't been approved by the secretary yet, and so it could not be presented as part of the Council's public comment. The inability of the Council to present its own ecosystem management plan for the NWHI fed into the no-take MPA proponents' portrayal of the NWHI as a fragile ecosystem that lacked adequate management. The Council would complete development of the CRE FMP in October 2001, and soon afterward, it received an unexpected long distance call from the Council on Environmental Quality asking for a copy of the plan.[32]

One of the fishermen who provided comments on the future of the NWHI was James Keliipio Mawae from Ho'olehua, Moloka'i, who wrote these words: "The people coming to Hawai'i from the mainland now want to stop us from fishing on the reefs our forefather's left undamaged. To fish is a birthright of Hawaiians. To stop us from fishing is to deny us the means to perpetuate our culture. Who can tell what the future will bring to Hawai'i or what role the NWHI will play in the future? What if natural or other disasters disrupt the imports of food into the islands? The Hawaiian people will once again need a large 'ice box.' There are good reasons for not putting up a 'fence' around the Northwestern Hawaiian Islands ..." (Mawae 2000).

A monument would likely have been declared in 2000 had it not been for forceful opposition from the Hawai'i Congressional Delegation, Hawai'i Governor Cayetano and Hawai'i former governors John D. Waihe'e III (1986–1994) and George R. Ariyoshi. On November 13, 2000, Cayetano met with President Clinton who was making a brief stop in Kona on Air Force One. Cayetano asked the president to refrain from proclaiming the NWHI as a national monument, which would effectively ban most commercial fishing, ecotourism and other recreational activities in approximately two-thirds of the Hawai'i archipelago. As reported the next day in the *Honolulu Advertiser*, Cayetano "urged a less restrictive approach that would allow state and federal agencies, the fishing industry and other parties to draft a memorandum about how to protect the reefs and endangered species in the islands." Cayetano said, "We want to work something out where we have some conservation provisions in place and yet allow people to fish there."

---

[32]   Mark Mitsuyasu in discussion with Michael Markrich, June 10, 2008.

Hawai'i Governor Benjamin J. Cayetano met with President Clinton in 2000 to express his opposition to a NWHI marine monument. *Pinterest/Culture Shocka photo.*

Cayetano opposed a monument designation on three points, based on information provided to him by the Hawai'i State attorneys. The first was the issue of the ceded lands. If the monument was declared, then the future of those lands would be decided by people in Washington, D.C., far away from Hawai'i. The action would set a precedent as to whether the ceded lands were only those from shore out to 3 nm, as opposed to the full 200 nm. The second issue was whether the monument process, including decisions on fishing policy, would be done with local input or by presidential fiat. The third reservation was whether the process protected fishing rights. For these reasons Clinton did not issue an executive order for an NWHI monument. This position was respected by the Clinton administration and was not altered until the Bush administration.[33]

---

[33]   Johns op. cit.

Babbitt objected to his plan for a NWHI monument being thwarted by Cayetano. He went to Senator Inouye to complain. Inouye told him that the Hawai'i fishing industry and the governor of Hawai'i were opposed to the designation.

Unwilling to go against the both the governor of Hawai'i and Senator Inouye, Clinton signed Executive Order 13178 on December 4, 2000, designating the NWHI as a Coral Reef Ecosystem Reserve (CRER), to be managed by the Department of Commerce under the National Marine Sanctuaries Act (NMSA). In order for the president to be able to make such a designation, the NMSA had been amended three weeks earlier. On November 13, 2000, Clinton signed the NMSA Amendments Act of 2000 (Public Law No. 106-513), which, coincidentally, was the same day that he had met with Cayetano. The NMSA Amendments law appropriated $4 million for each fiscal year 2001, 2002, 2003, 2004 and 2005 to take action to initiate the designation of the reserve as a national marine sanctuary, establish a NWHI CRER Advisory Council (to include at one representative from Native Hawaiian groups) and to manage the reserve until it was designated as a sanctuary.

The boundaries of the NWHI CRER mimicked those of the NWHI Protected Species Zone, which the Council had created in 1991 (see Chapter 8). The Council staff was in disbelief that the CRER executive order not only supplanted the Protected Zone with the reserve but also used language from the draft CRE FMP, which the White House's Council on Environmental Quality had prevented the WPRFMC from presenting during the visioning session in Washington, D.C., but later requested a copy of it. The reserve encompassed 137,000 square miles (355,000 square kilometers), making it 13 times the size of all the existing U.S. sanctuaries combined. Executive Order 13178 established no-take reserve preservation areas from the seaward boundary of

Hawai'i state waters and submerged lands to a mean depth of 100 fathoms around Nihoa, Necker, Gardner, Maro, Laysan, Lisianski, Pearl and Hermes, and Kure and 12 nm around various banks with some exemptions for bottomfish fishing. The NMSA Amendment stated that "no closure areas around the [NWHI] shall become permanent without adequate review and comment." The order immediately set in motion the debate as to whether the newly designated reserve and potential future sanctuary would allow fishing or not. There was a certainty among National Ocean Service employees that the federal government should not support any commercial fishing in the area. The bitter interagency rivalry continued.

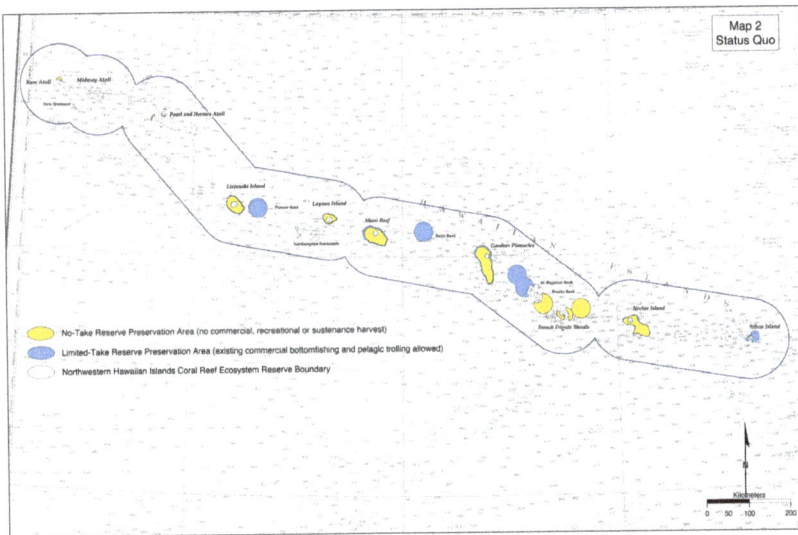

President Clinton's executive order of December 4, 2000, established the NWHI CRER and no-take and limited-take reserve preservation areas. The order allowed existing commercial bottomfish fishing and pelagic troll fishing and specified that a process begin to consider converting the area into a national marine sanctuary. Source: NOAA NMSP 2004, page C-50. *NOAA National Marine Sanctuary Program photo.*

There was particular objection from fishermen that the new Clinton executive order proposed the closure of 4% of the area as "no-take." This was considered too small an environmental protection effort by environmental activists and too much by fishermen. The fishermen said that the 4% that was proposed for closure represented 30% of the established fishing grounds for bottomfish (Wright 2000). Although there were local advocates for the proposed sanctuary, there was also a sense that this change was being imposed upon Hawai'i by federal regulators interested in taking more land and waters from Native Hawaiians.

On January 18, 2001, Clinton issued Executive Order 13196, which finalized the NWHI CRER (created by Executive Order 13178) by making the reserve preservation areas permanent and capping the number of permits, as well as the "annual aggregate take" for particular types of fishing. The annual aggregate take is based on historical levels of permit issuance and "take." For the lobster fishery, the annual aggregate level was to be the permittee's individual take in the year preceding December 4, 2000.

According to Babbitt, the order, while not exactly what he wanted, provided 80% of the protection needed to turn the area into a monument. There was no discussion of compensating fishermen or Native Hawaiians.

Executive Order 13178 created a newly designated NWHI CRER Advisory Council, which was to be run from the National Ocean Service's National Marine Sanctuary office in Washington, D.C., until the area went through a sanctuary designation process. Robert Smith transferred from the USFWS to the Department of Commerce to become the first coordinator of the NWHI CRER Advisory Council (Grover 2016, page 15). The new reserve created a major opportunity for the National Ocean Service to expand within NOAA and provided a large infusion of government funds. The Coral Reef Conservation Act, signed into law on December 23, 2000, authorized an appropriation of $16 million each year for fiscal years 2001, 2002, 2003 and 2004 for purposes of the Act. In 2002, Timothy Keeney, NOAA deputy assistant secretary for oceans and atmosphere, reported to Congress that that funding increased to $27 million and $28 million in fiscal years 2001 and 2002, respectively.

The complete closure of the NWHI fishery had been temporarily forestalled, but the bitter battle between state and federal control over the land and waters of the State of Hawai'i in the NWHI continued.

Environmentalists put the blame for Clinton's refusal to sign an executive order creating a NWHI monument on the WPRFMC and, more specifically, its executive director, Kitty M. Simonds, just as they had done for the collapse of the lobster fishery, the decline of the monk seal population and other environmental woes. Reflecting the strong emotions elicited by this issue was Ellen Athas, formerly of the Council on Environmental Quality, who was quoted in the *Cascadia Times* as saying that the action of continuing to allow fishing in NWHI was "perverse" (Koberstein 2003).

Lessons Learned

The NWHI lobster fishery was a fishery that developed at a time when scientific information about the stocks and their habitat was lacking and the nation's new fishery law focused on managing fisheries to produce optimum yield. While Council members and fishermen wanted to take early protective action, NMFS overrode them. Meanwhile, new traps proved extremely efficient, and the fishery skyrocketed due to soaring prices internationally for rock lobsters and fishermen following the market (Markrich 2020). As the catch per unit effort for the NWHI lobsters declined, the Council, NMFS Honolulu Lab and the fishing industry worked together to develop the first quota management approach in the Pacific Region. When the catch per unit effort fell below 1.0, the quota would be ratcheted down; when the catch per unit effort went above 1.0, the quota would be increased.

In 1989, an unforeseen oceanographic regime shift led to a major recruitment failure in the spiny lobster stock. The question to be asked was whether to manage the fishery based on a new optimum yield/maximum sustainable yield that recognizes the regime shift or as an overfished fishery with the aim to rebuild the stock to the original maximum sustainable yield.

In the end, the potential for a sustainable lobster fishery that would have benefited the Hawaiian people was closed as a result of a polarized political process.

5.2 Other Crustacean Fisheries in the Region

The commercial crustacean fishery in the main Hawaiian Islands has historically focused on lobster, Kona crab and deepwater shrimp. Lobster was a traditional source of food for Native Hawaiians and was sometimes used in early religious ceremonies. For a time after the arrival of Europeans, the lobster fishery was the most productive of Hawai'i's commercial shellfish fisheries. The reported catch in the main Hawaiian Islands was about 60 mt. By the early 1950s, the commercial catch of spiny lobsters had dropped by upward of 85%.

A small commercial fishery for Kona crab has existed since 1948, most of which is caught on Penguin Bank. An intermittent deepwater shrimp fishery began in 1967. Activity varies from year to year with an annual average of up to three vessels reporting the catch of deepwater shrimp to the State of Hawai'i.

In American Samoa, spiny lobsters constitute the bulk of the crustacean fishery. Lobsters are often present at important meals such as weddings, funerals and holidays. Formerly, they were harvested and consumed on the village/family level. Currently, they are primarily caught by commercial fishermen in territorial waters and purchased at the market. Fishing for lobster is labor intensive and not without some danger because they are typically taken by hand on the outer slope of the reef. Spiny lobsters are usually speared at night by free divers while hunting for finfish. Since 2006, boat-based fishermen have increasingly caught the majority of spiny lobsters. Crustaceans harvested in American Samoa are processed at sea on the vessel and marketed as fresh product or as frozen lobster tails. The domestic processing capacity and processing levels are equal to or exceed the harvest.

Little is known about Guam's crustacean fishery. Fishing for these species primarily occurs inshore with fishermen fishing recreationally or commercially on a small scale. Local fishermen assert that lobsters around Guam generally avoid traps, and it is illegal on Guam to spear spiny lobsters or take lobsters or crabs with eggs. Thus, lobsters are typically taken by hand on the outer slope of the reef. This method is labor intensive and not without some danger. Lobster can also be found in tidal pools along reef flats during low tides when there is a certain full moon.

A deepwater shrimp fishery was attempted in Guam, but no known operations have occurred since. Deepwater shrimping that might occur around Guam in the future would most likely use traps at depths generally greater than 1,000 feet (300 meters).

The CNMI crustacean fishery primarily targets spiny lobster and deepwater shrimp in nearshore waters with catches taken almost exclusively within 3 nm of the inhabited southern islands.

Beyond this boundary, lobster habitats become relatively small and difficult to access. Local fishermen assert that lobsters around CNMI generally avoid traps. Anecdotal information suggests bottomfish fishermen occasionally anchor to dive at night for lobsters, mostly for personal consumption.

Spiny lobster in the Mariana Archipelago are typically taken at night by divers hunting for finfish.

A deepwater, trap-based fishery emerged in the CNMI in the 1990s to harvest shrimp, mostly around Saipan and Tinian on flat areas near steep banks at depths greater than 1,000 feet (300 meters). Two fishing companies landed 26,808 pounds (about 12 mt) between May 1994 and February 1996. Shortly thereafter both companies exited the fishery, citing excessive gear loss as the primary reason. One company used an ovular trap made of plastic, weighing about 15 pounds (6 kilograms). It reported a trap loss of less than 4% per set when using an average of about 13 traps per string. Another company used a lightweight trap (5.5 pounds, or 2.5 kilograms) with nylon netting, which could tear away relatively easily and be recovered if it became ensnared on the bottom. Trap size was smaller, and catch per trap was on average 76% of the catch of plastic traps, but the company was able to deploy many more traps per string with less risk of gear loss.

The deepwater shrimp fishery in CNMI has faced a number of challenges. Gear loss has been a common problem and makes many ventures unprofitable. A second difficulty has been fluctuating market demand for the product due to the short shelf life and inconsistent quality of the catch. Lastly, these fisheries have generally experienced local depletion on known fishing grounds, which has led to a drop in catch rates. While other banks might have abundant stocks, fishermen are generally reluctant to explore these areas because unfamiliarity with them could lead to even greater rates of gear loss.

Deepwater shrimp. In the CNMI, the species has a short shelf life and an inconsistent quality. These factors, along with gear loss, led two companies targeting the species in the 1990s to leave the fishery. *NOAA photo/Office of Ocean Exploration and Research, 2015 Hohonu Moana.*

A few fishermen have expressed interest in fishing for lobsters in the Pacific Remote Island Areas, and at least two have attempted it.

However, tropical lobsters (green spiny, *Panulirus penicillatus*) do not enter traps readily. A lobster harvest exploration in 1999 in Palmyra and Kingman Reef waters was unsuccessful. This venture is also believed to have attempted to target the red crab (*Chaceon* spp.) and deepwater shrimp.

Under the Crustacean FMP, it is unlawful for any person to fish for, take or retain lobsters with explosives, poisons or electrical shocking devices, as this will kill other marine life in the area instead of just the targeted ones. The permitted methods for harvesting spiny lobster and deepwater shrimp ensure there is no to minimal bycatch. The fisheries are subject to an ACL to prevent overfishing, and a federal permit is required to harvest federally managed crustacean species in U.S. waters. Permit holders are required to provide catch and effort data. Fishermen have the option of using NMFS approved electronic logbooks in lieu of paper logbooks.

To support fishery monitoring, vessel operators must report their port, as well as the approximate date and time at which spiny and slipper lobsters will be landed, not less than 24 hours, but not more than 36 hours, before landing. They must also report the location and time that offloading spiny and slipper lobsters will begin, not less than six hours and not more than 12 hours before offloading. The NMFS regional administrator notifies permit holders of any change in the reporting method and schedule at least 30 days prior to the opening of the fishing season. To support fishery monitoring, when requested to do so by the NMFS regional administrator, all fishing vessels must carry an observer.

Amendment 11 to the Crustacean FMP, completed by the Council in 2002 and implemented by NMFS in 2004, established no-take marine protected areas (MPAs) around Kingman Reef, Jarvis, Howland and Baker.

TABLE 3: CRUSTACEAN FMP: SUMMARY OF COUNCIL ACTIONS		
DATE	ACTION	MEASURES
1983 Mar 9	FMP came into effect	Permit, data reporting and observer requirements within EEZ waters around the main Hawaiian Islands, American Samoa and Guam. Measures for the NWHI, including federal permit requirements, spiny lobster minimum size limit, gear restrictions, ban on harvest of egg-bearing female spiny lobsters, mandatory logbook program, requirement to carry a fishery observer if directed by NMFS and a ban on lobster fishing within 20 nm of Laysan Island, all NWHI waters shallower than 10 fathoms and all NWHI lagoons.
1983	Amendment 1	Adopted the State of Hawai'i's lobster fishing regulations for the federal waters around the main Hawaiian Islands.
1985	Amendment 2	Modified the allowable trap opening dimensions to minimize risk of harm to the Hawaiian monk seal while allowing flexibility in trap design.

1985	Amendment 3	Revised the minimum spiny lobster size specifications for the NWHI management area from a carapace length-based limit (7.7 centimeters) to a limit on tail width (5.0 centimeters).
1986 Oct 31	Amendment 4	Applied the NMFS emergency interim rule closing lobster fishing 20 nm of Laysan and within the fishery conservation zone landward of 10 fathoms in the NWHI to slipper lobsters.
1988 Jan 14	Amendment 5	Established minimum size for slipper lobster, required escape vents in all lobster traps and release of egg-bearing female slipper lobsters. Revised some permit application and reporting requirements. Changed the FMP name from "Spiny Lobster" to "Crustaceans."
1991 Jan 28	Amendment 6	Defined recruitment overfishing for lobster stocks in terms of spawning potential ratio. A stock would be considered recruitment overfished when the ratio minimum threshold is 20%.

1992 Mar 26	Amendment 7	Established a NWHI limited-access program (15 transferable permits based on historical and current participation and renewed contingent on minimum landing requirements), an adjustable fleet-wide NWHI annual harvest guideline, a closed season (January through June) and a 1,100 maximum limit on the number of traps per vessel. Revised reporting requirements and certain other provisions.
1994 Aug 9	Amendment 8	Eliminated the NWHI minimum landings requirements for permit renewal. Allowed the catch per unit effort target that is used to set the harvest guideline to be changed through the framework process. Modified reporting requirements.

1996 Jul 5	Amendment 9	Set the annual harvest guideline on a constant percent of the population (proportional to the estimated exploitable population size) based on a 10% risk of overfishing (associated with a constant harvest rate of 13% per year). Specified that annual harvest guidelines be published by NMFS no later than February 28 of each year. Eliminated earlier in-season adjustment procedures, minimum size limits and prohibitions on harvesting of egg bearing females (but they must be counted against the annual harvest guideline). Provided a mechanism for certain regulatory adjustments to be made through framework procedures.
1997 Jul 1	Regulatory Amendment 1	Implemented a VMS program for the NWHI crustacean fishery.
1998 Jul 29	Regulatory Amendment 2	Allocated the 1998 NWHI harvest guidelines on a bank-specific basis (Necker Island, Gardner Pinnacles, Maro Reef and all remaining NWHI lobster fishing grounds combined).

1998 Jul 8	Regulatory Amendment 3	Divided the NWHI into four fishing grounds across which harvest is allocated. Allowed fishing vessels with NMFS-certified VMS to be within those fishing grounds immediately after grounds closure, provided the vessel is steaming to port or other open fishing grounds.
1999 Feb 3	Amendment 10	Designated EFH. Described bycatch and fishing communities for American Samoa, Guam and the CNMI. The provisions for Hawaiʻi were approved later on August 5, 2003.
2002 Jun 14	Amendment 11	Prohibited the harvest of management unit species in the no-take MPAs established around Rose Atoll in American Samoa, Kingman Reef, Jarvis, Howland and Baker. NMFS final rule implemented Amendment 11 on February 24, 2004.
2006 Oct 26	Amendment 12	Permit and reporting requirements for federally managed crustaceans caught in EEZ waters around CNMI and the Pacific Remote Island Areas.

2008 Nov 21	Amendment 13	Added deepwater shrimp genus *Heterocarpus* as a management unit species and required federal permits and reporting for its harvest in federal waters of the region.

Chapter 6

Precious Coral Fisheries (1983–2009)

Richard Grigg was instrumental in the development of both the Precious Coral PMP and FMP, based on his seminal work in the 1970s. Much of the early exploratory work in Hawaiʻi relied on dredges as a sampling tool. In 1970, the State of Hawaiʻi marine affairs coordinator, John Craven, facilitated access to the *Star II*, a two-man submersible, which was then being used by Maui Divers. In 1971, the administration of the sub was moved to the University of Hawaiʻi and the sub was renamed the *Makalii*. Using this vehicle, Grigg was able to provide some of the first descriptions of the species of the Makapuʻu coral bed.

As a result of this background, Grigg was considered the ideal choice to head the Precious Coral Plan Team. He began his task by first looking at how other nations managed precious coral beds, using techniques such as bed rotations, closed seasons, size limits and weight quotas. But he thought none of them was appropriate for the circumstances in the Hawaiian Islands where, instead of wide swaths of precious coral on the seafloor, small individual patches are found isolated on seamounts. Since part of the task of the FCMA was to optimize fishing within sustainable limits, Grigg sought a means of managing the scarce precious coral resources on the basis of lifecycle and ecology—the way that many other fisheries are managed.

The FMP maintained the regulatory approach of the PMP, and the four categories of management units were continued: 1) Established Beds (defined as beds of known areas in which optimum yield had been determined; 2) Conditional Beds ("beds known to contain precious corals but in which the optimum yield is calculated based on the area of the conditional beds relative to the area of the Makapuʻu Bed"; 3) Refugia (places off limits to harvesting); and 4) Exploratory Beds (new areas to be explored).

The Precious Coral FMP was completed and implemented in September 1983. It established the management unit species and management areas and classified known beds. The intent of the FMP (as with the PMP) was to encourage or facilitate precious coral fishing in an ecologically sound manner. The plan set an annual harvest limit (a proxy for maximum sustainable yield) at 1,000 kilograms (~2,200 pounds) for all managed precious coral species in areas classified as "exploratory." Half the allotment was designated for foreign fisheries and the other half to domestic fishermen. The Council intended to encourage U.S. vessels to explore and harvest corals in areas that might otherwise be dominated by foreign vessels.

Unfortunately, the FMP did not end the damage caused by poachers. It was estimated that 90,000 kilograms (~200,000 pounds) of precious coral were taken illegally from U.S. waters in the Western Pacific Region in 1985 (WPRMFC 1987a). Council Executive Director Kitty M. Simonds voiced the frustration that many felt at the time about not being able to do more to protect a resource from being obliterated by foreign greed and what seemed to be an inexhaustible U.S. market for precious coral jewelry. "The foreigners are out there poaching, and they are selling our corals back to us," Simonds said in frustration in 1988 (Tenbruggencate 1988). Foreign poaching of precious corals from Taiwan and Japan would end in the early 1990s as a result of increased

U.S. Coast Guard presence and a collapse in the world market for precious coral jewelry.

The Precious Coral FMP established the Wespac Bed, located between Nihoa and Necker Islands, as refugia. The objectives of the refugia were to a) preserve coral beds as natural areas for purposes of research; b) establish control areas that could be used in the future to measure environmental impacts of coral harvesting; and c) establish possible reproductive reserves for enhancement of recruitment into adjacent waters.

Over the years, amendments to the FMP enhanced coverage and protection of precious coral in the EEZ to include the Pacific Remote Island Areas and expanded the managed species to include all of the genus *Corallium* (initially only the known commercial species of that genus were included). Provisions for the issuance of experimental fishing permits were designed to stimulate the domestic fishery. Domestic fishermen were encouraged to use submarines rather than dredges to avoid unnecessary damage to coral beds, and a "selective harvest only" rule was passed for the established beds.

However, domestic harvesting of deepwater precious corals was coming to an end. Maui Divers stopped harvesting precious coral with its submarines after an accidental death during the course of operation caused it to lose its insurance. Another company, American Deepwater Engineering, operated for a year and a half, ceasing operations after 1999 and leaving the precious coral business following President Clinton's executive orders creating the NWHI CRER, which reduced by 75% the potential areas with the most accessible precious coral beds in Hawai'i.

Precious Coral Plan Team Chair Richard Grigg (left) and HDAR Administrator Bill Devick visit an American Deepwater Engineering submersible in Honolulu harbor (circa 2000). The Council's Precious Coral FMP encouraged domestic fishermen to use submarines rather than dredges to avoid damage to coral beds and required selective harvest only for established beds. *WPRFMC photo/Sylvia Spalding.*

Although possibilities exist to harvest deepwater precious corals throughout the Western Pacific Region, no commercial operations are presently engaged in doing so. One of two operators expressed interest, using remotely operated vehicles, but none did so in the EEZ except in a few places in the main Hawaiian Islands. Nonetheless, in 2004, the Council prohibited the harvest of precious coral in the no-take MPAs it had created under the CRE FMP (see Chapter 9), including areas around Rose Atoll in American Samoa and the Pacific Remote Island Areas. Additionally, based on new information on the life history of gold coral, the Council recommended a five-year moratorium on the harvest of gold coral in 2008, and then it renewed that moratorium in 2013 for an additional five years and again in 2018 for another five years (NMFS 2018).

The only significant harvesting effort for precious coral today by U.S. fishermen is of black coral, which grows in relatively shallower waters than other precious corals. In an effort to protect black coral stocks in Hawai'i, in 2006 the Precious Coral Plan Team recommended a decrease in the maximum sustainable yield of black coral, and a Council amendment of the FMP (implemented in 2008) set a quota of 5,000 kilograms (~11,000 pounds) for every two-year period in the 'Au'au Channel. The amendment covers both federal and state waters combined as a precautionary measure.

The effort initiated by the Council to protect and manage precious corals stands was a remarkable achievement, indicative of the Council's conservative management approach. According to Grigg, "It is important to recognize that the Council was a pioneer in developing rules and regulations to successfully manage precious coral resources. The Precious Coral FMP could in fact be used as a template for other precious coral fisheries in the world (Italy, France, Japan and Taiwan) that are in urgent need of comparable management programs" (Grigg 2010).

TABLE 4: SUMMARY OF CURRENT KEY PRECIOUS CORAL MEASURES FOR THE HAWAI'I ARCHIPELAGO

- Federal permit and logbook reporting.
- Use of only selective gear that can discriminate or differentiate between type, size, quality or characteristics of living or dead corals.
- Bed-specific quotas.
- Closed areas.
- Minimum height 10 inches for live pink coral.
- Minimum stem diameter 1 inch or minimum height 48 inches for live black coral.
- Moratorium on gold coral through June 30, 2023.

TABLE 5: PRECIOUS CORAL FMP: SUMMARY OF COUNCIL ACTIONS		
DATE	ACTION	MEASURES
1983 Sep 29	FMP came into effect	Established the plan's management unit species, management area, closed areas, and permit and reporting requirements. Classified several known beds.
1988 Jul 21	Amendment 1	Applied the FMP management measures to the Pacific Remote Island Areas by incorporating them into a single Exploratory Permit Area. Expanded the managed species to include all species of the genus *Corallium*. Outlined provisions for the issuance of experimental fishing permits designed to stimulate the domestic fishery.
1991 Jan 22	Amendment 2	Defined overfished for Established Beds as when the total spawning biomass (all species combined) has been reduced to 20% of its unfished condition. This definition applies to all species of precious corals and is based on cohort analysis of the pink coral, *Corallium secundum*.

1998 Oct 19	Amendment 3	Established a framework procedure for adjusting management measures in the fishery.
3 Feb 1999	Amendment 4	Defined overfishing. Described bycatch and fishing communities. The provisions regarding the Hawai'i fishing communities became effective later, on August 5, 2003.
2002 Apr 17	Framework Measure 1	Revised the definitions of "live coral" and "dead coral." Suspended the harvest of gold coral at Makapu'u Bed. Applied minimum size restrictions only to live precious corals and the minimum size restrictions for pink coral to all permit areas. Prohibited the harvest of black coral with a stem diameter of less than 1 inch or a height of less than 48 inches (with certain exceptions) and the use of non-selective fishing gear to harvest precious corals.

2002 Jun 14	Amendment 5	Prohibited the harvest of managed species in the no-take MPAs around Rose Atoll in American Samoa, Kingman Reef, Jarvis, Howland and Baker. NMFS final rule implemented on February 24, 2004.
2007 Nov 14	Regulatory Amendment 1	Removed an exemption allowing fishermen who reported black coral harvest to the State of Hawai'i within five years prior to April 17, 2002, to harvest black coral at a minimum base diameter of 3/4 inch.
2008 Aug 13	Amendment 7	Designated the 'Au'au Channel bed as an established bed with a harvest quota for black coral of 5,000 kilograms every two years for federal and state waters combined. Implemented a five-year gold coral harvest moratorium for the entire region while research on life history is conducted.

Chapter 7

Bottomfish and Seamount Groundfish Fisheries (1986–2009)

7.1 Hawaiian Islands Bottomfish Fishery Overview

In 1973, the *Taihei Maru*, captained by Heisei "Bill" Shinsato, commercially fished bottomfish in the NWHI at a time when there were no federal regulations. Shinsato was a forward-looking fisherman and concerned that the federal government might close fishing in the NWHI. Shinsato appealed to commonsense and self-interest, saying, "If we overfish an area to a point where the fish cannot replenish, we have lost our fishing grounds" (Shinsato 1973).

Most bottomfish live 30 years or longer and start reproducing only after five or more years. Depending on the species, they will feed either during the day or the night. They live in relatively specific habitats, and, as a result, it is difficult for even skilled fishermen using hook-and-line techniques to remove so much as to threaten the stock. However, if overharvested, bottomfish stocks could take years to recover to healthy levels. To prevent this from happening, for many years Hawaiʻi commercial fishermen would rotate among bottomfish grounds, periodically resting some areas while shifting effort to others. The length of the rest period was dependent on the individual fisherman's accumulated knowledge of his catches from each individual area. In some cases the period was short; for others, it could be longer than 10 years.[34]

The most commercially valuable bottomfish species in Hawai'i waters are the shallow- to mid-water uku (gray jobfish, *Aprion virescens*) and seven deepwater species, known as the "Deep 7": 'opakapaka (pink snapper, *Pristipomoides filamentosus*), onaga (long-tail red snapper, *Etelis coruscans*), ehu (short-tail red snapper, *E. carbunculus*), hapu'uupu'u (sea bass, *E. quernus*), lehi (silver jaw jobfish, *Aphareus rutilans*), kalekale (yellowtail snapper, *P. auricilla*, and Von Siebold snapper, *P. zonatus*) and gindai (flower snapper, *P. zonatus*). Bottomfish are found throughout the Hawaiian archipelago, stretching from the main Hawaiian Islands up through the NWHI. These fish are found living in depths of about 300 to 1,200 feet (90 to 400 meters) and have long been regarded as a delicacy. Some of the earliest known writings depicting Native Hawaiians fishing describe the techniques used to catch bottomfish.

Ancient Hawaiian fishing gear, including line made from the fiber of olona (*Touchardia latifolia*) and hook made of turtle shell. Olona fishing lines measuring 300 feet (about 90 meters) in length could be tied together to catch deepwater bottomfish. *Bishop Museum photo.*

---

[34]   Kurt Kawamoto in discussion with Michael Markrich, September 8, 2016.

Long before electronic fish finders were invented, ancient Hawaiians could find bottomfish sites hundreds of feet deep off the coast. They looked carefully for tiny organisms living in nutrient streams appearing on the dark surface of the sea. These nutrient upwellings indicated that the bottomfish existed below, usually along a narrow band of steep undersea ledges and ridges, typically no more than a hundred yards wide, known as a "deep slope." Once found, these areas were memorized by fishermen, who would mentally note three or more landmarks on shore that would line up with the area. Another method of locating fishing grounds was to carefully observe the surrounding land topography. Each valley or stream was usually reflected in the undersea. During the sampan era of bottomfish fishing, additional grounds were serendipitously found after an anchor broke loose from a known spot and then got stuck at a newly discovered spot.[35]

The special equipment and skills needed to catch bottomfish—a large enough vessel and the ability to locate the fishing grounds and to catch the fish with special baits and hooks—limited the number of entrants to the fishery for many years. However, as Hawai'i's economy grew during the 1960s and the technology of fiberglass construction brought the price of vessels down, many more people were able to afford small boats and become the serious, noncommercial fishermen known in Hawai'i as "weekend warriors." Their new vessels had affordable electronic fish finders, and the low price of fuel gave them access to the fishing grounds that had once been the exclusive domain of full-time commercial fishermen. Although, under the law, fishermen were not supposed to sell their catches without a commercial license, demand for fresh bottomfish by small restaurants and stores—fueled by a growing

---

[35]   Kawamoto ibid.

tourism industry—was met by intermittent cash purchases of fish brought in by many otherwise recreational fishermen.

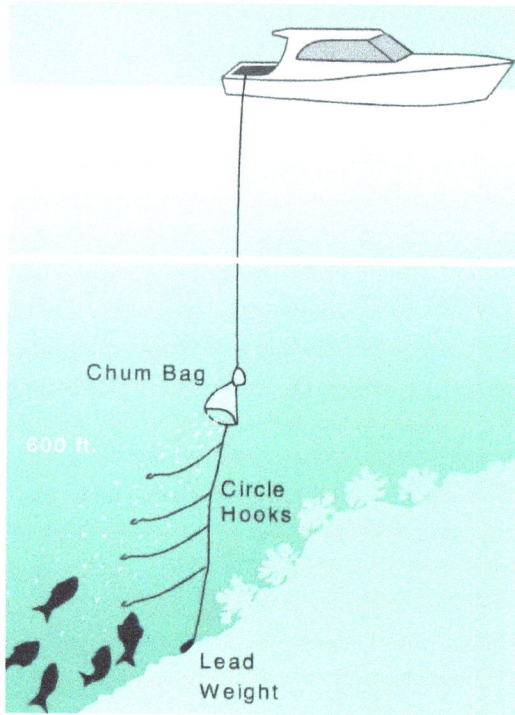

Modern bottomfish gear in Hawai'i consists of a hand or electric reel, *palu* (chum) bag, mainline and branch lines, circle hooks and lead weight.

As more noncommercial fishermen entered the fishery, the commercial fishermen recognized that these activities were impacting their livelihoods. They saw the noncommercial fishermen not only as economic competition but also as a danger to the long-term sustainability of the bottomfish stock. Since some of the most valuable bottomfish fishing grounds—such as Penguin Banks off O'ahu and Ka'ula Rock off the coast of Kaua'i—were in federal waters, the commercial fishermen came to the Council for help.

7.2 The Northwestern Hawaiian Islands Bottomfish Fishery

Two Council advisors, Kuni Sakamoto and Charlie Yamamoto, thought that something should be done about the increased fishing pressure on bottomfish, recalls former Council biologist Paul Bartram. Sakamoto and Yamamoto were some of the last old time fishermen who caught bottomfish in the NWHI using the old style Hawaiian sampans. Yamamoto came to early Council meetings and said, "Bottomfishing is changing. It used to be the domain of professionals who all knew each other, and now it's changing to weekend warriors." Sakamoto and Yamamoto thought there ought to be some kind of plan to regulate the fishery. They weren't the only ones, but they were the most outspoken.[36] They asked that the Council begin work on a Bottomfish FMP.

In 1978, University of Hawai'i doctoral candidate Steve Ralston was hired to analyze the lifecycle of bottomfish. The evidence gathered was the basis for the creation of a Bottomfish FMP. The Council, NMFS and Hawai'i Department of Land and Natural Resources, Division of Aquatic Resources (HDAR), informally agreed that the federal government would manage the NWHI bottomfish fishery and the State would manage the main Hawaiian Islands bottomfish fishery. This agreement would later come to an end.

The Council's bottomfish monitoring efforts started in 1984 and were the first time that information on the bottomfish fishery had been collected in detail. One of the first Council-sponsored research projects involved staff members going to the United Fishing Agency fish auction in Honolulu to find out who was selling bottomfish and how much was being landed. This started the collection of fishery auction data, which is used to this day.[37] In addition, the Council contracted Brooks

---

[36] Paul Bartram in discussion with Michael Markrich, October 17, 2012.

[37] Kawamoto op. cit.

Takenaka, who was then assistant manager of the fish auction, to study the characteristics of the fishermen. He found many of the new sellers were not part of the group of the longtime commercial bottomfish fishermen but were the weekend warriors about whom the commercial fishermen had complained. While he did that study, Council staff members interviewed fishermen about their catches and established the first baseline of typical species and size distributions per fishing trip.

By 1986, so much progress had been made that the joint Council–Honolulu Laboratory research effort resulted in obtaining 98% to 99% of the available information on the size of bottomfish passing through the auction. The information was collected to determine bottomfish population dynamics (WPRFMC 1986b).

The first draft of the FMP for Bottomfish and Seamount Groundfish Fisheries in the Western Pacific Region was completed in 1984. An important issue was the harmful effects of harvesting bottomfish with bottom trawl nets, which severely damaged the ocean habitat. Another concern was the status of Hancock Seamount where foreign fleets had previously overfished the groundfish stocks.

In 1986, the Bottomfish FMP was approved and implemented. It included the following major components:

1) A prohibition of certain destructive fishing techniques, including explosives, poisons, trawl nets and bottom-set gillnets.

2) A moratorium on the commercial harvest of seamount groundfish stocks at the Hancock Seamount.

3) A permit system for fishing for bottomfish in the waters of the EEZ around the NWHI.

The plan also established a management framework that provided for regulatory adjustments to be made, such as catch limits, size limits, area or seasonal closures, fishing effort limitations, fishing gear restrictions, access limitations, and permit and/or catch reporting requirements, as well as a rules-related notice system.

The new Bottomfish and Seamount Groundfish Fishery FMP covered stocks in the federal waters in the NWHI. The plan team for the FMP recommended that no regulations be enacted in Guam and American Samoa, where overfishing for bottomfish was at the time not a concern. (CNMI was not included in the plan at this stage.)

"The Council proposed logbooks because past experience with the Pelagic FMP showed they provided fishermen with a way to provide their catch data in an organized way," explained Council bottomfish specialist Mark Mitsuyasu. "People [in NMFS] asked why we would want to do that, saying the fishermen were not going to tell us the truth anyway! Well, eventually the fishery was subject to 100% observer coverage on the NWHI bottomfish fishing boats, and, when the catch report data was put together with the observer data, the observers confirmed what the fishermen were reporting. In the end, we saw very similar results."[38]

Eventually the logbook program was implemented, and, as a result, the Council was able to build a reliable database of catch and effort for managing the NWHI commercial bottomfish fishery.

Developing the Nation's First Limited-Entry Program for Finfish

During this time period, the NWHI bottomfish fleet grew from three vessels in 1978 to 40 permitted vessels in 1988. Several of the new vessels were high seas albacore fishing vessels that participated seasonally when albacore tuna were not available. Additional vessels from the Alaska salmon gill net fleet also came to Hawaiʻi to try their

---

[38]   Mitsuyasu op. cit.

hand at bottomfishing.[39] Landings from the NWHI during this 10-year period ranged from 143,000 pounds (about 65 mt) in 1978 to a high of 1 million pounds (about 454 mt) in 1987, and the Council began to consider limiting entry to the NWHI.

The NWHI fishery continued to attract new fishermen, despite relatively moribund prices (WPRFMC 1994). This at first mystified the Council fishery planners until it was discovered that the Hawai'i bottomfish fleet was being touted nationally as a legal tax deferral investment for high-income professionals.[40]

The idea surfaced that one way to protect the fishery from large numbers of off-island fishermen would be to create a limited-entry program, with special opportunities for Native Hawaiians and others who had historical ties with the local fishery. This would be modeled on programs that protect the cultural fishing rights of Alaskan Natives and Native American tribes. The limited-entry program would have two impacts: it would restrict the overall number of vessels allowed in the fishery and it would provide a special opportunity to Native Hawaiians who, it was thought, were due some kind of compensation for loss of their lands and waters to the federal government.

Simonds recalls, "We wanted to find a way to help Native Hawaiian families get back into this kind of business, so it wouldn't be just people from the mainland benefiting from Hawai'i resources."[41]

There was no precedent for an FMP limited-entry program providing Native Hawaiian fishing preference rights. However, Council members felt that a case could be made based on anthropological and historical

---

[39]   Kawamoto op. cit.

[40]   Bartram op. cit.

[41]   Simonds op. cit.

evidence. On the U.S. mainland these were considered acceptable means of providing proof of fishing rights for Native American tribes.

Phil Meyer, an economic consultant experienced with Native American fishing rights, was contracted by the Council to develop the bottomfish limited-entry system. He believed that a limited-entry preference for Native Hawaiians could be implemented. One problem, however, was how to define a Native Hawaiian. There was disagreement with the Office of Hawaiian Affairs over the definition, and Council Chair Yee said it was the Council's responsibility to come up with a working definition, acceptable to both that office and island fishermen (WPRFMC 1986b).

This discussion led to a proposal for a special clause in the Bottomfish FMP dealing with Native Hawaiians. The issue was significant because, as NOAA attorney Hochman pointed out, Native Hawaiians were not covered under a treaty in the same way as were other Native Americans. It was not enough to simply establish criteria that identified Native Hawaiians and to assert sociocultural claims for preference. In order for limited-entry program for Native Hawaiians to exist, it would be necessary to justify preferential treatment in terms of the MSA statutes (ibid).

Native Hawaiian Council member and commercial fisherman Buzzy Agard, HDAR biologist Alvin Katekaru and attorney Gerald A. Sumida were appointed by Yee to work with the Office of Hawaiian Affairs to develop a proposal for Native Hawaiian rights for inclusion in the Bottomfish FMP. To support this initiative, a letter from the Office of Hawaiian Affairs administrator dated August 7, 1986, stated that ample evidence existed to establish a legal and moral claim for special consideration of Native Hawaiian fishing rights. The new project came to be called the Native Hawaiian Rights Project.

Meetings were then held with the Office of Hawaiian Affairs, which agreed to pay 50% from its ceded lands income toward the cost of the proposed Native Hawaiian Rights Project. Questions were raised during the meetings as to the legality of a special allocation to Native Hawaiians. NOAA attorney Hochman and consulting attorney Sumida said that an argument for the Native Hawaiian limited-entry proposal could be based only on aboriginal law because there is no treaty between Native Hawaiians and the federal government. Meanwhile, the Council decided to also support the rights of other indigenous peoples in the Western Pacific Region, specifically the American Samoans, Chamorro and Carolinians. Paul Stevenson, one of the early Council members from American Samoa, recommended the formation of a Standing Committee on Fishing Rights of Indigenous People (Ka'ai'ai 2016). The new committee supplanted the Native Hawaiian Rights Project with Guam Council member Rufo Lujan as its chair.

Guam Council member Rufo Lujan (circa 1981) served as the inaugural chair of the Standing Committee on Fishing Rights of Indigenous People and, in 1994, as chair of the Council. The committee and its predecessor, the Native Hawaiian Rights Project, explored the possibility of preferential fishery allocations for indigenous fishing communities in the Western Pacific Region. *WPRFMC photo.*

Limited-entry programs, where fishing permits were held at a specific number, were controversial at this time because there were clear winners and losers. Those who received permits were often able to make a good living because of reduced competition. However, those who were left out of a fishery complained of economic discrimination. At the same time, some Council members and NMFS felt limited-entry programs could add unnecessary expense to fishery management because they required extra staff to monitor them. The preference was that fishermen should not be managed until the catch per unit effort went down, after which restrictions should be put on the fishery. However, several Council members came to believe that a fishery without control on the number of entrants would soon be overcapitalized with too many vessels competing at low prices for ever dwindling stocks.

The Council bottomfish planners were also aware that, if the fishing regulations for bottomfish in the NWHI were too rigid, pressure would shift back to bottomfish populations in the main Hawaiian Islands. Regulations would have to balance maintaining future access with limiting entry of new participants. The door to the more abundant fishing grounds of the NWHI would be left open for fishermen, but the point would be made clear that access could be limited at any time.

The Council then began work on one of the nation's first limited-entry schemes in an effort to control participation in the NWHI bottomfish fishery. The 1988 amendment to the Bottomfish FMP divided the NWHI into two parts. The smaller southern Mau Zone was open access but required a federal permit to fish in the waters that surround Necker and Nihoa Islands. The second and larger Ho'omalu Zone offered more exclusive fishing opportunities for bottomfish but required a special limited-entry permit—this applied to the area from 165 degrees west longitude to the western end of the EEZ around the NWHI, roughly from French Frigate Shoals to Kure Atoll (WPRFMC 1986b).

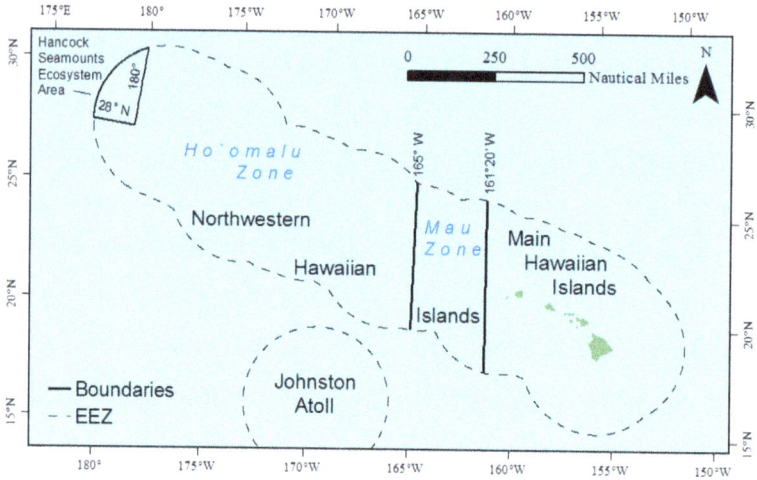

The 1988 amendment to the Bottomfish FMP divided the NWHI into two parts, creating one of the nation's first limited-entry schemes. This was done in an effort to control participation in a rapidly escalating fishery. A moratorium on seamount groundfish harvesting at Hancock Seamount, established in 1986, remained in effect. *WPRFMC illustration.*

The Mau Zone was created to allow all fishermen to travel north from the main Hawaiian Islands to catch bottomfish, where they could gain experience and earn "history of fishing effort" to possibly qualify for a Hoʻomalu Zone permit. However, access to the potentially more lucrative Hoʻomalu Zone would be tightly controlled. The area was limited to operators of vessels no longer than 60 feet (18.3 meters) who had participated in the bottomfish fishery or who had demonstrated financial commitment to the area before the control date of August 7, 1985. Because it was common knowledge that fishing was better in the Hoʻomalu Zone, this control date was fixed with the intent of preventing speculators from obtaining the permits and reselling them. In addition, permits for the Hoʻomalu Zone would be subject to renewal each year, based on a fixed fishing performance standard, i.e., "use it or lose it." No new permits would be issued unless it was determined that there were sufficient fish stocks to sustain an increase.

In addition, limited-entry permits were non-transferrable, meaning they could not be sold or otherwise transferred to new entrants.

From a fisherman's point-of-view, there were some negative outcomes from the strict regulations put in place in the NWHI. Bottomfish fisherman Timm Timoney recalled, "At least five of the boats during the 1978 to 1988 period were high seas albacore boats that only fished three to four months during the winter and then lost their permits due to the inflexibility of the "use it or lose it" [provision]. They had qualified for permits and lived in Hawai'i, but NMFS and local managers could not see that participating in multiple fisheries helped ensure economic survival and reduced pressure on specific fisheries. However, the strict requirements did get rid of the folks who weren't serious and/or using it as a tax dodge." Timoney also recalls that the non-transferability of permits "was a huge contentious issue with qualified permit holders." [42]

Many people who came into the fishery were not serious about fishing, affirmed bottomfish specialist Mitsuyasu. After the establishment of strict requirements on who could fish and on what basis, people began dropping out of the fishery. [43]

In 1989, NMFS, consistent with the requirements of the FMP, issued six limited access permits for the Ho'omalu Zone. Landings peaked at about 1 million pounds (about 454 mt) in 1986 and dropped significantly after the limited-entry program was implemented to about 400,000 pounds (about 180 mt) a year. This was expected as vessels left the fishery.

In 1991 the plan was again amended allowing the placement of observers on vessels (at the request of the NMFS) if the vessel intended to fish within 50 nm of the NWHI, an area the Council had delineated as a Protected Species Zone (see Chapter 8).

---

[42] Timm Timoney in discussion with Sylvia Spalding, September 21, 2010.

[43] Mitsuyasu op. cit.

Native Hawaiian Preferential Permits

The Standing Committee on Fishing Rights of Indigenous People recommended that studies be conducted on native fishing rights in the region. As a result of this recommendation, the Council published five studies in 1989 and 1990 to determine if sufficient evidence existed to support a legal basis for preferential rights that could become a part of limited-entry systems. Robert Iversen of Pacific Fisheries Consultants, archaeologist Tom Dye and attorney Linda Paul were hired to do an in-depth study of Native Hawaiian fishing rights and practices (Iversen et al. 1990a, 1990b). Similarly, several anthropologists and historians under Council contract would compile histories on the lives and cultures of indigenous fishing communities of American Samoan, Guam (Chamorro) and the CNMI (Chamorro and Carolinian) (Amesbury and Hunter-Anderson 1989a, 1989b; Severance and Franco 1989). The purpose of this research was to lay the foundation for provisions in limited-entry programs that would give preferential treatment to indigenous fishermen in Hawai'i and the U.S. Pacific territories.

Legal research commissioned by the Council indicated that native peoples could historically assert rights to high seas resources under two legal theories: 1) effective exercise of sovereign control and 2) proof of long and continuous usage. After the forceful overthrow of the Hawaiian monarchy in 1893 and America's subsequent annexation of Hawai'i, Native Hawaiians were unable to exercise sovereignty. The difficulty for the Native Hawaiian position was that, lacking a treaty, Native Hawaiians had no mechanism other than the courts and the Hawai'i State Legislature to petition federal or state government agencies for participation in management of or access to fishery resources.

The Council's historical studies were unable to verify bottomfish fishing in the NWHI prior to the 1930s. According to Robert Iversen, the research group tried but was unable to find surviving Native Hawaiians who could provide knowledge of such activity. Ultimately, the Standing Committee on Fishing Rights of Indigenous People adopted the Office of Hawaiian Affairs definition of a Native Hawaiian fisherman as a person who can document that his or her relatives resided in Hawai'i prior to 1778.

As in the past, when confronted with an unprecedented situation, Council Executive Director Kitty M. Simonds researched similar examples from other countries. In 1991 she wrote a letter to the Ministry of Agriculture and Fisheries in New Zealand inquiring about its preferential programs for the Maori people. The response from the ministry's director of policy discussed *"taiapure"* (translated literally as "a coastal patch"), a local management tool established in an area that has customarily been of special significance to Maori people as a source of food or for spiritual or cultural reasons.

The Council's studies and research established the justification and initiative for the Council to seek remedies for shortcomings in the MSA regarding recognition of traditional aboriginal fishing rights and access to resources. They also formed the basis for the Council's efforts to enhance inclusion of traditional ecological knowledge and participation of indigenous peoples in the federal fisheries decision-making process.

Backed with the knowledge from these studies, the Council worked with federal agencies and Congressional representatives to ensure the 1996 amendment to the MSA recognized the importance of traditional knowledge and fishery practices in the Western Pacific Region. For example, the 1996 reauthorization (the SFA) included provisions [Section 305(i)(2)] for a Community Development Program for the Western Pacific Region. This provision was modeled on the Alaska Community Development Quota program. However, unlike the Alaska program, the Western Pacific Community Development Program allows benefits to be delivered to the community beyond just quotas. Specially, the Council sought to provide preferential fishing rights in the form a limited access permits and exceptions for training purposes.

In 1999, a limited-entry program became effective in the Mau Zone (Amendment 5). Using the new Community Development Program authority under the SFA, the Council set aside 20% of the new Mau Zone limited-entry permits for qualifying Native Hawaiian fishermen. The Mau Zone limited-entry program was implemented with a cap of 10 total permits, two of which were Community Development Program permits.

Native Hawaiian fishermen were never able to take advantage of this long-researched indigenous preference program. The NWHI CRER would be proclaimed in 2000, and the nation's first marine national monument, which would supplant it, would take away all fishing rights (except sustenance fishing, i.e., catching to eat while within in the monument) and phase out all existing bottomfish fishing permits.

A Sustainable Fishery

One of the novel features of the Bottomfish FMP was the indicators developed by the Bottomfish Plan Team and managers that allowed them to determine whether too much pressure was being put on the fishery. Under the FMP rules, the following stress indicators were established for the bottomfish fishery:

- If the catch per unit effort for any species were to decline to 50% or less of the average aggregate catch per unit effort for the first three years of available data, it would be considered a cause for concern and possibly action pending analysis on the effects of the fishery and the need for adjustment in management.

- If more than 50% of the catch of a species were below the size at first maturity, the stock could be stressed by fishing activity. Further analysis would be conducted to determine what action if any would be appropriate to conserve the stock.

- If the spawning stock biomass per recruit were less than or equal to 20% (a critical threshold in recruitment) for any bottomfish species, overfishing would be declared and the fishery for that species would be shut down.

- Any combination of these indicators would put regulators on the alert for a "yellow light situation" when analysis would be needed to determine whether new management measures were necessary. A "red light" situation would exist if emergency action were required.

In 1996, just prior to the establishment of the Mau Zone limited-entry program, NMFS issued a high of 27 open access permits. By 1999, through the implementation of the Mau Zone limited-entry program and the performance standard, the number of active vessels was reduced to 10 for the Mau Zone and seven for the Hoʻomalu Zone. Landings dropped, and the reduced effort significantly lowered the risk of overfishing.

Information from the logbooks indicated that the catch rates remained fairly consistent between 1986 and 2002, going from 2,206 pounds (about 1 mt) per trip to 2,496 pounds (about 1.1 mt) per trip in the Mau Zone and from 5,301 pounds (about 2.4 mt) to 4,651 pounds (about 2.1 mt) per trip in the Hoʻomalu Zone (WPRFMC 2004a, page 3-51).

TABLE 6: TOTAL COMMERCIAL BOTTOMFISH
LANDING FROM THE NORTHWESTERN AND MAIN
HAWAIIAN ISLANDS, 1984–2004 (1,000 pounds)

(Note: Amendment 2 in 1988 imposed landing requirements that caused several boats to lose their permits and leave the fishery.)

Source: NMFS and HDAR (In: WPRFMC 2007b, page 3-41).

YEAR	MAU ZONE	HOʻOMALU ZONE	TOTAL NWHI	MAIN HAWAIIAN ISLANDS
1984	NA	NA	661	807
1985	NA	NA	922	763
1986	NA	NA	869	810
1987	NA	NA	1,015	783
1988	NA	NA	625	1,164
1989	118	184	303	1,006
1990	249	173	421	646
1991	103	283	387	548
1992	71	353	424	587
1993	98	287	385	348
1994	160	283	443	458
1995	166	202	369	440
1996	133	176	309	440
1997	105	241	346	513
1998	66	266	332	479
1999	54	269	323	455
2000	49	213	262	497
2001	50	236	286	367
2002	112	127	239	351
2003	99	152	251	334
2004	97	169	266	366

Losing the Fishery

Legal challenges against the fishery began in 2000 with the Earthjustice Legal Defense Fund filing a complaint on behalf of the Greenpeace Foundation, Center for Biological Diversity and Turtle Island Restoration Network, alleging that the U.S. Department of Commerce and NMFS, in connection with their authorization of the bottomfish and crustacean fisheries in the NWHI, violated the ESA, National Environmental Policy Act and the Administrative Procedure Act (Greenpeace Foundation, et al. v. William M. Daley, et al. (D. Haw.) Civ. No. 00-00068 (SPK) (FIY)). Concern focused on bottomfishing interactions with the endangered Hawaiian monk seals in the NWHI. The *Honolulu Star-Bulletin* reported on November 18, 2000, that Samuel King (senior judge of the U.S. District Court for the District of Hawai'i) would hold a hearing on Earthjustice's request to close the bottomfish fishery, where about a dozen bottomfish vessels operating in the NWHI harvest snappers—onaga and 'opakapaka—and jacks, such as ulua and papio (Barayuga 2000). King requested and met with active NWHI bottomfish fishermen to better understand the fishery to inform his decision. In March 2001, King denied the plaintiffs motion for a permanent injunction of the NWHI bottomfish fishery.

Although the fishery was not closed, NMFS in 2001 stopped issuing new bottomfish permits in the NWHI following Clinton's Executive Order 13196, which finalized the establishment of the NWHI CRER. That order said, "there shall be no increase in the number of permits of any particular type of fishing (such as for bottomfishing) beyond the number of permits of that type in effect the year preceding the date of this order."

The annual level of aggregate take under all permits of any particular type of fishing was also capped to that taken in the preceding years. The order also made permanent a suite of reserve preservation areas that banned fishing from the seaward boundary of Hawai'i state waters and submerged lands to a mean depth of 100 fathoms around Nihoa, Necker, Gardner, Maro, Laysan, Lisianski, Pearl and Hermes, and Kure and 12 nm around various banks. Bottomfishing was provided some exemption and could continue seaward of a mean depth of 10 fathoms around Nihoa and Gardner; seaward of 20 fathoms at Necker and Maro; seaward of 50 fathoms at Laysan and Lisianski. The fishery could also continue for five years at Raita Bank and the first bank west of Saint Rogation. After five years, fishing around these banks would be allowed only if it was determined that it did not have an adverse impact on the resources at those banks.

NMFS estimated that the effects of these closures, using just the depth contours, would affect 30% to 35% of the fishery's catch and revenue (Pooley 2000). Eventually the number of NWHI bottomfish vessels would fall to eight in 2006, with four active in the Ho'omalu Zone and four active in the Mau Zone.

Between 2001 and 2004, the CRER Advisory Council office attempted to get a draft environmental impact statement through the National Environmental Policy Act process needed to transition the NWHI CRER into a sanctuary designation, but it was not successful in answering scientific questions as to why the NWHI should be closed to fishing. Lacking scientific data, the local CRER Advisory Council depended upon three arguments for a proposed no-take sanctuary in the NWHI:

1) There was interaction between bottomfish fishermen and monk seals that was negative to the seals' health.

2) Over-harvesting of lobsters had resulted in failing diets of monk seals.

3) Interactions between fishermen and protected species had resulted in an unacceptable level of monk seal and seabird deaths.

The CRER Advisory Council was unable to provide documented proof of these allegations, which were essentially the USFWS concerns from 1991, all of which the Council believed had long since been addressed.

Although most sanctuaries allow fishing, many in the National Ocean Service and USFWS felt strongly that it should not be allowed in NWHI. As Robert Smith, coordinator of the NWHI Reserve, said at an April 2001 CRER Advisory Council meeting, "The train has left the station. We intend to make some decisions on the Reserve Operations Plan ... in five to six weeks" (CRER Advisory Council 2001).[44]

At the same meeting Simonds responded to offensive characterizations of the Council and insinuations that the outcome was predetermined regardless of the public process. She said, "The Council supports the sanctuary. The Council does not support an executive order making decisions on fisheries." This was a basic point. Under the National Environmental Policy Act process, the sanctuary designation takes place using the best available science and is open to public input. The Act's guidelines were put in place to prevent rule-making on natural resources without public input.

Under the NMSA, the drafting of fishery regulations for a national marine sanctuary is assigned to the appropriate regional fishery management council. In preparing the draft regulations, the council is to use as guidance the national standards of the MSA to the extent that they are consistent and compatible with the goals and objectives of the proposed sanctuary designation.

Council staff members found it odd, therefore, when on September 20, 2004, the NMSP provided it with a document titled "Proposed Northwestern Hawaiian Islands National Marine Sanctuary: Advice and

---

[44] The final NWHI CRER Operations Plan would be published in 2005 under Acting Reserve Coordinator 'Aulani Wilhelm.

Recommendations on Development of the Draft Fishing Regulations under the National Marine Sanctuaries Act Section 304(a)(5)." The document, written "to assist" the Council in drafting fishing regulations, included model fishing regulations for the fishing alternatives the NMSP considered most consistent with the goals of the proposed sanctuary; results of the fishing alternative analysis; and resource and statistics used to evaluate these alternatives. The NMSP recommended that fishing be allowed by permit but restricted through zoning and other means for six types of fishing, including bottomfish fishing. It recommended prohibiting four other types of fishing—commercial pelagic longlining (which was already prohibited in the area through the Council's Pelagic FMP), precious coral, coral reef species and crustacean.

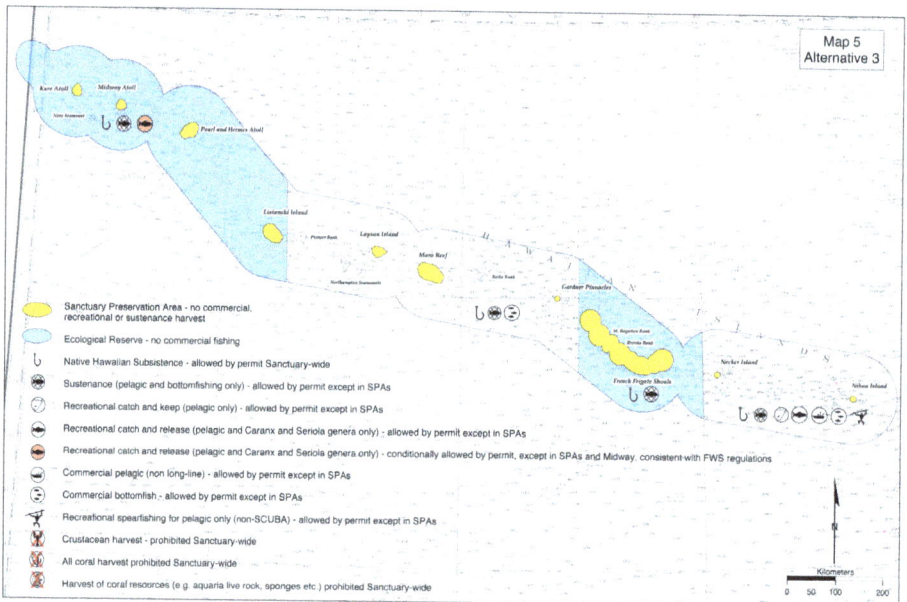

The fishing alternative that the NMSP considered most consistent with the NMSA and with the goals and objective statement for the proposed NWHI sanctuary was provided in a 2004 document to the Council. The NMSP proposed measures were more restrictive than the executive order that created the NWHI CRER. *NOAA NMSP 2004, page map C-53.*

In January 2005, the Council held public hearings on Maui, Kaua'i, O'ahu and the island of Hawai'i to solicit input and comments on a range of alternatives and associated draft fishing regulations for the proposed NWHI sanctuary. Additionally, the Council provided opportunities to submit written comments on this topic between January 19 and February 18, 2005, via mail, email and fax. While a thousand comments were received from local respondents with the majority in favor of continued protection through Council management, more than 14,000 comments were from U.S. mainland and foreign sources (the majority in form emails) in favor of the strongest protection possible (WPRFMC 2005a). This demonstrated the power of an orchestrated marketing campaign outside the region over the MSA-intended local management process.

The Council submitted its recommendations for draft fishing regulations for the proposed sanctuary for NOAA review and consideration on April 14, 2005. The draft regulations proposed that bottomfish fishing be allowed for up to 17 vessels maximum (10 in the Mau Zone and seven in the Ho'omalu Zone); vessel size be limited to 60 feet (18.3 meters) in length overall; use of bottom trawls, explosives, poisons and other destructive gear be prohibited; and vessels be subject to federal fishery observers if requested by NMFS. The Council also recommended that fishing be allowed by Native Hawaiian communities with preferential Native Hawaiian participation through the issuance of two of the 10 Mau Zone bottomfishing permits under the Community Development Program. These were essentially the regulations already in place for the fishery.

While the Council worked within the processes of the NMSA and MSA to support sustainable fishing in the NWHI based on science, other efforts were being undertaken by politicians and environmentalists to close the fisheries.

On May 16, 2005, Congressman Ed Case (D-Hawaiʻi) introduced H.R. 2376 Northwestern Hawaiian Islands National Marine Refuge Act of 2005, which emphasized the end of fishing in the NWHI. He described the area as an ocean equivalent of "Yellowstone National Park" (Case 2005). Case's bill was followed by newspapers columns attacking the Council. As before, much of the opposition by the local no-take MPA activists was directed not so much towards the Council but towards Simonds. Published personal attacks accused her of manipulating the process (Koberstein 2006). As the politics of this orchestrated public relations campaign intensified in local and national media, they had the effect of creating a negative image of the Council that would affect future policy decisions.

Bobby Gomes, a Native Hawaiian fisherman from Maui and holder of a Hoʻomalu Zone bottomfish permit, and his crew show onaga they caught in the NWHI in June 2005, at a time when the future of the fishery and their fishing careers were at stake. *Bobby Gomes photo.*

The NWHI commercial fishing vessel *Laysan,* owned and operated
by the husband-and-wife fishing team of Tim and Timm Timoney,
measures under 60 feet in length as required by the Bottomfish
FMP. Timm (the wife) served as a Council member 1992–1995.

Meanwhile, plans were made by Governor Lingle to proactively end
commercial fishing in state waters in the NWHI. On September 29,
2005, Lingle signed a bill creating the world's largest marine refuge in
the NWHI. The rules set forth that extractive use would be banned in
100% of state waters (out to 3 nm) except for use by Native Hawaiians
who would not be able to bring back the fish and sell them. Peter Young,
then chair of the Hawai'i Board of Land and Natural Resources, said,
in words that directly impacted the Council, that the Lingle declaration
of the refuge would "support an organized phase-out of commercial
fishing in federal waters to make state and federal waters closed to
fishing." Young went on to say, "We're going to change our position
and officially encourage the sanctuary program to follow the lead of the
State and prohibit fishing in the federal waters as well" (Lewis 2005).

Despite Lingle's declaration and Young's statement that they
planned to close Hawai'i waters, the Council was being assured that

National Environmental Policy Act rules were being honored and that the process for regulating the NWHI was still open. Thus the Council continued to work on the sanctuary plan in good faith with the National Ocean Service and NMFS. In January 2006, NOAA Administrator and Under Secretary of Commerce Conrad Lautenbacher, a retired admiral, wrote to the Council stating that NOAA planned to publish a draft environmental impact statement and draft fishing regulations for the proposed NWHI sanctuary in June 2006. He said NOAA was considering three alternatives: the first would allow limited fishing indefinitely; the second would end fishing by 2025; and the third would end fishing five years after sanctuary designation. The Under Secretary went on to say that NOAA believed that there was a credible basis for moving forward with proposing catch and permit limits through amendments to the Council's existing FMPs rather than as regulations under the NMSA, but, to meet NOAA's draft environmental impact statement timeline, the Council would have to transmit its amendment package(s) to NOAA no later than May 1, 2006. In response, the Council completed draft amendments to the Bottomfish and other FMPs regarding fishing activities in the proposed sanctuary on April 3, 2006.

Proclaiming the First Marine National Monument

Brown and others who had proposed the NWHI monument during the Clinton administration did not believe that they would get traction with the plan during the Bush administration, which had shown little interest in environmental causes. In fact, the Bush administration had initially publicly questioned the legitimacy of a monument designation. Still, carryover employees within the Council on Environmental Quality maintained their determination to create a monument in the NWHI.

On June 15, 2006, the Council was convening its 133rd meeting in Pago Pago, American Samoa, when the news came that President George W. Bush had established the NWHI as the nation's first marine national monument through Presidential Proclamation 8031 under the authority of the Antiquities Act. The monument comprised the emergent state and federal land and waters out to approximately 50 nm around the NWHI, i.e., essentially the same boundaries of the NWHI Reserve and the Council's Protected Species Zone before that. The proclamation and its implementing regulations required case-by-case permits for access to the monument, imposed a zero annual harvest guideline for NWHI lobsters and prohibited commercial fishing for bottomfish and associated pelagic species after June 15, 2011. Commercial fishing was also prohibited in ecological reserves, special preservation areas and the Midway Atoll Special Management Area established by the proclamation. Sustenance fishing (in which all fish harvested within the monument are consumed within the monument) could be allowed as a term or condition of a monument access permit. Access could also be granted for the conduct of Native Hawaiian cultural practices. All domestic vessels entering the monument were required to notify authorities in advance, as well as carry an active vessel monitoring system (VMS), an enforcement method that the Council had pioneered 15 years earlier (see Chapter 8). The above restrictions were not applicable to persons who were not citizens, nationals or resident aliens of the United States (including foreign flagged vessels) unless in accordance with international law.

Like the Council, persons at the NMFS headquarters in Silver Spring, Maryland, were surprised by the announcement. One experienced senior Honolulu NMFS executive said that he was convinced it was done on the spur of the moment "because of all the mistakes."

In November 2006, the Council's executive director, Hawai'i vice chair and coral reef ecosystem/habitat coordinator met with Dinah Bear, general counsel for the White House's Counsel on Environmental Quality, in Washington, D.C., to discuss four concerns about the NWHI monument proclamation. Chief among them was the provision that prohibited bottomfish vessels to deploy an anchor, which is an essential component for the fishery in the NWHI, where ocean currents are very strong. The WPRFMC contingent noted the incongruity of banning the use of 80-pound rebar anchors by the bottomfish vessels while allowing NOAA research vessels to use 1,000-pound anchors in the same area. This concern would eventually be addressed in 2007 with the allowance of a management permit to be issued by the monument managers to NMFS authorizing it to allow NWHI fishery permit holders to use anchors in non-coral areas.

Other concerns the WPRFMC presented to Bear included the impact of the no-fishing areas (i.e., ecological reserves and special preservation areas) on the bottomfishfishery. Nearly 100% of the uku (grey snapper) representing 15% of the fishery's total landings had been historically caught in these areas. The WPRFMC asked that consideration be given to opening these areas to the bottomfish fishery, noting that there was no impact to any ecosystem parameters. It also asked that the bottomfish fishery be allowed to operate on a limited basis in the future to ensure availability of local bottomfish since the main Hawaiian Islands was experiencing local depletion of the species and increasing closed areas.

The WPRMC representatives also brought to Bear's attention Native Hawaiian issues. It noted that Native Hawaiians viewed the NWHI as "kupuna" (ancestral) islands and the monument as "taking" and "withdrawal" of the islands and surrounding waters in the context of the 1893 overthrow of the Kingdom of Hawai'i. The monument proclamation also was requiring Native Hawaiians to obtain "permits" and "approval" from the federal government to conduct their cultural activities in these ancestral areas. The WPRMC asked the Council on Environmental Quality, which advises the president, to consider allowing the taking of resources by Native Hawaiians for community sharing and the setting aside of some islands for a future Native Hawaiian entity, similar to what had been done with the island of Kahoʻolawe in the main Hawaiian Islands.

On February 28, 2007, Bush amended his initial proclamation through Proclamation 8112, giving the NWHI monument a Native Hawaiian name, Papahanaumokuakea. Proclamation 8112 also specified that, under Native Hawaiian Practice Permits, "any living monument resource harvested from the monument will be consumed or utilized in the monument" and replaced the word "Sanctuary" with "Monument" on the map accompanying the original proclamation.

The president's proclamation significantly changed how the monument was to be administered. When the monument was first proposed under the Clinton administration, it was to be managed by the Interior Department. When Bush established the monument, management was changed to include three entities: the State of Hawaiʻi and the U.S. Departments of Commerce and the Interior. Overall the monument is housed in the NMSP under the Department of Commerce. Even so, each entity was to have its own jurisdiction and would give up its authority only if there were collective approval.[45]

In May 2006, the State of Hawaiʻi Department of Land and Natural Resources and federal officials (USFWS, NMSP and NMFS) signed a memorandum of agreement on NWHI management (NOAA

---

[45] Johns op. cit.

et al. 2006).[46] In the news media it was presented as a triumph of a few activists against an all-powerful fishing Council that was excluded from the agreement.

The legality of the agreement was controversial. From the beginning John Craven, the former marine affairs coordinator for the State of Hawai'i, said that the monument was unconstitutional because the Hawai'i State Constitution declared that no sovereignty of Hawai'i can be conceded without ratification by the State Legislature. By signing an agreement with the federal government for the submerged lands from the islands to 3 nm from shore, Hawai'i had essentially given up sole sovereignty over the NWHI without asking the will of the people. He believed, were this to ever be challenged in court, the legality of the monument would fail and fishing would be permitted because the Antiquities Act cannot extend over state sovereign waters.[47] Craven said that the declaration was also in violation of the international law of the sea and would one day be challenged in court.

Division of the Native Hawaiian Community

A Hawaiian name, Papahanaumokuakea, for the NWHI Monument was announced by First Lady Laura Bush. It raised questions from the Council as to why the ancestral name Moku Papapa, which the people

---

[46]   On December 8, 2006, a memorandum of agreement was signed by Secretary of Commerce Carlos M. Gutierrez, Secretary of the Interior Dirk Kempthorne and Governor Linda Lingle of Hawai'i to provide for coordinated administration of all the federal and state lands and waters within the boundaries of the monument.

[47]   Craven op. cit.

of the adjoining island of Niʻihau had used for centuries for the Leeward Islands, had not been chosen.[48]

There had been little interest during the CRER Advisory Council sessions of finding evidence that would link Native Hawaiians to habitation sites in the NWHI. During the 2000 NWHI scoping sessions, the Council's indigenous coordinator, Charles Kaʻaiʻai, sought to have the federal government seek out and recognize Native Hawaiian burials found in the NWHI under the North American Graves Protection and Repatriation Act, but his suggestion was marginalized. Kaʻaiʻai said, "They put it on the list and went on to other things. The State as trustee for the Native Hawaiian people should be asking why this work is not being done."[49]

Meanwhile, long-time Native Hawaiian fishermen such as Leo Ohai spoke out forthrightly about the monument and the damage it would do to their lives and the lives of their children.

In 2010, the Koani Foundation and Na Koa Ikaika KaLuhui Hawaiʻi, two indigenous Hawaiian organizations, would unsuccessfully petition the United Nations Educational, Scientific and Cultural Organization against the nomination by the United States of the Papahanaumokuakea Marine National Monument (MNM) as a World Heritage Site. The petitioners raised objections to the nomination in two respects: "As indigenous peoples who have owned, used and occupied the lands, territories and resources of the NWHI from time immemorial and whose human rights elucidated in the U.N. Declaration on the Rights of Indigenous Peoples are being violated" and "as beneficiaries of the

---

[48]   Interestingly, the NOAA Marine Sanctuary's interpretive center for the NWHI, established in 2003 and located in the southernmost island of the archipelago, is named the Mokupapapa Discovery Center, reflecting the traditional name for the Leeward Islands.

[49]   Charles Kaʻaiʻai in discussion with Michael Markrich, January 25, 2021.

Ceded Lands Trust established pursuant to Section 5(f) of the State of Hawai'i Admissions Act." The petition noted that "[t]he territories and resources, which are subject to the U.S. nomination, are trust assets of the Native Hawaiian peoples bringing these objections" (Bandarin 2010, Attachment, p. 2).

Transfer of Effort

Bush's overlay of the Reserve with the Papahanaumokuakea MNM essentially ended all fisheries in the NWHI in 2009 (the NMFS deadline for fishermen to surrender their permits so as to be compensated for closure of the fisheries) and any hope for preferential NWHI bottomfish permits for Native Hawaiians. In the final year, the total landing of bottomfish from the few remaining vessels was less than 50,000 pounds (about 23 mt), only a fraction of what was sustainably produced from the area through the 1990s.

The closing of the NWHI bottomfish fishery put additional pressure on main Hawaiian Islands bottomfish stocks which were already stressed. Since bottomfish such as 'opakapaka remain the signature dish for many restaurants whose focus is Pacific Rim Cuisine, closing the NWHI fishery resulted in a shortage of fresh local, Hawai'i-caught fish for hotels and restaurants. Whereas in 1990 virtually all of the bottomfish sold in Hawai'i was locally sourced, by 2008 imports from Tonga, New Zealand, Indonesia, Fiji and Australia made up more than half of the fresh snapper and grouper sold in the state (WPRFMC 2010).

As these countries' waters were heavily fished for bottomfish to satisfy the Hawai'i market, there could be long-term negative impacts on their limited but valuable deep-slope fisheries. The Bush monument was created with no thought of what it might do to bottomfish stocks in Pacific Islands. Many of these island countries lack the resources to police their fisheries (Govan et al. 2008).

In March 2006 at Honolulu harbor, Frank Goto, general manager of the Honolulu fish auction and inaugural Council member (2nd from left), joins NWHI bottomfish fishermen including Oʻahu-based Tim Timoney, Kauaʻi-based Troy Lanning and Maui-based Bobby Gomes in an effort to keep the sustainable fishery open. The fishermen display six of the seven deepwater bottomfish featured in Hawaiʻi's signature Pacific cuisine. *WPRFMC photo*.

How the Monument Came to Be

No full explanation has been given for the change of heart that encouraged President Bush to sign the proclamation establishing the NWHI monument. Some people who followed the issue, such as Craven, believed that there was a strategic purpose and the action was actually taken for national security reasons as a "buffer against terrorism."[50]

But the action was portrayed in the press as evidence of Bush's interest in environmentalism, leading others to believe that it was done solely for its public relations value because at the time the Bush administration was at record lows in opinion polls. It was considered an

[50] Craven op. cit.

easy way to win favor with environmental groups at low political cost since it affected only eight fishing boats.[51]

Political cartoonists had a heyday with tongue-in-cheek speculations for the designation, such as Gary Hoff's (*Hawaii Tribune Herald*) suggestion that Bush planned to train monk seals for national security monitoring. Other cartoonists shed light on other facets of the government's action. For example, Drew Sherman juxtaposed the government's using the Hawaiian Islands to create the world's largest marine refuge and the U.S. Senate's refusal to debate the Akaka Bill to establish a federal process for U.S. recognition of Native Hawaiians.

This 2006 political cartoon explores possible national security reasons for establishing the NWHI marine national monument. *Drew Sheneman illustration.*

---

[51]   Johns op. cit.

This political cartoon examines how the 2006 NWHI marine monument proclamation overlooked the welfare of the people of Hawai'i, especially Native Hawaiians. *Hawaii Tribune-Herald/* Gary Hoff illustration.

Heather Elizabeth McDowell Ward provided this interpretation in her PhD dissertation, Creating the Papahanaumokuakea Marine National Monument: Discourse, Media, Place-Making and Policy Entrepreneurs:

> The rise of the Northwestern Hawaiian Islands to the top of a full 2006 political agenda resulted from a combination of factors and complicated interactions, all achieved through orchestrated communication efforts employing evocative media. The collective efforts of an elite network 'made place' by envisioning the Northwestern Hawaiian Islands as a unique, fragile ecosystem—distinct geographic space inscribed with particular characteristics and meanings worthy of territorial boundaries and policy protections. The National Oceanographic and Atmospheric Administration's Office of National Marine Sanctuaries perhaps exercised the most control because its director

and staff understood the power of persuasive media and managed communication between interested parties. (Ward 2010, abstract, para. 3)

Among the media used was the book *Archipelago: Portraits of Life in the World's Most Remote Island Sanctuary* by award-winning nature photographers Susan Middleton and David Liittschwager, published in October 2005 by the National Geographic Society. The NWHI CRER would co-present exhibits of photographs from the book beginning in February 2006 throughout the main Hawaiian Islands, describing the exhibit as "Forty dramatic portraits of marine and terrestrial flora and fauna, landscapes and seascapes, and interpretive imagery and information collectively express the biological exuberance of the NWHI." The announcement said the "reserve is co-presenting the exhibit in collaboration with community host venues on the various islands to raise awareness about NWHI and support for its long-term protection ...."

Some attribute Bush's decision to First Lady Laura Bush's viewing of *Voyage to Kure*, a documentary film by Jean-Michel Cousteau. In it, Cousteau focuses on a boat in the NWHI, which he describes as a suspected illegal fishing vessel. Council staff called the U.S. Coast Guard about the vessel to learn that it was legally operating, contrary to Cousteau's dramatic portrayal otherwise (Campagnoni 2006). The Council further learned that it was a Hawaiʻi bottomfish vessel operated single-handedly by an 80-year-old fisherman.

In contrast to the emotive messages about the NWHI from environmental advocates, the Council's outreach on the NWHI appealed to reason and was based on the best available science. For example, the NMFS Pacific Islands Fisheries Science Center director had stated that the NWHI bottomfish stocks were in good condition based on 25 years of monitoring, research and stock assessments and that the integrity of the NWHI ecosystem "continues to exist and indeed thrive, despite decades of fishing in this region" (Pooley 2005).

The Council produced two videos—*Living the Legacy: The Northwestern Hawaiian Islands* (2000) and *The Northwestern Hawaiian Islands* (2005)—with interviews from an array of scientists, fishermen and chefs.[52] Council efforts to have its videos aired on Hawai'i Public Television were rebuffed because the producer regarded the Council as a special interest group, even though it is a government instrumentality and nonprofit organization, similar to the Smithsonian Institution. Interestingly, environmental NGOs were not considered a special interest group. The Council's videos were, therefore, aired on 'Olelo community television. Additionally, the Council was confined by law to what it could say and to whom. It could not lobby Congress or engage in grass roots lobbying. Individuals and organizations for a no-take NWHI designation receiving funding from Pew Charitable Trusts (formerly, Pew Charitable Foundation) appealed to the Department of Commerce's Inspector General in 2005 and to the U.S. Government Accountability Office (through U.S. Representative Henry Waxman [D-California]) in 2007 to have the Council investigated for lobbying and other matters. While the Council was cleared of wrongdoing in both cases, it gave occasion for these NGOs and their representatives to characterize the Council as being under investigation. In their press releases and other outreach and lobbying efforts, they called for the resignation of Simonds as the Council's executive director (Aila et al. 2007, Bonk et al. 2007).

---

[52] A third video, produced in partnership with the Teaching Learning Network, was completed but could not be aired because it featured sport-fishing and ecotourism on Midway, which became prohibited activities under the monument designation.

Western Pacific Regional Fishery Management Council
History of the Fisheries

Living the Legacy

The Northwestern Hawaiian Islands

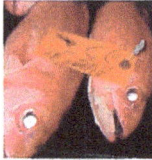

The *Living the Legacy* video, produced by the Council, explores the history of the fisheries in the NWHI, a place often described as "pristine" by marine monument proponents.

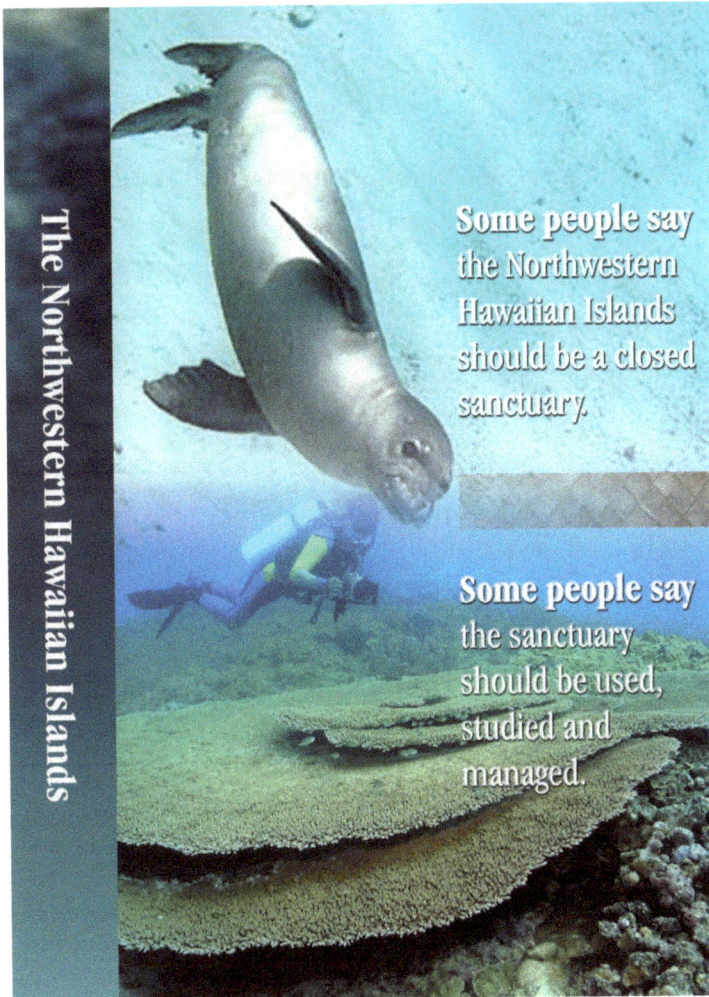

The Northwestern Hawaiian Islands, the second video about the Leeward Islands produced by the Council, examines the scientific research and management of the NWHI fisheries.

After the monument designation, several persons and organizations claimed that it was their actions that led to Bush's decision. Among them was the Pew Charitable Trusts, which said they "were able to amplify the efforts of (local Hawai'i activist organizations) in helping create the conditions that greatly facilitated the president's decision" (Pew 2007, pages 10–11).

Pew is one of the largest philanthropic trusts in the United States. According to Pew NWHI project manager Jay Nelson, Pew environmental campaigners are always on the lookout for possibilities to implement significant changes in policy in terms of greater environmental protection.[53] When approached by environmental activists in the late 1990s, Pew responded with financial support and organized a comprehensive campaign for the NWHI monument. According to a 2007 Pew report, the organization played a significant role in challenging the U.S. fisheries management institutions including the Council system. Pew began paying activists in Hawai'i to help promote the concept of a no-take MPA in the NWHI but eventually decided that additional efforts were needed.

In a letter, found on the website of the public relations firm Scott-Foster and Associates, Nelson wrote: "One of the challenges of creating and sustaining a conservation campaign is crafting and disseminating a compelling story. In the case of the NWHI and Western Pacific Fishery Management Council (Wespac) it is a multifaceted complicated story, which significantly increases the difficulty for conservation advocates" (Nelson n.d.).

---

[53]   Jay Nelson in discussion with Michael Markrich by telephone, October 1, 2007.

Council Executive Director Kitty M. Simonds and staff wear "Bad Kitty" T-shirts at the 131st Council meeting in Honolulu. The shirts were distributed outside the meeting at a small anti-Council rally attended by the Pew's NWHI Director Jay Nelson. *WPRFMC photo.*

Pew wrote in its 2007 annual report: "The decision to direct our efforts towards this particular island chain was deliberate. ... The Northwestern Hawaiian Islands offered an unparalleled opportunity to establish an ecosystem-scale reserve, one in which extractive activities could be prohibited and natural biological processes could occur unaltered by direct human interference. From the late 1990s prior to the Trust's involvement, local conservation groups in Hawaiʻi engaged in important work to protect the area. Yet, as the result of an internal analysis conducted in early 2005, Pew staff concluded that the efforts of these organizations were not likely to be sufficient in themselves to accomplish the goal of ending commercial fishing in this area."

From January 2005, Pew took the following steps:

1) Hired an experienced advocate/biologist to become the Pew's NWHI project director, to help coordinate and focus existing environmental efforts to ensure that the public supported—and federal agencies adopted—the most protective and conservation-oriented management measures possible including an end to commercial fishing at the site.

2) Provided two full-time professionals to assist Hawaiian organizations in building public support in Hawai'i for NWHI protection.

3) Engaged a respected communications firm in Hawai'i to reach out to businesses and organizations that did not understand the conservation concerns.

4) Hired a media consultant to assist conservation groups in their effort to create a network of organizations in support of protection and to prepare and disseminate press releases and conduct other outreach to media.

5) Launched an effort to engage Hawaiian chefs, who represent an economically significant sector of the island economy, in support of fully protecting the Hawaiian archipelago.

6) Encouraged a study at the University of Hawai'i looking at the economic impact of closing commercial fishing on restaurants in the state, consumers and the state itself, with the findings to be distributed to policy leaders and the public.

7) Engaged a prominent retired judge to lead a professional team to begin outreach and buyout negotiations with the eight bottomfish permit holders.

8) Contracted with several legal experts to investigate specific fisheries-related issues, including fishing permits, agency actions and fisheries regulations.

9) Informed a number of policymakers, who, in turn, helped to educate others.

10) Worked with Washington, D.C.–based conservation partners to ensure that recreational fishing organizations were fully briefed on the [NWHI] issue and that their questions were fully addressed in advance of the president's decision.

In Hawai'i, a small state with only one million people, the presence of such a well-financed media and lobbying campaign, spending reportedly hundreds of thousands of dollars, was soon felt. The intervention of a mainland-based philanthropic organization with vast financial resources—in what would otherwise have been a purely local regional dispute centering on interagency rivalry and native sovereignty—tipped the scales in favor of the environmental activists. The Pew public relations campaign contributed to lack of understanding of the immense value of the submerged lands the greater Native Hawaiian community was being asked to surrender in favor of what was presented as a "greater good." The Council was portrayed in the media as unjust usurpers of the public process and as thoughtless managers in the thrall of commercial fishing interests and indifferent to the well-being of the natural world (Koberstein 2006).

The NWHI became the subject of a very high profile multi-million-dollar media campaign coordinated by the NMSP to influence public opinion, involving the National Geographic Society, National Public Radio and Jean-Michel Cousteau's Ocean Futures Society, among others. The area became a symbol of national celebration (Ward 2010).

Jay Nelson said, "If the Native Hawaiians hadn't said they wanted this, we would have never supported it."[54] On Pew's website would be the claim that Native Hawaiians were nearly unanimously "in favor of protecting the NWHI from commercial activity" (Pew 2006).

William Aila Jr. was among the divided Native Hawaiian community members who supported Pew's efforts to turn the NWHI into a no-take marine preserve. Aila would later serve as a Council member and as its chair in 2012 and 2014. *WPRFMC photo.*

In contrast to 30 years earlier, when State Senator T. C. Yim spoke loudly against the colonial attitude of the federal government in taking Hawai'i land and waters without offering compensation, the response in Hawai'i to the monument designation was generally favorable. Even the Office of Hawaiian Affairs was represented on the new monument advisory board. The only protest was muted and subtle. U.S. Senator Inouye, the leader of the Hawai'i Democrats, did not attend the signing at the White House.

---

[54]   Nelson ibid.

7.3 Main Hawaiian Islands Bottomfish Fishery

One of the complications in development of the Bottomfish and Seamount Groundfish FMP was that, when the Council was established in 1976, there were serious questions about the completeness and quality of the HDAR bottomfish catch reports. Noncommercial fishermen operating in state waters were not required to report their catch, and commercial fishing data were often incomplete.

It was estimated that the majority of the main Hawaiian Islands bottomfish grounds are found in state waters rather than federal waters. Therefore, the Council deferred management lead in the main Hawaiian Islands to the State of Hawai'i.

From 1986 to 2004, the Pacific Islands Fisheries Science Center and the Council assessed the stocks of the federally managed bottomfish by, among other methods, calculating their spawning potential ratio from annual commercial catch data provided by HDAR. As explained by Drazen et al. (2014), an annual spawning potential ratio of 20% was established as the critical threshold for designating a stock as recruitment overfished. While the spawning potential ratios for all species of bottomfish in the NWHI were consistently above this level, those for onaga and ehu in the main Hawaiian Islands had been below 20% for well over a decade as reported by the Bottomfish Plan Team.

In response to the chronic low spawning potential ratios of onaga and ehu, the State of Hawai'i in June 1998 implemented several management measures for bottomfish in the state waters of the main Hawaiian Islands for Deep 7 bottomfish. They included a bag limit of five for the seven most commonly caught bottomfish per noncommercial fisherman, gear restrictions, closed areas and mandatory registration of all bottomfishing vessels with the state. The closed areas included 19 bottomfish restricted fishing areas (BRFAs) throughout the main

217

Hawaiian Islands, with much of the areas in federal waters. HDAR was required by law to review the effectiveness of the closures after five years and report back to the legislature.

The creation of the NWHI CRER in 2000 led to the reduction of the NWHI catch, which was supplying approximately half of the local bottomfish. This encouraged the approximately 400 fishermen who caught bottomfish around the main Hawaiian Islands to increase their catch and fish wholesalers to source bottomfish to import.

The 1996 MSA reauthorization required the regional fishery management councils and NMFS to manage fisheries based on maximum sustainable yield. The spawning potential ratio method would no longer suffice to meet Congressional requirements for monitoring stock status. In 2003, the Pacific Islands Fisheries Science Center produced a report on the status of the main Hawaiian Islands bottomfish stock that met the new biomass-based requirements of the MSA. The Plan Team reviewed the report findings and flagged potential "overfishing" in the main Hawaiian Islands based on the new control rule. The spawning potential ratio for onaga and ehu were also still well below the 20% threshold. Immediately the Council requested NMFS to reinitiate the stock assessment for the main Hawaiian Islands bottomfish fishery. It also established a working group to assist the State in assessing its management regime by July 1, 2003. Joint Council and State scoping meetings were held throughout the archipelago to discuss BRFA performance and the new federal overfishing/overfished control rules with the fishermen.

In 2004, the Council convened a Bottomfish Stock Assessment Workshop, bringing together international and national stock assessment experts on demersal fisheries. Workshop experts found that substantial information existed on the biology, distribution and abundance of bottomfish stocks. However, much of the information was fragmented and, as a consequence, could not be used to effectively manage the resource. A slew of recommendations to improve the data situation were made, including the following:

a) Reporting catch and effort on finer scales,

b) Supporting hydroacoustic research to assess biomass,

c) Sensitivity analysis on effects of MPAs on fishing mortality and catch per unit effort,

d) Pacific Islands Fisheries Science Center using stock assessment funds to collect size frequency data to determine age at maturity, and

e) Establishing a group to develop action plans and budgets to implement workshop recommendations.

The Council also encouraged the State to analyze its existing 'opakapaka tagging data so they could be used to inform the next assessment.

The Council was notified on May 27, 2005, that overfishing was occurring in the main Hawaiian Islands Deep 7 species complex and that as a consequence bottomfish fishing had to be reduced by 15%. The following year the federal mandate was to further reduce bottomfish fishing in the main Hawaiian Islands by 24%.

In response to bottomfish overfishing, at its 127th meeting on June 2, 2005, the Council immediately recommended that a working group of Plan Team members convene to develop options to reduce effort in EEZ waters and that a targeted survey of main Hawaiian Islands bottomfish fishermen be conducted in coordination with HDAR to better understand fishery participation and gather socioeconomic information. A control date was established, and the Council again requested that the Pacific Islands Fisheries Science Center use stock assessment funds to collect bottomfish size and frequency information.

Before the end of 2005, several rounds of public meetings were held throughout the state by the Council in coordination with HDAR to solicit input on management options. The community largely supported seasonal closures with options for quotas for full-time fishermen. Clear opposition was heard on the State's existing and proposed area-based closures.

By March 2006, the Council took action to develop an amendment for an annual bottomfish seasonal closure between the months of May 1 and August 31 in all main Hawaiian Islands waters, when peak bottomfish spawning takes place. Should the State not support coordinating consistent regulations, the Council would move to close federal waters at Penguin Bank and Middle Bank. The State was at first reluctant to agree to seasonal closures, largely because its policy was to trust in the new BRFAs that it had created to deal with this issue. At this time, the original 19 BRFAs had been replaced by 12, which took effect in 2007. The new BRFAs were based on a review conducted in 2005, which concluded that the original system did not protect adequate amount of preferred habitat (Parke 2007).

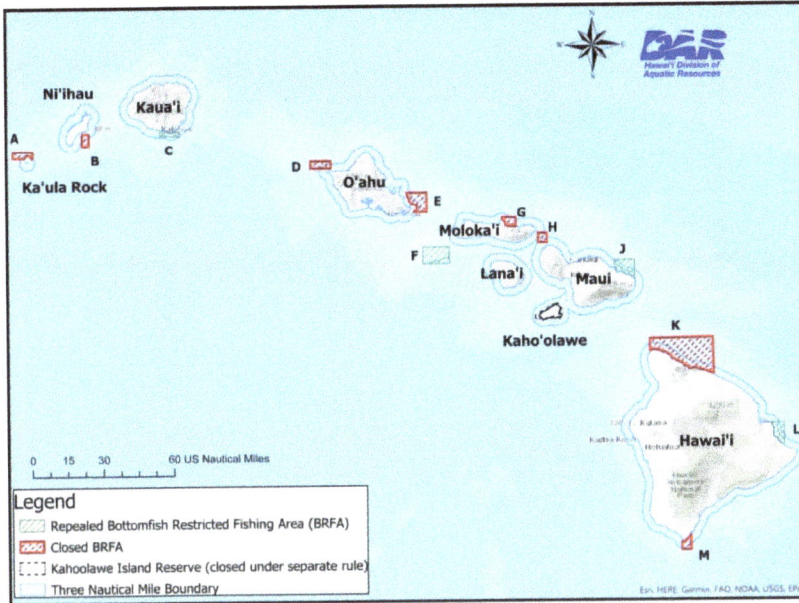

The State of Hawai'i has relied on the BRFAs it created as a key management tool for the main Hawaiian Islands bottomfish fishery. The 2007 restructuring of the BRFA system reduced the number of BRFAs from 19 to 12 and was meant to better align the closed areas with bottomfish habitat. *HDAR illustration.*

The State eventually agreed to coordinate management efforts after the 2006 reauthorization of the MSA included a provision for federal preemption of state rules when fisheries are determined to be experiencing overfishing or are determined to be overfished and resources straddle federal/state jurisdiction. However, despite widespread objections from fishermen and Council recommendations, the State of Hawai'i would maintain the BRFAs, many of which extend into federal waters (Hospital and Beavers 2011).

In 2007, the Council, HDAR and the NMFS Pacific Islands Regional Office jointly announced that both commercial and noncommercial fishing would be prohibited for the main Hawaiian Islands Deep 7 bottomfish in both state and federal waters from May 1 to September 3, 2007. Exempted from the seasonal ban were the eight fishing boats allowed to fish in the NWHI, grandfathered under existing permits until 2011—although Congress, through NMFS, was moving to get them out earlier. The closure of the main Hawaiian Islands bottomfish fishery while the NWHI fishery remained opened caused some confusion among restaurant owners and their patrons, which the Council worked to alleviate through outreach and education, such as the production and distribution of tent cards to be placed on dining tables.

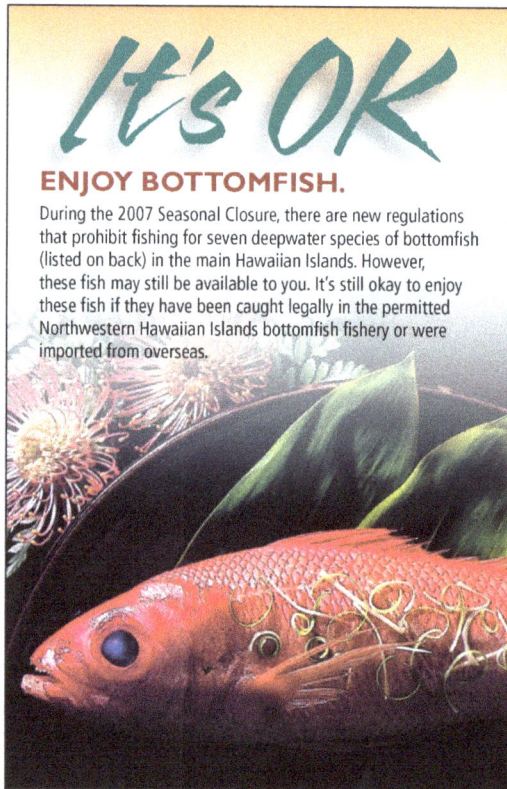

It's OK

ENJOY BOTTOMFISH.

During the 2007 Seasonal Closure, there are new regulations that prohibit fishing for seven deepwater species of bottomfish (listed on back) in the main Hawaiian Islands. However, these fish may still be available to you. It's still okay to enjoy these fish if they have been caught legally in the permitted Northwestern Hawaiian Islands bottomfish fishery or were imported from overseas.

Tent cards on restaurant tables and counters helped to clarify confusing messages as to the legality of enjoying a meal of bottomfish from the NWHI when catching and selling bottomfish from the main Hawaiian Islands was banned. *WPRFMC photo.*

The seasonal bottomfish closures in the main Hawaiian Islands were implemented as a precautionary measure, meant to remain in place until a total allowable catch (TAC) could be established for the fishery. In 2008, the Council implemented a fully integrated bottomfish management regime in coordination with the State. The centerpiece of the regime was built around the annual fleet wide quota, the first quota-based management system used in the small boat fishery. If the quota were reached based on commercial data (as reporting is not required by the noncommercial fishery), the commercial and noncommercial fisheries in both state and federal waters would be closed until the season reopened. The fishing year for the Hawai'i bottomfish fishery was established as September 1 to August 31 the following year. The State changed its commercial marine license reporting from monthly to trip-based for Deep 7 bottomfish and required bottomfish vessel owners to annually renew their registration. The intent was to improve accuracy of forecasting any needed closure to ensure the TAC would not be exceeded.

The Council implemented noncommercial bottomfish permit and reporting requirements and a daily bag limit of five total Deep 7 species per day per person. Landings were tracked weekly and monitored against the annual quota by HDAR. When total landings approached the annual quota, the Pacific Islands Fisheries Science Center would forecast when the quota would be reached. The Council executive director, HDAR administrator and NMFS Pacific Islands Regional Office regional administrator would meet to select the closure date for the fishery based on the Pacific Islands Fisheries Science Center projection. Because the fishery was monitored near real time through trip reports, the forecast window to close the fishery was shortened to just three weeks.

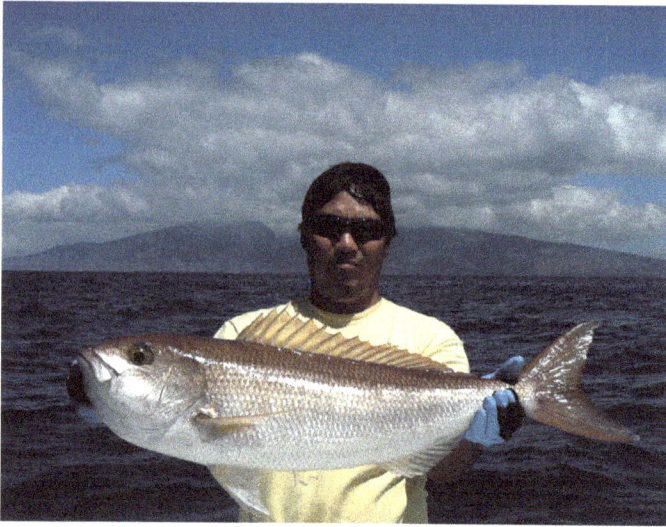

Maui-based commercial fisherman Layne Nakagawa shows a recently caught 'opakapaka (pink snapper), one of the most prized deepwater bottomfish in Hawai'i. *Layne Nakagawa photo.*

The quota-based management regime was new to Hawai'i's small boat fishery, so significant effort was put into outreach and education to support the implementation of the new rules. The Council, NMFS and State created and hosted a new "bottomfish.info" website to share information with the public. Joint statewide workshops were held to train fishermen on new federal noncommercial permit and reporting requirements, noncommercial bag limits and changes to the state bottomfish registration program and trip reporting. The Pacific Islands Fisheries Science Center staff provided training to U.S. Coast Guard staff on species identification. NOAA Office of Law Enforcement ramped up patrols at the local harbors to educate fishermen on the new requirements.

Because the SSC did not support the NMFS-produced stock assessments, it used the conservative 25th percentile of the 25-year running catch average to establish the TAC for the main Hawaiian

Islands Deep 7 bottomfish. The first TAC for the 2007–2008 fishing year was set at 178,000 pounds (about 80 mt) and was overreached by 10%. At this time, the bottomfish vessels were required to report their catches monthly, which hindered the accuracy of calculating when the TAC would be reached for a fishery that was dependent on market demand and calm weather. The 2008–2009 TAC was set at 241,000 pounds (about 110 mt) and was exceeded by 8%, with 259,194 pounds (about 120 mt) caught. The 2009–2010 TAC was set at 254,050 pounds (about 115 mt). Anticipating that the TAC would be reached, the fishery was closed on April 20, 2010. When the reports were tallied, only 208,412 pounds (about 95 mt) of the TAC had been caught, with an underage of 18% of the TAC.

TABLE 7: MAIN HAWAIIAN ISLANDS BOTTOMFISH TOTAL ALLOWABLE CATCH, FISHING YEARS 2007–2010				
FISHING YEAR	CATCH LIMIT	ANNUAL CATCH	OVERAGE / UNDERAGE	% OF TARGET
2007–2008	178,000	196,147	18,147	10%
2008–2009	241,000	259,194	18,194	8%
2009–2010	254,050	208,412	-45,638	-18%

Eventually, the TAC would be replaced by an ACL (see Chapter 10, Section 10.2, and Chapter 12, Section 12.1).

7.4 American Samoa Bottomfish Fishery

In the 1950s, bottomfish fishing in American Samoa remained largely a subsistence practice. However, many of the small boats were equipped with outboard engines and used hooks made of steel instead of pearl shells, and fishing lines fabricated from monofilament instead of hand woven sennit. In this hook and line fishery, one or several hooks were attached to a mainline weighted with a sinker and lowered to a desired depth to target one or several species of grouper, snapper and emperor.

It was not until the early 1970s that the American Samoa bottomfish fishery evolved into a commercial venture. In 1972, a government subsidized boat building program was initiated to provide local fishermen with larger, more powerful wooden dories for bottomfish fishing. However, mechanical problems and other difficulties took their toll, leading to the demise of the dory fleet.

Today, most fishermen who participate in the American Samoa bottomfish fishery use alia vessels—locally built, twin-hulled, aluminum or wood/fiberglass boats about 30 feet (about 9 meters) long, powered by self-contained 40-horsepower, gasoline outboard engines. These boats typically have not been outfitted with depth sounders, electric/hydraulic reels, global positioning system (GPS) or chilling capabilities, though the American Samoa government and the Council have both recently provided financial support to improve safety and efficiency. In addition, larger vessels with greater technological capabilities and fishing ranges have been entering the fishery in recent years. Fishermen still generally stay within 20 nm of land, though the better equipped and larger vessels do venture further.

The performance of the America Samoa bottomfish fishery varies according to the need for bottomfish at government and cultural events. The fishery's performance is also routinely disturbed by natural disasters. In 1987, a hurricane destroyed half of American Samoa's small-boat fishing fleet, including all the local vessels in Manu'a Harbor. American Samoan fishermen turned to the Council for help in getting financial assistance to restart their fishery. The Council did not provide them with funds for this purpose but acted as an interface with the Federal Emergency Management Agency and other federal agencies. The Council, as it often does, served a purpose beyond fisheries management, in this case acting as a source of local government assistance.

Total landings mostly declined in the 1990s, a trend that continued through the 2000s as a result of the territory being hit by two consecutive hurricanes in 2004 and 2005. Catches were in somewhat of a rebound when a tsunami in 2009 caused a commercial failure of the fishery, with revenue declining by about 80%. Seventeen bottomfish vessels were damaged or destroyed, accounting for as much as 75% of the fleet. Of the vessels known to have sustained damage, approximately 35% were total losses and an additional 42% were in need of major repairs (Markrich and Hawkins 2016, p. 24).

While the gear types and fishing strategies used in the fishery tend to be relatively selective for desired species and sizes, measures in the Bottomfish and Groundfish FMP served to further reduce bycatch and minimize bycatch mortality. They included prohibitions against the use of bottom trawls and bottom set gillnets and against the possession or use of any poisons, explosives or intoxicating substances to harvest bottomfish. Five types of non-regulatory measures aimed at reducing bycatch and bycatch mortality and improving bycatch reporting were also implemented: (1) outreach to fishermen and engagement of fishermen in management, including research and monitoring activities, to increase awareness of bycatch issues and to aid in development of bycatch reduction methods; (2) research into fishing gear and method modifications to reduce bycatch quantity and mortality; (3) research into the development of markets for discard species; (4) improvement of data collection and analysis systems to better quantify bycatch; and (5) outreach and training of fishermen in methods to reduce barotrauma in fish that are to be released.

In 1987, the Council established a system to allow implementation of limited access for the American Samoa bottomfish fishery. The fishing year for American Samoa fishery begins on January 1 and ends December 31. Since March 2004, bottomfish fishermen have been prohibited from operating within the 50-fathom isobaths around Rose Atoll, a no-take MPA developed by the Council.

The fishery was subsequently constrained by additional MPAs established by President Bush and by the Office of National Marine Sanctuaries. On January 6, 2009, under the authority of the Antiquities Act, Bush established the Rose Atoll Marine National Monument through Proclamation 8337, prohibiting fishing within 50 nm of the atoll. The monument would be jointly managed by the USFWS and NOAA to protect the coral reef ecosystem, seabirds and sea turtles. In 2012, the Office of National Marine Sanctuaries expanded its 0.25-square-mile (0.65-square-kilometer) Fagatele Bay National Marine Sanctuary to comprise six sites covering 13,581 square miles (35,175 square kilometers) and renamed it the National Marine Sanctuary of American Samoa. Zone B of the Aunuʻu site is reserved for research and does not permit bottomfish fishing even for sustenance, subsistence, traditional or recreational purposes. Fishing was also prohibited within 12 nm of Rose Atoll (Muliava), which now was now triply protected as an MPA under the MSA, a monument under the Antiquities Act and a sanctuary unit under the NMSA. Noncommercial fishing outside of the 12 nm Muliava sanctuary unit was prohibited unless authorized by a permit.

U.S. Coast Guard Petty Officer Frank Thompson inspects the alia bottomfish vessel of Elvin "Eo" Mokoma in American Samoa (circa 2002). The alia is a small, double-hulled vessel with a short range and limited capacity. *David Hamm photo.*

Many alia fishermen use the traditional, wooden hand-reel (right) to catch bottomfish as they do not own a modern electric reel (left). *WPRFMC photo.*

The American Samoa bottomfish fishery would face additional hardships with the implementation of ACLs (see Chapter 10, Section 10.2, and Chapter 12, Section 12.2).

7.5 Guam Bottomfish Fishery

Bottomfishing on Guam is a combination of recreational, subsistence and small-scale commercial fishing. Handlines, home-fabricated hand reels and small electric reels are the commonly used gear for small-scale fishing operations, whereas electric reels and hydraulics are the commonly used gear for the larger operations in this fishery.

The fishery can be divided into a shallow-water complex and a deepwater complex. The shallow-water target species are primarily reef-dwelling snappers, groupers and jacks, while the deepwater complex consists primarily of deepwater snappers. More people participate in shallow-water bottom fishing than deepwater fishing because the costs are lower and calm fishing grounds are more accessible. Shallow-water fishermen seldom sell, and they fish primarily for recreational or subsistence purposes using vessels measuring under 25 feet (7.6 meters) in length overall. Deepwater fishermen fish primarily for commercial purposes; they utilize the offshore banks and include a few vessels measuring more than 25 feet (7.6 meters) in length overall.

Participation in the Guam bottomfish fishery peaked in the early 2000s, with nearly 500 vessels participating. Since then, participation has fluctuated between 250 and 350 vessels. Guam's bottomfish fishery is highly seasonal. Most offshore banks have high shark densities and are deep, remote and subject to strong currents. Generally, these banks are accessible only during calm weather in the summer months (May to August/September). During this time of year, bottomfishing activity increases on the offshore banks in federal waters and on the east side of the island in the territorial waters. Both these locations are productive fishing areas that are less accessible during most of the year due to rough seas.

James Borja shows an example of shark depredation on Guam bottomfish. *James Borja photo.*

Charter bottomfishing has been a substantial component of the fishery, accounting for about 15% to 20% of all bottomfish trips from 1995 through 2004. Charter vessels typically make multiple two- to four-hour trips on a daily basis. The charter fleet includes vessels that engage in trolling and bottomfish fishing. However, larger bottomfish-only vessels can accommodate as many as 35 patrons per trip. These vessels generally fish in the same area and release most of their catch, primarily small triggerfish, grouper and goatfish. They occasionally keep larger fish, and a portion of the catch may be prepared as sashimi for their guests.

In 1987, the Council established a system to allow implementation of limited access systems for bottomfish fisheries in EEZ waters around Guam within the framework measures of the FMP.

In 2006, federal permitting and reporting requirements for bottomfish fishing in Guam became effective, as did measures prohibiting vessels greater than 50 feet (15.25 meters) in length from targeting bottomfish species within 50 nm of Guam. The use of poisons, explosives, intoxicating substances, bottom trawls and bottom set gillnets is prohibited. Due to concerns over habitat impacts, vessels larger than 50 feet (15.25 meters) are prohibited from anchoring at Guam's Southern Banks or in the U.S. EEZ seaward of Guam west of 144 degrees 30 minutes east longitude, except in the event of an emergency caused by ocean conditions or by a vessel malfunction that can be documented.

Stephen Meno, a former member of the Council's Guam Advisory Panel, proudly displays his bottomfish catch (circa 2018–2019). Guam fishermen typically use small vessels to catch bottomfish and pelagic species. *James Borja photo.*

Despite the federal reporting requirements, the data collection programs in place in Guam would prove to be inadequate leading to a pessimistic assessment of the bottomfish fishery (see Chapter 12, Section 12.3).

7.6 Northern Mariana Islands Bottomfish Fishery

Like Guam, the CNMI bottomfish fishery can be divided into a 100- to 500-foot (approximately 30- to 150-meter) shallow-water sector and a greater than 500-foot (about 150-meter) deepwater sector. The shallow-water harvest comprises primarily reef-dwelling snappers, groupers and jacks, such as the redgill emperor, while the deepwater catch is primarily snappers. The fishery is predominantly small boats engaged in small-scale commercial and subsistence fishing within a 50-nm radius of the island of Saipan, though some venture a bit further and there are a few Rota-based boats. A few larger vessels sporadically participate in the deepwater bottomfish fishery in the islands north of Saipan.

Fishermen largely ceased deepwater bottomfishing in early 1990s, resulting in a drop in overall bottomfish landings. Consistent fishing activity resumed in 1994, and the number of bottomfish trips more than doubled from 2000 to 2001 to the highest levels in 18 years. The number of commercial bottomfish trips then declined until 2005, when troll fishermen began conducting more bottomfishing as an alternate activity (Markrich and Hawkins 2016, p. 29). Maintenance and repair costs, as well as challenges with crew retention, apparently make it difficult to sustain a CNMI bottomfish operation for more than a few years. The participation of individual fishermen in the bottomfish fishery tends to be less than four years, with length of participation in the deepwater fishery slightly longer, which probably reflects the greater investment required to participate in the deepwater fishery. Deepwater trips tend to be longer, and vessels measuring 30 feet (9 meters) and longer have better chilling capacity and so tend to be more efficient. Successful bottomfish fishing depends on having a skilled captain and experienced fishermen who have knowledge of the location of specific underwater features. CNMI bottomfish fishermen rely on land features to guide them to fishing areas.

CNMI fisherman Lino Tenorio shows off his shark-depredated bottomfish. *Lino Tenorio photo.*

In 2008, the Council developed measures that prohibited commercial fishing vessels greater than 40 feet (12 meters) from fishing within 10 nm of Alamagan Island and within approximately 40 nm of the southern islands of the CNMI, i.e., Rota, Aguijan, Tinian, Saipan and Farallon de Medinilla. It also required them to carry active VMS units. The Council would reopen these areas in 2016 after assessments of the multi-stock complex determined the bottomfish fishery was not achieving optimum yield and that the closure was discouraging fishermen from upgrading to larger vessels.

CNMI bottomfish fishermen have voiced concerns about the incremental loss of access to prime fishing grounds around Farallon de Medinilla, due to military use of the island for bombing and military training practices. Additional loss of fishing grounds occurred with establishment of Marianas Trench MNM by President George W. Bush on January 6, 2009, through Proclamation 8335 under the authority of the Antiquities Act. The monument encompasses 95,000 square miles (250,000 square kilometers) and allows only sustenance, recreational and traditional indigenous fishing to occur around the islands of Asuncion, Maug and Uracas at the northern end of the archipelago. Commercial fishing is prohibited in the area.

Establishment of the monument was supported by the Pew Charitable Trusts, which was always on the lookout for new environmental protection possibilities, and was opposed by many CNMI residents. Approximately 100 persons from Saipan, principally from the indigenous Carolinian community, held two demonstrations opposing the monument when James L. Connaughton, chair of the White House's Council on Environmental Quality, visited the island on October 19, 2008. Community concerns included permanent fishing prohibitions; permanent oil, gas and mineral extraction prohibitions; and the requirement to obtain written permission from the federal government to enter ancestral waters that have been under indigenous control for thousands of years (Marianas Conservation 2008). After greeting the protestors, Connaughton proceeded to the Hyatt hotel where he was met by about 65 Friend of the Monument supporters wearing Marianas Trench Monument T-shirts provided by the Pew Environment Group, a division of the Pew Charitable Trusts. In the view of many CNMI residents and the CNMI government, promises made by Connaughton during his visit would be broken (see Chapter 12, Section 12.3).

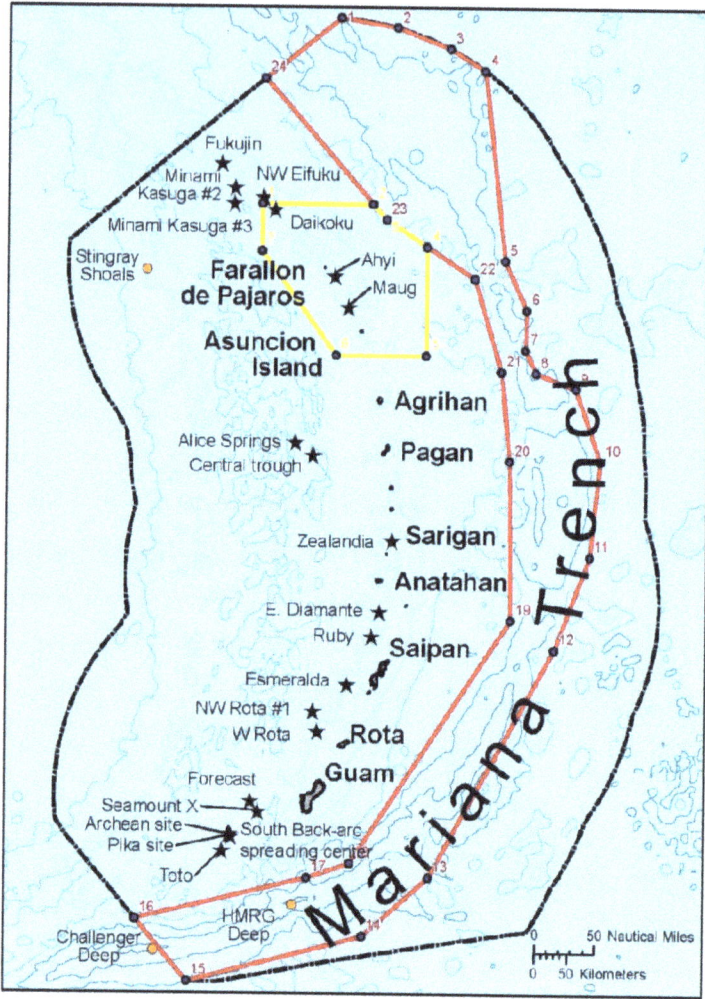

-10000 m	★	Active Hydrothermal Submarine Volcanoes
-8000 m		Trench Unit (59,732 nm²)
-6000 m		Islands Unit (12,388 nm²)
-4000 m		EEZ
-2000 m		

The Marianas Trench Marine National Monument includes three units: Island, Trench and Volcanic. Commercial fishing, which is federally defined as selling even a single fish, is prohibited in the Islands Unit. *NOAA illustration.*

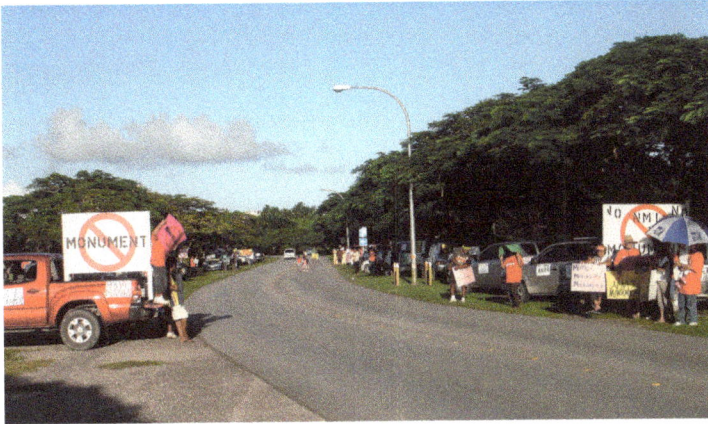

Approximately 100 persons from Saipan, principally from the indigenous Carolinian community, demonstrate against a proposed marine national monument in the CNMI. The protest coincided with the arrival of James L. Connaughton, chair of the White House's Council on Environmental Quality, to the island on October 19, 2008. *MarianasConservation.org photo.*

The CNMI bottomfish fishery would continue to face hardships due to dwindling access to fishing grounds (see Chapter 12, Section 12.3).

7.7 Pacific Remote Islands Bottomfish Fishery

Bottomfish harvesting in the Pacific Remote Island Areas occurred at a relatively low level due to their remote isolation, jurisdictional status and existing federal regulations. In 1998, two Hawai'i-based troll and handline vessels fished in the U.S. EEZ around Palmyra and Kingman Reef. These vessels targeted both pelagic and bottomfish species, including deep-slope snappers and sharks. Trips in 1999 targeted the two-spot snapper (*Lutjanus bohar*) at Kingman Reef. Fishing stopped after results of a single specimen submitted for testing showed slight traces of ciguatera (an illness caused by eating fish containing toxins produced by the marine microalgae *Gambierdiscus toxicus*). Hawai'i-based vessels have been reported to make sporadic commercial fishing

237

trips to Palmyra and Kingman Reef to harvest bottomfish. Commercial data held by the State of Hawai'i for the years 1988–2007 indicates that over this period 19,095 pounds (8,661 kilograms) of bottomfish and reef fish were caught at Palmyra, Kingman Reef and Johnston. This is equivalent to 477 pounds (216 kilograms) annually of bottomfish and reef fish.

In March 2004, the Council prohibited the harvest of bottomfish and seamount groundfish management unit species in the no-take MPAs it had established in waters out to 50 fathoms around Kingman Reef, Jarvis, Howland and Baker. In 2006, federal permitting and reporting requirements for bottomfish fishing in the Pacific Remote Island Areas became effective.

On January 6, 2009, President Bush, again through authority of the Antiquities Act, proclaimed establishment of the Pacific Remote Islands MNM. The monument included coral reefs, pinnacles, seamounts, islands and surrounding waters within 50 nm of Johnston, Howland, Baker, Jarvis, Kingman Reef, Palmyra and Wake, encompassing a total area of 87,000 square miles (225,000 square kilometers). The monument proclamation prohibited commercial fishing but allowed the possibility of recreational fishing.

TABLE 8: BOTTOMFISH AND SEAMOUNT GROUNDFISH FMP: SUMMARY OF COUNCIL ACTIONS

DATE	ACTION	MEASURES
1986 Aug 27	FMP came into effect	Prohibited the use of explosives, poisons, trawl nets and bottom-set gillnets. Established a moratorium on the commercial harvest of seamount groundfish stocks at the Hancock Seamounts. Implemented a permit system for fishing for bottomfish in the waters of the EEZ around the NWHI.
1987 Nov 11	Amendment 1	Established framework measures to implement limited access systems for bottomfish fisheries in EEZ waters around American Samoa and Guam.
1988 Sep 6	Amendment 2	Divided the NWHI EEZ into the Ho'omalu and Mau Zones. Established a limited access system for the Ho'omalu Zone, with non-transferable permits and landing requirements for permit renewal and for new entry into the fishery. Left access to the Mau Zone unrestricted, except for vessels permitted to fish in the Ho'omalu Zone. Required new entrants to both fisheries to complete a protected species workshop before receiving permits.

1991 Jan 16	Amendment 3	Defined recruitment overfishing and delineated a process by which overfishing would be monitored and evaluated.
1991 May 30	Amendment 4	Requirements for vessel owners to notify NMFS at least 72 hours before leaving port if they intended to fish in a protected species study zone that extended 50 nm around the NWHI and to allow federal observer to be placed on the vessels to record protected species interactions if deemed necessary.
1999 Feb 3	Amendment 6	Designated EFH and described bycatch and fishing communities. Provisions regarding Hawai'i fishing communities, overfishing definitions and bycatch were approved later on August 5, 2003.

1999 May 28	Amendment 5	Established a limited-entry program for the Mau Zone in the NWHI with non-transferable permits and landing requirements for permit renewal; required attendance by the primary vessel operator at a protected species workshop; established a Community Development Program under which 20% of Mau Zone permits were reserved for Native Hawaiians; instituted a maximum vessel length of 60 feet (18.3 meters) for replacement vessels in the Hoʻomalu or Mau Zones.
2004 Mar 25	Amendment 7	Prohibited the harvest of bottomfish and seamount groundfish MUS in the no-take MPAs around Rose Atoll in American Samoa, Kingman Reef, Jarvis, Howland and Baker.
2004 Aug 19	Framework Measure 1	Extended the moratorium on harvesting seamount groundfish from the Hancock Seamount by another six years until August 31, 2010.
2006 Sep 12	Amendment 8	Established new permitting and reporting requirements for vessel operators targeting bottomfish species around the Pacific Remote Island Areas.

2006 Dec 4	Amendment 9	Prohibited vessels greater than 50 feet (15.25 meters) in length overall from targeting bottomfish MUS within 50 nm of Guam; required these vessels to obtain federal permits and to submit federal logbooks.
2009 Jan 12	Amendment 10	Prohibited commercial bottomfish vessels greater than 40 feet (~12 meters) in length in waters 0 to 10 nm around the southern islands of CNMI and 0 to 10 nm around the Alamagan; required these vessels to have federal permits and logbooks and to carry active VMS units.
2008 Apr 4	Amendment 14	Implemented a TAC, federal noncommercial permits and reporting requirements, noncommercial bag limits and a closed season for fishing for main Hawaiian Island Deep 7 species. Defined the fishing year for the fishery as September 1 to August 31 the following year. Required permits and reporting.

Chapter 8

Pelagic Fisheries—Overcoming Challenges to Support Sustainable U.S. Fisheries (1987–2009)

Completion and implementation of the Billfish PMP in 1980 did not end the controversy over management of fisheries that targeted tuna but incidentally caught substantial amounts of other pelagic species. The PMP dealt only with prospective foreign longline fishing and was principally driven by a desire to protect billfish stocks or at least the availability of billfish in waters used by local commercial and noncommercial fishermen in Hawai'i and by sportfishing boats out of Kona. It was known that foreign longline vessels had fished extensively in and around the U.S. EEZ, taking substantial catches of billfish; control of this fishing was the first priority of the PMP and the Council's developing FMP.

Even after the PMP was approved, disagreements still existed about the extent of closures needed to ensure protection from potential foreign harvests of locally important billfish and other species, such as tuna, mahimahi and wahoo. The Council noted that the effect of the PMP had been to eliminate foreign longline fishing. This went beyond the intent of the Council and its member governments, some of which had interests in engaging in joint venture arrangements with foreign industries to generate funds and investments for much needed fishery development in the islands. The foreign vessels provided the territories with revenue through their purchases of food, fuel and repair services when they called into port.

At the same time, the PMP did not provide any certainty about the future conditions in the fishery. For example, the number of foreign fishing applications had been increasing over time, even as there was no foreign fishing occurring under the PMP. Also, the PMP did not address tuna stocks or fishing gear types. In this context, the Council was especially concerned about drift gillnets and the lack of controls over them. In addition, the PMP established non-retention of billfish as a principal management measure, which was not only difficult to enforce but also likely to lead to waste of important fishery resources. Perhaps equally important, however, the PMP did not provide any basis for considering measures to control domestic fisheries, a weakness that needed to be resolved. This would require reliable fishery monitoring and data collection programs.

The development of the Billfish FMP was ultimately a collaborative effort involving the Council, the NMFS Southwest Regional Office and scientists from the NMFS Honolulu Laboratory. The elements of the FMP included the following:

- Designation of a broader mix of species in the management unit (adding mahimahi and wahoo to the billfish and oceanic sharks under the PMP);

- Extensive area closures for foreign longline fishing around Hawai'i, Guam and American Samoa;

- A total prohibition of drift gillnet fishing by foreign and domestic vessels in the U.S. EEZ, except domestic vessels that obtained experimental fishing permits subject to reporting catch and effort, including interactions with sea turtles and marine mammals;

- Permit and reporting requirements for foreign fishing vessels (as under the PMP); and

- An annual report and five-year review of the FMP to determine if the objectives were being met and if any management changes were needed.

The Billfish FMP was completed in 1986 and submitted to the Commerce Secretary for approval; the final rule to implement the FMP was published on March 23, 1987.

8.1 Saving North Pacific Albacore from Drift Gillnets

One of the Council's first actions under its Pelagic FMP was prohibiting drift gillnet fishing throughout the region's EEZ waters. This important and precedent-setting action may ultimately have contributed to international action to address the damaging potential of this gear.

In 1988, Michael Bailey of Earth Trust, an environmental NGO, warned the Council that Asian-based fishing boats were setting approximately 30,000 miles (48,280 kilometers) of drift gillnets in the North Pacific. He expressed concern that, if left unregulated, these miles of untended floating drift nets could lead to the loss of migratory albacore stocks and incidentally harm other significant living marine resources such as porpoises and humpback whales (WPRFMC 1988a, pp. 13–14). Council members, with help and advice from the Pacific Basin Development Corporation (PBDC), deliberated on ways to protect the albacore and other resources at risk by these unregulated drift nets.[55]

---

[55] The PBDC is comprised of the governors of Hawai'i, American Samoa, Guam and the CNMI. Established in 1980, its mission is to advance economic and social development in the U.S. Pacific Islands.

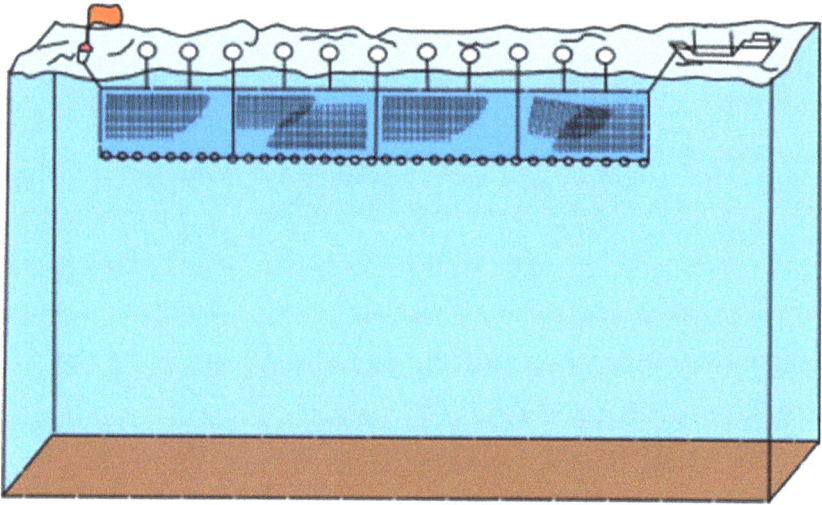

Drift nets hang vertically in the water column through the use of floats attached to a rope along the top of the net and weights attached to another rope along the bottom of the net. These nets usually target schools of pelagic fish, but bycatch of non-target species led the Council to ban the gear in the Western Pacific Region in 1987 and the United Nations to restrict its use in 1992. *Wikimedia Commons illustration/Jose Ramon Garcia Ares.*

The principal fishing nations of Asia—including Japan, China and Thailand—had invested heavily in miles of nets so were reluctant to stop drift net fishing. The U.S. territories conducted large amounts of business with these foreign fishing fleets through sales of fuel, supplies and shore-side provisions, such as repair facilities. To pressure the fishing nations to stop the practice, the Council mobilized its members to refuse service in state and territorial ports to fishing vessels using floating drift nets.

In 1989, American Samoa Governor Peter Tali Coleman spoke to the Council about the devastating impacts that drift gillnets were having on albacore. He said that he had asked the tuna canneries in American Samoa not to do business refueling fishing boats with drift nets because the fishing method was decimating local albacore stocks.

For the leader of a territory whose economy was based on tuna, this was an act of significant political courage.

The Council's drift gillnet ban along with the refusal of American Samoa and, subsequently, Guam and CNMI to refuel or sell provisions to boats involved in drift gillnetting caused the Asian fisheries in the region to discontinue the practice. In 1991, four years after the Council's action was implemented, the United Nations banned driftnet fishing on the high seas. The United States followed with legislation prohibiting U.S. vessels from using driftnet gear on the high seas. While there are still reported and suspected fishing vessels illegally using driftnets on the high seas, the scale of the fishery has dramatically decreased since the United Nations ban went into effect in 1992.

The work with Earth Trust is an example of the many times the Council has collaborated with environmental NGOs to safeguard vulnerable species.

8.2 Hawai'i Pelagic Fishery

Arrival of Vessels from the Atlantic and Gulf of Mexico

When the Pelagic FMP was approved, the need to manage domestic fisheries in the near future was not considered likely. Noncommercial and commercial fisheries seemed to coexist peacefully. Indeed, the distinction between the fisheries was not always clear. As with some other Hawai'i fisheries, many in the pelagic fishery who fished for recreation would sometimes sell their catch to cover the costs of the fishing experience. This was simply an accepted practice that exacerbated the problem of getting accurate catch data.

There were no conflicts between major gear types, either. While purse-seine fishing and high seas albacore trolling occurred in various parts of the Pacific, they did not seem to pose the risk of any adverse effects on the local domestic fisheries, so the need to try to control those gear types was not apparent. Besides, they were targeting tuna, and tuna were still exempt from management under the FCMA. Thus, domestic management of pelagic fisheries was not significant at the time of the FMP development. This changed dramatically and became critical when the arrival of fishing vessels from the Atlantic and the Gulf of Mexico caused the local longline fishery to grow exponentially in the late 1980s.

Prior to this, during the early 1980s the local longline fleet consisted of 37 vessels that set their gear 900 to 1,500 feet (275 to 450 meters) below the surface to catch yellowfin and bigeye tuna. This fleet was descended from the flag-line vessels that were established in Hawai'i by Okinawan immigrants in 1917 and was augmented by Korean-American fishermen after World War II. The fleet fished with what was known as basket-style gear, consisting of tarred rope and floats marked with flag-topped bamboo poles.

Council member Thomas Webster coils tarred lines aboard his vessel, one of the last vessels of the Hawai'i flagline fleet. With the arrival of longline vessels from the U.S. mainland, the fishery would evolve to the use of monofilament fishing line. *WPRFMC photo.*

One of the early fishermen to migrate from the Atlantic to Hawai'i was Capt. Bruce Mounier aboard the F/V *Magic Dragon*. His initial interest was to harvest lobster. However, when intense competition between European and U.S. Atlantic fishing fleets caused excessive harvests and declines in the stocks of Atlantic swordfish and bluefin tuna in the mid- to late-1980s, Mounier turned his eye toward broadbill swordfish. Known locally in Hawai'i as shutome, swordfish was regarded as an incidental catch of limited value. Realizing that there was a large market for it on the mainland and that swordfish existed in waters surrounding Hawai'i, Mounier began searching for the stock in a systematic way. After a number of experimental sets, it became clear there was a substantial swordfish in waters near the NWHI offshore from where he had fished for lobster and to the east.

Hawai'i is a major provider of swordfish for domestic consumers, accounting for 55% of the American harvest and about 14% of the swordfish Americans consume. The fishery began in the 1980s with the discovery of nearby swordfish grounds and an influx of longline vessels from the U.S. mainland. *Hawaii Seafood Council photo.*

Word of Mounier's success quickly reached fishermen throughout the continental United States, who began moving to Hawai'i in the fishery equivalent of a gold rush. Among them were the "black boats," so called because of the color of their hulls. Six of these vessels were the first group to enter the Hawai'i swordfish fishery. They came together from the East Coast and had fished swordfish up and down the North Atlantic. They were larger than almost any, if not all, existing longline boats in Hawai'i. They had more sophisticated electronics and were configured for longer trips.

At about the same time, celebrity chef Paul Prudhomme responded to concerns regarding the overharvesting of red drum (*Sciaenops ocellatus*) for his signature Cajun style blackened fish dish by using yellowfin tuna as a substitute. As a consequence, the price of yellowfin tuna on the mainland began to rise. Locally, the demand for yellowfin tuna in Hawai'i prior to 1980 was small and mostly limited to Japanese restaurants and small markets. Known as 'ahi, these tuna were primarily caught by small longline, troll and handline vessels. That changed after a Vietnamese-American fishing boat captain on the Gulf Coast heard about the Hawai'i fishing grounds from family members in the fish business in Honolulu. On their advice he brought his vessel from the Texas Gulf Coast through the Panama Canal to Hawai'i. It was through his influence that large numbers of Vietnamese-American fishermen began migrating to Hawai'i. Many of the vessels in this fishing fleet, which originated in Texas and Louisiana, had been made eligible for special U.S. government fishing vessel financing programs as a part of a post–Vietnam War settlement. Some converted their shrimp trawlers to longliners so that they could participate. As opposed to the East Coast longliners, the Vietnamese-American vessels used a mixed fishing strategy that initially caught shallow tunas (yellowfin and albacore) and miscellaneous pelagic species such as mahimahi and wahoo, known locally in Hawai'i as ono.

There were several reasons the vessels from the Gulf and East Coast began moving to Hawai'i. The State of Hawai'i, trying to diversify its economy, began offering subsidies, while at the same time the federal government was offering low-interest loans to fishermen. Hawai'i waters also looked inviting because federal and state regulators on the Gulf and East Coast had increased pressure on commercial swordfish fishermen in response to claims by big-game sportfishermen that recreational fishing was being adversely affected.

The Pelagic FMP at the time had no regulatory limit on the number of fishing vessels that could operate, and the FCMA prohibited discriminatory allocation of fishing rights based on residence or area of origin. All U.S. fishermen had equal opportunity to fish out of Hawai'i (subject to licensing and reporting requirements) and could not be turned away. The number of vessels from the Atlantic and Gulf converging on Hawai'i resulted in a quadrupling in the size of the Honolulu-based longline fleet.

Between 1987 and 1991, the number of longline fishing vessels in Honolulu Harbor increased from 37 to 156 (Boggs and Ito 1993). The increase took place as the Hawai'i longliners, which were geared for yellowfin and bigeye tuna, were going through a technical evolution, changing from traditional Japanese-style basket gear to new spooled, monofilament mainline stored on a mechanical drum capable of setting many more hooks. Consequently, the number of longline hooks set by the Hawai'i fishery quadrupled from an estimated 3.2 million in 1987 to more than 12.1 million set in 1991. It was a huge increase in industrial fishing effort throughout the EEZ surrounding Hawai'i and on the high seas, which had a ripple effect for many years on both the economy and the environment.

The total catch of the swordfish longline fishery in Hawai'i increased from 600,000 pounds (272 mt) in 1989 to 4.2 million pounds (1,905 mt) in 1990 (WPRFMC 1992a).

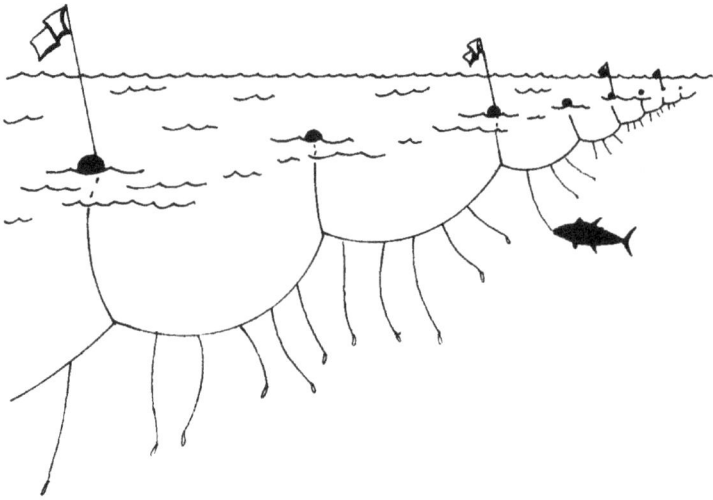

Longline fishing has been the main source of tuna and other fish caught by Hawai'i's commercial pelagic fishery since the early 1980s. About 125 vessels set gear deep in the water column during the day to target sashimi-quality bigeye tuna, and about 30 vessels shallow-set at night part of the year to catch swordfish, a premium grilling fish.

As the number of longline vessels in Hawai'i continued to increase, social problems began to develop. The most serious altercations involved Vietnamese-American fishermen who had arrived from the Gulf and local small-boat fishermen.

The fishermen from the Gulf Coast were accustomed to fishing near shore. Many began setting their 35-mile (56-kilometer) longlines near FADs that had been placed offshore by the State of Hawai'i and were popular with local Hawai'i fishermen. This caused an immediate overlap between the effort of the newly arrived longliners and the Hawai'i-based noncommercial troll and handline fishermen who were accustomed to fishing in these same areas. Conflicts that broke out between the groups soon had to be addressed by the Council. Troll fishermen who had their gear entangled with the longlines became angry and vandalized the longlines. Shots were fired!

The destruction of their gear infuriated the longline fishermen, who had to bear the cost of retrieval and repair or replacement of the ruined gear. The longline fishermen began threatening to defend their property with guns. The local Hawaiʻi fishermen responded with counter threats. The issue arose in angry outbursts at Council meetings. The arguments grew so heated that William "Bill" Paty Jr., who had succeeded Yee as Council chair, had to appeal for peace and the end of violent threats.[56]

Hilda Diaz-Soltero, NMFS Southwest Region administrator (at table, center, turning page), listens intently to the discussion at the 87th Council meeting, held August 8–10, 1995, at the Sheraton Makaha on the Waiʻanae coast of Oʻahu. *WPRFMC photo.*

The rapid increase in fishing pressure was negatively impacting the ability of local handline and troll fishermen to make a good living

---

56   William "Bill" Paty Jr. in discussion with Michael Markrich, 2008.

catching yellowfin and other pelagic species at the FADs and in nearshore waters.

One of the problems for the Council, however, was that the MSA in the late 1980s did not include the management of tuna under the Council's purview.

The Push to Include Tuna as a Managed Species

[The following text on tuna inclusion is freely adapted from Carr (2004) with permission given to the Council for its use.]

The Western Pacific Council had advocated tuna inclusion since its creation in 1976. However, in the mid-1980s, criticism of its fisheries management and questioning of its necessity gave the Council reason to advocate tuna inclusion with increased urgency. In 1985, a draft report of the U.S. Department of Commerce's Office of the Inspector General recommended elimination of the Western Pacific Council and the transfer of its responsibilities to the Pacific Council, on the grounds that there were not meaningful fishery resources to be managed in the Western Pacific Region. The Western Pacific Council responded that most of the catch from fisheries in the region occurred in federal waters, regulated by the Council, and that substantial tuna fishing within the FCZ surrounding the U.S.-flagged Pacific Islands had to be taken into account. Shortly thereafter, in 1986, a blue ribbon study of fishery management commissioned by NOAA recommended tuna inclusion. However, the same study recommended elimination of the Western Pacific and Caribbean Fishery Management Councils on the grounds that most fisheries within their regions were conducted within state, commonwealth and territorial boundaries, rendering those Councils unnecessary.

Further impetus was given to the Western Pacific Council's efforts for tuna inclusion in the MFCMA in the late 1980s by the tremendous growth in the Hawai'i-based longline fishery for tuna and swordfish. This rapid increase in the longline fleet gave rise to serious conflicts with recreational fishermen, who pushed the Council to exclude the commercial fishermen from certain fishing grounds. Even later, established commercial longline fishermen would advocate that the Council limit new entrants to the fishery.

In early 1989, the Council persuaded the PBDC to support tuna inclusion. The support of Hawai'i Governor John D. Waihee III, American Samoa Governor Peter Tali Coleman, Guam Governor Joseph F. Ada and CNMI Governor Pedro P. Tenorio gave tuna inclusion a self-determination cachet that would be usefully set against the ambitions of the U.S. distant water tuna fleet. More significantly, the support of American Samoa's governor for tuna inclusion would serve to counter the opposition of the distant water fleet and the two processors with canneries in American Samoa, on the grounds that tuna inclusion threatened the economic viability of the canneries.

Hawai'i Governor John D. Waihe'e III (1986–1990) chaired the Pacific Basin Development Council. *Wikimedia Commons/Michi Moore photo.*

American Samoa Governor Peter Tali Coleman (1956–1961, 1978–1985, 1989–1993). *National Governors Association photo.*

Guam Governor Joseph F. Ada (1987–1995).
National Governors Association photo.

CNMI Governor Pedro P. Tenorio (1982–1990). *WPRFMC photo.*

In the U.S. Congress, Senator Inouye was the ranking majority member of the Senate Commerce Committee, the committee with primary jurisdiction over fisheries issues. He also sat on the Committee's National Ocean Policy Study, a subcommittee that dealt specifically with fisheries and oceans issues. Inouye's support of the Council was long-standing, and members of the Council, its staff and other advocates of tuna inclusion from Hawai'i were personally acquainted with him.

On the House side, Republican Congresswoman Pat Saiki, one of two members of the House representing Hawai'i, would act as the leading advocate for tuna inclusion. Saiki sat on the House Merchant Marine and Fisheries Committee, the committee with primary jurisdiction over fisheries issues, and on its Subcommittee on Fisheries and Wildlife Conservation and the Environment. Although a relatively junior member of Congress, having been first elected to the House in 1986, Saiki proved to be a persistent and surprisingly successful advocate for tuna inclusion.

Congresswoman Patricia Saiki (1987–1990) would act as the leading advocate for tuna inclusion in the House of Representatives. *U.S. House of Representatives photo.*

In their respective houses, Inouye and Saiki would act as the torchbearers for tuna inclusion. Their staffs collaborated closely with the Council and relied heavily upon it in developing arguments and materials to support tuna inclusion and in moving tuna inclusion amendments through Congress.

In the summer of 1989, Congress held hearings on MFCMA reauthorization, as well as separate hearings devoted solely to the tuna inclusion issue in both chambers. The House Fisheries Subcommittee hearing overwhelmingly featured opponents of tuna inclusion. The Subcommittee declined to invite representatives of any councils or governors to testify. Of the seven witnesses who testified before the subcommittee, only one, the representative of a recreational fishing conservation organization, testified in favor of tuna inclusion. Testimony against tuna inclusion was presented by the Departments of State and Commerce, the U.S. Tuna Foundation, the America Tunaboat Association, the East Coast Tuna Association and a representative of Atlantic Coast swordfish fishermen.

In the Senate, the proponents of tuna inclusion enjoyed a far more receptive hearing before the Commerce Committee's National Ocean Policy Study. Indeed, as the witness list for the House hearing had been loaded against tuna inclusion, the witness list for the Senate hearing was loaded in favor of tuna inclusion. The witness list for the July 20, 1989, hearing of the Committee on Commerce, Science and Technology included Council Chair Paty, who testified on behalf of the five regional fishery management councils, which, at the time, favored tuna inclusion. Also testifying in favor of tuna inclusion were the Honorable Peter Tali Coleman, governor of American Samoa; U.S. Admiral Ronald J. Hays (retired), former commander in chief of the U.S. Pacific Command; Richard Shomura, former director of the NMFS Honolulu Lab; Peter Fithian, chairman of the Hawaiian International Billfish Association;

Trudy Nishihara, Hawaii Fishing Coalition; and Fritz Amtsberg, Hawai'i commercial fisherman.

Testimonies against tuna inclusion were presented by U.S. Senator Pete Wilson (R-California); U.S. Representative Eni Faleomavaega (D-American Samoa); Edward E. Wolfe Jr., U.S. State Department deputy assistant secretary for oceans and fisheries affairs (accompanied by Carmen Blondin, NOAA deputy assistant secretary of commerce for international interests); James Walsh, U.S. Tuna Foundation (accompanied by August Felando, American Tunaboat Association president); James Joseph, Inter-American Tropical Tuna Commission director; and Michael Franks, Coastal Seafood Processors president.

During the July 20, 1989, Senate hearing in Washington, D.C., William "Bill" Paty Jr. (chair of the Council and of the Hawai'i Board of Land and Natural Resources) provided testimony on behalf of five regional fishery management councils, which at the time favored including tuna as a federally managed species. *WPRFMC photo.*

Margaret Cummisky, who served in Senator Inouye's personal office and in his Senate Commerce and Appropriations Committee offices, worked closely with the Council on the issue of including tuna in the MFCMA. *WPRFMC photo.*

In early August 1989, Senator William Roth (R-Delaware), noting the support of five regional fishery management councils and numerous sportfishing and conservation associations, introduced legislation to repeal the tuna exclusion provision of the MFCMA. In response to Saiki's request for specific language to accomplish tuna inclusion, the chairman of the Hawaii International Billfish Association proposed the straightforward language of the Roth bill terminating the MFCMA's tuna exclusion provisions. This would become known as the Saiki Amendment. While its language was simple, securing House passage of a tuna inclusion amendment would prove complex.

In the early fall of 1989, the staff of the House Fisheries Subcommittee developed legislation designed to maintain the tuna exclusion and preserve the juridical position but, at the same time, authorize the councils to collect data from tuna fishermen. Saiki offered her amendment to repeal the tuna exclusion from the MFCMA, but it was defeated by a voice vote.

Undeterred by the defeat of her amendment in the Subcommittee markup, Saiki made plans to again introduce her amendment at the full Committee markup scheduled for October 5, 1989. In support of this effort, she solicited individual letters from the PBDC governors tailored to each governor's area. Saiki's tuna inclusion amendment was adopted during the full Committee markup by voice vote.

On January 8 and 9, 1990, during the Congressional recess, U.S. Representative Walter B. Jones Sr. (D-North Carolina), the powerful chairman of the Merchant Marine and Fisheries Committee from 1981 to 1992, presided over a hearing of the full Committee in Honolulu on various ocean issues, including fisheries. Although the Committee had already reported MFCMA amendments, including Saiki's tuna inclusion amendment, tuna inclusion supporters from the region used the hearing as an opportunity to reiterate their commitment to it.

A key role was played by the Council during the hearing. At the hearing, Congressman Jones heard U.S. Navy personnel testify that, wherever they went in the Pacific Islands, they were often hated because U.S. tuna boats could fish in the waters of the Pacific Island countries with impunity and without paying. Concerned about U.S. security in the region, Jones became an unexpected ally of the efforts of the Council to change the law and include tuna in the FCMA.[57]

On February 6, 1990, the House approved the MFCMA reauthorization amendments, which included the Saiki tuna inclusion amendment, by an overwhelming vote of 396 to 21 (with 14 abstentions). Representative Glenn Anderson (D-California) insisted on a recorded vote rather than a voice vote, saying, "When the domestic tuna industry, a $26 billion industry, collapses at the weight of this provision, I can say I had no part in it."

---

[57] Simonds op. cit.

As they had in the House, the tuna inclusion forces would also have to overcome advocacy by Senate committee staff of "compromise" legislation supported by the U.S. tuna industry that would maintain the MFCMA's tuna exclusion and preserve the juridical position. In fall 1989, the U.S. distant-water tuna industry successfully lobbied the National Ocean Policy Study staff to develop and support such compromise legislation. Under the compromise legislation, the regional councils would be divested of the authority they had to manage highly migratory species (viz., billfish) and such management authority would be given to a "Highly Migratory Council" consisting of one representative from each of the regional fishery management councils. This Highly Migratory Council would not have the authority to manage tuna but would act as an advisory body to U.S. delegations to international organizations, such as the Inter-American Tropical Tuna Commission and the International Commission for the Conservation of Atlantic Tunas. In response to the staff "compromise" legislation, the Western Pacific Council met and rejected the proposal for a Highly Migratory Council.

The distant-water tuna industry was not the only segment of the U.S. fishing industry opposed to tuna inclusion. Commercial swordfish fishermen on the East Coast had avoided regulation by the fishery management councils because billfish management plans had regularly been rejected by NMFS, pursuant to the 1979 NOAA legal opinion, for failing to allow a reasonable opportunity to fish for tuna. These swordfish fishermen feared that tuna inclusion would remove this impediment to billfish management and that, if they were subject to fishery management council measures, "the recreational fishermen would regulate them out of business." The concerns of commercial swordfish fishermen led Senator John Kerry (D-Massachusetts), a member of the Commerce Committee and the National Ocean Policy Study vice chairman, to oppose tuna inclusion.

The Western Pacific Council communicated its objection to Inouye. After collaborating with the senator's fisheries staffer, the Council's executive director informed the executive directors of the other regional councils supporting tuna inclusion that the senator wanted to hear from the councils about their views regarding tuna inclusion. In addition, the Western Pacific Council worked closely with the Pacific Island governors and the PBDC to maintain their continued support for tuna inclusion and to ensure that such support was communicated to Inouye.

In April 1990, tuna inclusion received a boost when the three largest U.S. canneries—StarKist, Bumble Bee and Van Camp—announced they would no longer purchase or market tuna caught by purse-seine vessels that encircled dolphins. Because encircling dolphins swimming above tuna was the predominant method of fishing by the U.S. purse-seine fleet fishing in the Eastern Tropical Pacific Ocean, the processors' moratorium had the effect of mooting the issue of negotiating satisfactory access for such vessels to the EEZs of Latin American countries. The need for leverage for such negotiations had long been one of the primary arguments of the U.S. distant-water tuna industry and the State Department against tuna inclusion.

Before the full Commerce Committee markup of the MFCMA amendments, the Western Pacific Council sought to fortify Inouye's commitment to tuna inclusion. Already in March 1990, Inouye had written the Committee's chairman to make clear his intention to introduce a tuna inclusion amendment. Inouye provided several reasons for tuna inclusion, including the need for more effective tuna management than was being provided under the auspices of the International Commission for the Conservation of Atlantic Tunas and the need for the Western Pacific Council to be empowered to effectively manage the rapidly growing longline fleet in its region. In an effort to galvanize Inouye for the Committee markup, Kitty M. Simonds, the Western Pacific Council executive director, wrote Inouye's fisheries staffer that management of tuna by the Secretary of Commerce, rather than the regional councils, "would again take away from the U.S. EEZ in the Pacific management of its only renewable resource" (Simonds 1990 May 20). Simonds made it clear that the councils needed to have the authority to "restrict the number of boats in the fishery" because the Western Pacific Council wanted to limit the influx of longliners into its region (Simonds 1990 May 14).

Going into the markup, the likely success of Inouye's tuna inclusion amendment was uncertain. When the full Committee met on May 22, 1990, it accepted Inouye's tuna inclusion amendment by a vote of 11 to 8. As an indication of the controversy surrounding the issue and the significance attached to it, the vote taken by the Committee on the tuna inclusion amendment was the only roll call vote taken on several amendments and on the bill as amended; all other votes were by voice vote.

The bill reported out of the Commerce Committee redefined "highly migratory species" to include tuna and gave the Secretary of Commerce the authority to manage highly migratory species on the Atlantic and Gulf Coasts and in the Caribbean but assigned such management authority in the Pacific to the fishery management councils. In addition, the bill's highly migratory species provisions prohibited secretarial management plans for highly migratory species from establishing domestic quotas for highly migratory species that differed from the allocations or quotas established for U.S. fishermen by international organizations.

The Western Pacific Council expressed concern about the possible implications of the limitation on domestic quotas for its management authority over highly migratory species. At the time, the provision tying domestic quotas to international quotas had practical effect only on fisheries in the Atlantic and the Gulf because of the International Commission for the Conservation of Atlantic Tunas' activities. Nonetheless, it appears the Western Pacific Council—and supporters of tuna inclusion in the region—feared that the U.S. distant-water tuna industry might be inspired to urge the Department of State to advocate for tuna quotas by the Inter-American Tropical Tuna Commission, in the hopes of restricting tuna management by the Western Pacific Council and the U.S. flag Pacific Islands.

At the time, the ability of the U.S. tuna industry to use the Inter-American Tropical Tuna Commission in that way seemed farfetched because the organization at that point did not include Hawai'i and other U.S. flagged Pacific Islands within its jurisdiction and for many years had not been establishing quotas but had been only collecting data. Moreover, although the U.S. flagged Pacific Islands hoped significant tuna fisheries would be developed in their waters, the U.S. distant-water industry had not shown a real interest in fishing those waters.

On October 11, 1990, the Senate debated and passed MFCMA reauthorization legislation asserting U.S. jurisdiction over tuna within the EEZ. While the proponents of tuna inclusion prevailed on this most important issue, some of them were not completely happy with the bill passed by the Senate. In particular, the Senate bill retained the provision requiring secretarial management of highly migratory species on the Atlantic and Gulf Coasts. It also included provisions crafted by the National Ocean Policy Study staff that appeared to temper, but not altogether eliminate, the requirement that secretarial plans for highly migratory species could not establish different quotas than those established for U.S. vessels by the International Commission for the Conservation of Atlantic Tunas. The bill passed by the Senate also added a provision to delay the effective date of the tuna inclusion amendments until January 1, 1992, offered by Inouye as a "compromise," which had not been included in the bill reported by the Commerce Committee. While not limited to highly migratory species, the bill passed by the Senate also included a provision, similar to that in the bill reported by the Commerce Committee, allowing the Secretary of Commerce to impose moratoria on new entrants to fisheries in order to prevent overfishing.

The legislation next returned to the House for its consideration of the Senate amendments. When it took up the Senate bill on October 23, 1990, the House concurred with the Senate's amendments, except for the provision authorizing the Secretary to impose temporary moratoria on new entrants to fisheries. The House's deletion of the temporary moratorium provision appears to have been spearheaded by Representative Donald Young (R-Alaska), who described the provision as one "which would have allowed the Secretary of Commerce to determine who could fish and who could not fish in our Nation's waters." Young said the provision "is contrary to every action taken by this House in the past 14 years that the Magnuson Act has been in effect and therefore has been removed from the bill."

On November 28, 1990, President George H. W. Bush (an avid sportfisherman) signed the Fishery Conservation Amendments of 1990 into law. In a signing statement, the president, although acknowledging that the legislation delayed elimination of tuna exclusion until January 1, 1992, declared that "as a matter of international law, effective immediately the United States will recognize similar assertions by coastal nations regarding their exclusive economic zones." The announcement removed a long-standing irritant in U.S. foreign relations with many nations. The U.S. juridical position that emphasized the need for coastal and fishing states to cooperate in the conservation and management of tuna throughout their range (as mandated by Article 64 of UNCLOS) would live on as an important element of U.S. fisheries diplomacy.

With tuna now included in the MFCMA, in 1991 the Council proposed to solve the conflict between the growing longline fleet and Hawai'i's small-scale fisheries by imposing an emergency moratorium on any new longline entrants to the fleet for three years, a measure approved by the Secretary of Commerce. This led to a limited-entry program that was implemented in 1994. Additionally, the Council set strict fishing boundaries, closing waters out to 25 to 75 nm offshore (depending on the area) to longline fishing to physically separate the two fleets. During the winter, the large commercial longline vessels were allowed to fish as close as 25 nm from shore on the windward (northern) sides of the main Hawaiian Islands to take advantage of a seasonal winter run of bigeye tuna. This action was possible because the small local trollers and handliners rarely went offshore in winter as yellowfin availability is lower and ocean conditions are more unsettled, creating a greater risk to small boats. In 1992, this created a permanent buffer zone between the local Hawai'i fishing fleets and the newly enlarged commercial swordfish and tuna longline fleet. The intent of these regulations was to protect the range of mobility to which local fishermen had become accustomed while still allowing the large commercial fishermen access to good swordfish and tuna grounds.

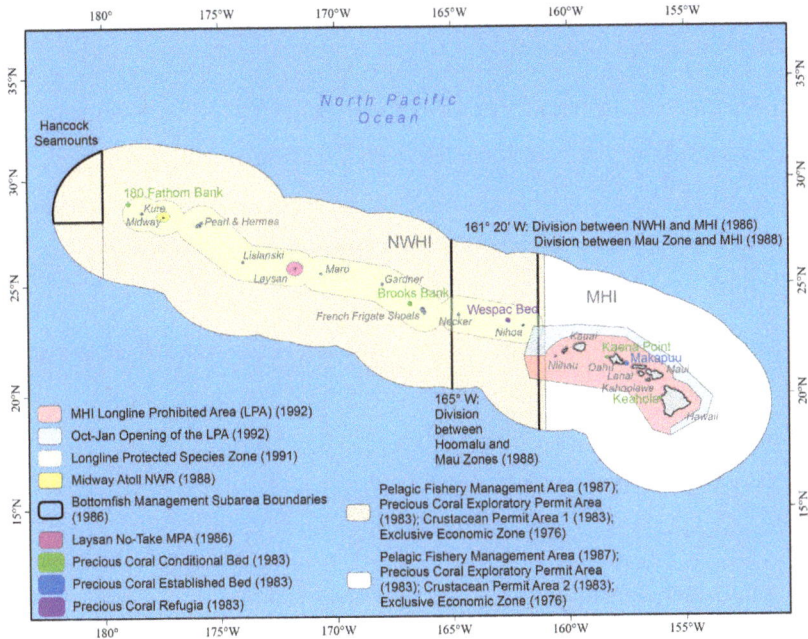

The Council has managed fisheries in the Hawai'i archipelago
through a variety of spatial management areas, from the precious
coral refugia in 1983 to the NWHI and main Hawaiian Islands
longline prohibited areas in 1991 and 1992. *WPRFMC illustration.*

At the same time the Council was creating the buffer space between
the two groups, it sought to limit participation based on objective criteria,
such as historic participation or long-term investment in the fishery.
New entrants could be prohibited from entering the fishery if they had
not participated in the fishery prior to 1991 or had not demonstrated
proof of investment by that year with intent to enter the fishery. NMFS
took permit applications and made decisions about permits based on the
records available in the State of Hawai'i or other sources. Ultimately,
the number of longliners allowed under this program was capped at 164
vessels, a limit which has remained in effect since.

Incidental Take Concerns: Monk Seals and the Protected Species Zone

The dramatic increase of the longline fishery in Hawai'i also created new issues for the Council in the NWHI. There was nothing in the Council's history that prepared it for the sudden deployment of more than 12.3 million hooks by U.S. fishermen in the upper ocean trophic layer. The Council would be forced virtually overnight to deal with the potential for unprecedented numbers of interactions with monk seals, seabirds, turtles and sharks. Action on monk seals was also spurred by appeals from Hawai'i longline fishermen who were aware of the consequences of interactions with endangered and threatened species without any mitigation measures in place.

These events, however, were just the beginning of an ongoing process to resolve real and imagined problems in the management of the longline fishery over the next three decades. The Council has probably spent more time and energy on the longline fishery—the largest and most valuable fishery in Hawai'i and the Western Pacific Region—than on any other fishery under its jurisdiction, primarily to deal with bycatch issues.

In the late 1980s, the expanding Hawai'i-based longline fleet began fishing for swordfish in the waters around NWHI. Some of the vessel captains used the same techniques they had been accustomed to using in Florida, laying their shallow-set hooks near the shore lines. As they moved up the chain of atolls and low-lying islands of the NWHI, contacts with monk seals increased. In 1990, the first official reports became available that the newly arrived swordfish longliners were interacting with monk seals. Head injuries on several live seals were observed. Light sticks used in swordfish fishing gear were also found on the shores of French Frigate Shoals.

Mitigating protected species interactions by the longline fishery in the NWHI was an industry-driven initiative, explained Council Executive Director Kitty M. Simonds.[58] Fisherman Jim Cook approached the Council to prohibit longline fishing from shore out to 20 nm in the NWHI. The NMFS regional administrator opposed the closure, Simonds said. However, Council discussions on the proposed closure continued and expanded to 30 nm around the islands and eventually to 50 nm.

At the 70th Council meeting from September 26 to 28, 1990, the Council approved Amendment 2 to the Pelagic FMP with the following provisions:

- Requirement for operators of domestic pelagic longline fishing and transshipment vessels to have a federal permit, maintain a federal logbook and, when fishing within 50 nm of the NWHI, have an observer on board if directed by NMFS;

- Requirement for longline gear to be marked with the official number of the permitted vessel; and

- Incorporation of waters of the EEZ around CNMI into the area managed under the FMP.

The creation of what would be known as the Protected Species Zone became the most sweeping preventive action to protect an endangered species taken by any Council up to that time. Typically such changes took six months or more to effectuate. In this case, however, the Council requested emergency action in November 1990 before processing the amendment to implement the mandatory observer coverage requirement and the Protected Species Zone, and the Secretary agreed. The emergency rule went into effect in the same month. Injuries to monk seals observed

---

[58]   Kitty M. Simonds in discussion with Sylvia Spalding, May 11, 2021.

in the first six months of 1991 spurred further Council action to request emergency closure of all waters within 50 nm of the NWHI to longline fishing. This was approved, and an emergency rule went into effect in April 1991. The final rule to implement Amendment 2 was then published in May 1991, with the Protected Species Zone formally coming into effect in July 1991, before the expiration of the emergency rule. The Protected Species Zone was set at 50 nm from the shore of most of the NWHI, with no longline fishing in those waters without an observer on board. The rule also authorized necessary changes to the zone by regulatory action rather than a full FMP amendment.

This was followed by Amendment 3 to the Pelagic FMP, implemented in October 1991, which created a permanent 50-nm longline exclusion zone around the NWHI to protect monk seals. It also implemented framework provisions for establishing a mandatory observer program to collect information on interactions between longline fishing and sea turtles. With the 50-nm Protected Species Zone in place, interactions with monk seals stopped.[59]

To monitor and enforce the area, the Council next pioneered the required use of GPS units on fishing vessels. British companies had used satellite tracking devices on top of their trucks so that the companies could keep track of their location. The Council had funded a pilot project in 1985 to deploy these devices on the Hawai'i longline fleet where it proved to be an exemplary success.[60]

This method of tracking fishing vessels, which came to be known as VMS, uses the GPS satellite constellation to record the exact coordinates of a fishing boat's position. Instead of having the U.S. Coast Guard conduct expensive monitoring flights, an enforcement officer needed

---

[59]   Paty op. cit.

[60]   Simonds 2008 op. cit.

only to view on screen the position of each participating longline fishing vessel, relative to the Protected Species Zone. The system proved so successful that it has become national and even international policy to mandate VMS for use in fisheries throughout the United States and the WCPO.

Addressing Marine Debris

Preventing fishery interactions in the NWHI may have played a part in improving the prospects for monk seal recovery. However, other causes of mortality remained, including shark predation, mobbing by aggressive males within the monk seal population and the loss of habitat. In addition, dozens of monk seals apparently died from handling by scientists and federal employees (Carretta et al. 2009). Another source of monk seal mortality and injury was, and continues to be, marine debris.

A large gyre of water, approximately 800 nm north of Hawai'i, acts as a collecting point for millions of tons of floating garbage including derelict fishing and cargo nets, other discarded fishing gear and plastic debris of all kinds that drift toward Hawai'i from high latitudes of the North Pacific. This debris presents a particular problem for animals in this area. Entanglement in the lost or abandoned cargo nets and fishing gear can cause injury or death of endangered monk seals and threatened turtles. In addition, the ingestion of small plastic debris particles kills or sickens seabirds. Although lost fishing gear is estimated to account for only 5% of the total marine debris, it has a disproportionate and negative impact on marine life.

While the marine debris impacting the populations of NWHI monk seals and other marine species, including coral, is not from Hawai'i fisheries, the Council has made concerted and collaborative efforts to address this issue, including the participation in and support of international workshops to find solutions to this problem.

The first international workshop on marine debris was held in Hawai'i in 1984 and entitled the Workshop on the Fate and Impact of Marine Debris. At the workshop it was determined that much of the derelict netting was constructed of synthetic fibers, which do not break down easily over time. This particular type of marine debris presents a long-term threat to fish and other marine organisms because it continues to "ghost fish" long after it is lost or abandoned. Among the species that suffered disproportionately were Hawaiian monk seals, this being especially problematic given the endangered status of the population. As more and more reports of marine life entanglements in the NWHI became apparent, the problem came to the forefront of public attention. In 1987, the Hawai'i State legislature passed Act 66 to make it unlawful to discard or dispose of any fishing net, trap or gear with netting in state waters. Hawai'i does not have significant net fisheries, and the most problematic derelict fishing gear was originating outside of Hawai'i waters. The solution would require international cooperation.

The Council co-sponsored a special seminar, held October 13 to 16, 1987, entitled the North Pacific Rim Fisherman's Conference on Marine Debris. The conference was attended by fishing industry representatives from Japan, China, Korea and the United States. At the conference, the attendees recognized that synthetic fishing gear constituted a growing threat to marine life and adopted the "Fishermen's Pledge for a Clean Ocean." The Council funded printing of the conference proceedings (Alverson and June 1988).

Addressing the impacts of plastic and synthetic fishing on the ocean is a focus of the Fishermen's Pledge for a Clean Ocean, adopted by participants at the 1987 North Pacific Rim Fisherman's Conference on Marine Debris. *WPRFMC illustration.*

Over the years, the Council sponsored and participated in a number of marine debris forums that took the problem directly to governments and fishermen, including a series of five International Marine Debris Conferences that were held in Hawai'i. As Hawai'i was host of these major conferences, it was said to be the cradle of the global marine debris movement. Although these marine debris symposia have contributed to the knowledge of marine debris issues, few of the ideas generated for solving the problem have been implemented. As the origin of the debris cannot easily be identified, and international agreements prohibiting discarding of pollution at sea are not well enforced, there is little ability to punish or deter activity that results in the accumulation of debris. At the same time, there is little or no incentive to collect the debris because it has no value and because no authority is specifically charged with its collection. On the local scale, since the debris does not originate from Hawai'i fisheries, Hawai'i fishermen and policymakers have little incentive to fund recovery of the debris.

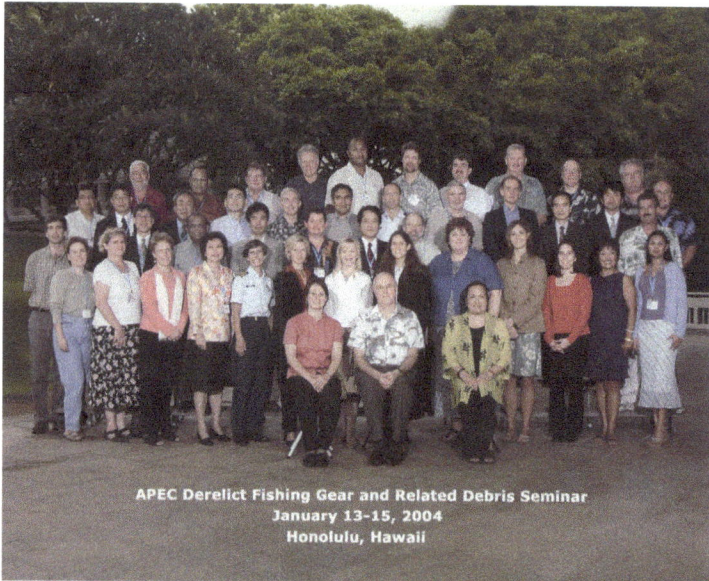

APEC Derelict Fishing Gear and Related Debris Seminar
January 13-15, 2004
Honolulu, Hawaii

Executive Director Kitty M. Simonds (seated front row, right) joins participants of the Asia Pacific Economic Cooperation's Derelict Fishing Gear and Related Debris Seminar, January 13 to 15, 2004, in Honolulu. Simonds provided the welcoming remarks, emphasizing the need to work together to address the problem in the Pacific.

NMFS initiated a marine debris collection effort in 1997 to remove lost nets, ropes and other marine debris in the NWHI. Under this program, a contractor was hired to collect the materials at a cost of approximately $25,000 per ton. The derelict fish nets recovered from the NWHI were chopped up at a local steel recycling company and incinerated at H-POWER, a waste burning facility that produces electricity.

As a follow up to this arrangement, the Council spearheaded a port collection program. Under this "Net to Energy" program, Hawai'i longline fishermen brought back derelict fishing nets they encountered at sea for incineration at Honolulu's waste energy recovery facility. The Council obtained funding for a roll-off bin to be located in Honolulu Harbor where Hawai'i longline fishermen could place debris and spent monofilament line. The bin, once full, was taken to Schnitzer Steel where the debris and monofilament were cut up for ease of handling before delivery to H-POWER for incineration. The amount of debris brought back by Hawai'i longline fishermen was enough to power 50 homes for a year. The Honolulu Harbor derelict fishing gear project was globally recognized as a public/private partnership that promotes marine stewardship.

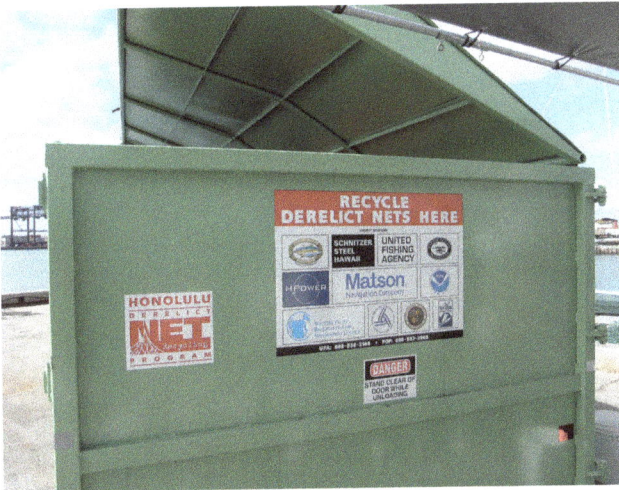

The Nets to Energy receptacle at Pier 38, Honolulu, is where longline fishermen deposit debris and spent monofilament line for incineration and conversion to energy (circa 2006). *WPRFMC photo.*

The Council additionally partnered with the Secretariat of the Pacific Community to print and distribute a series of "Think! Don't Throw!" posters. One poster provides a message for all those who go to sea on why they should care about marine debris and what they can do. Another includes a table comparing the amount of time it takes for various substances to break down, from an orange peel (6 months) to monofilament lines/nets (600 years).

Reducing Sea Turtle Interactions

After the initial Hawaiian monk seal interaction issue was resolved, the Council focused on the shallow-set longline fishery for swordfish and its interactions with sea turtles, seabirds and false killer whales (*Pseudorca crassidens*). Several measures were introduced to protect these species from being incidentally taken in the fishery.

Five species of turtles are found in the waters under the Council's jurisdiction: green sea turtle (*Chelonia mydas*), olive ridley (*Lepidochelys olivacea*), hawksbill (*Eretmochelys imbricata*), leatherback (*Dermochelys coriacea*) and loggerhead (*Caretta caretta*). All are listed as endangered or threatened under the ESA. The growth rate varies between species, but they are generally long-lived and slow to mature and many populations have historically experienced declines from overexploitation. They are at risk for many reasons. Humans harvest their eggs and eat their meat. Coastal development destroys their nesting sites. They are subject to entanglement and suffocation in marine debris. Disease, exposure to pollution and predation on land and at sea are constant dangers. Ocean warming may affect their reproduction and food supply. And there is the possibility of being caught incidentally by fishing gear.

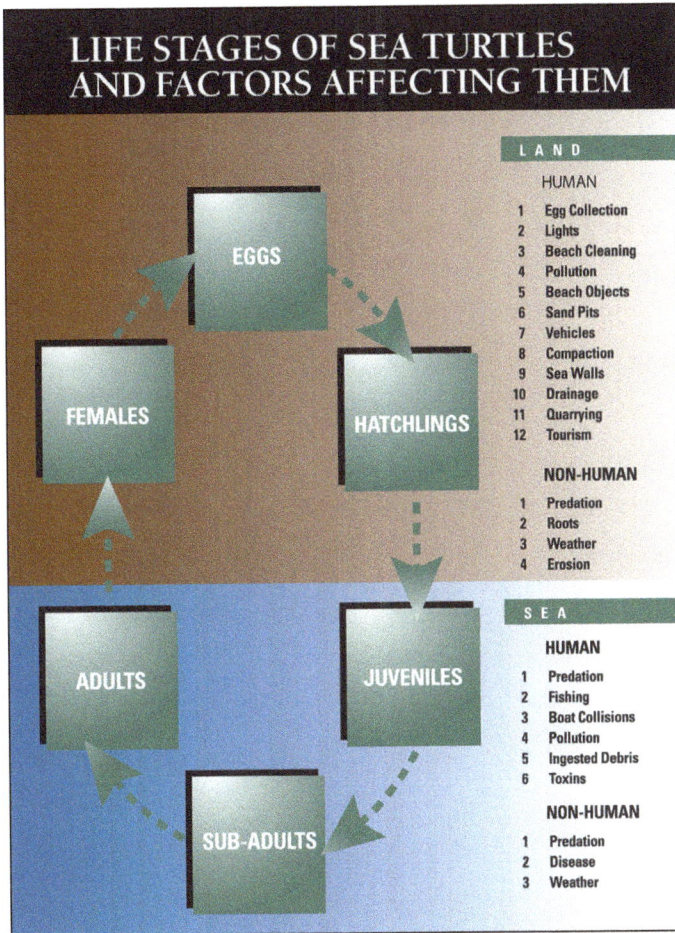

LIFE STAGES OF SEA TURTLES AND FACTORS AFFECTING THEM

EGGS

FEMALES

HATCHLINGS

ADULTS

JUVENILES

SUB-ADULTS

L A N D

HUMAN

1 Egg Collection
2 Lights
3 Beach Cleaning
4 Pollution
5 Beach Objects
6 Sand Pits
7 Vehicles
8 Compaction
9 Sea Walls
10 Drainage
11 Quarrying
12 Tourism

NON-HUMAN

1 Predation
2 Roots
3 Weather
4 Erosion

S E A

HUMAN

1 Predation
2 Fishing
3 Boat Collisions
4 Pollution
5 Ingested Debris
6 Toxins

NON-HUMAN

1 Predation
2 Disease
3 Weather

The poster "Life Stages of Sea Turtles and Factors Affecting Them" exemplifies the Council's efforts to provide outreach and education about the multiple factors that impact sea turtles at all stages of their lives. *WPRFMC illustration.*

The Council's fishery management measures have mostly addressed interactions with leatherback and loggerhead turtles. There was little knowledge and documentation of fishery interactions with sea turtles until observer programs recorded such events, especially in the Hawai'i longline fishery targeting swordfish. As that fishery developed and expanded into new fishing grounds, the vessels began setting their hooks in areas in which leatherback and loggerheads occurred. Shallow-set longlines used to catch swordfish began hooking turtles more frequently. Most turtles caught were released alive but with varying degrees of injury. It was estimated that the Hawai'i longline fishery's annual interaction rates averaged 112 leatherbacks, 418 loggerheads, 146 olive ridleys and 40 green sea turtles during the 1994–1999 period, producing an estimated mortality of 136 animals per year of all species combined (Simonds 2011).

This issue was raised at Council meetings, and discussions ensued about efforts to attach electronic tracking devices to the endangered and threatened turtles in order to better understand their migration patterns. At the April 1998 Council meeting, it was determined that the incidental catch of turtles exceeded federal limits in 1995 and 1996 (WPRFMC 1998). As a result NMFS would have to conduct or reinitiate consultations under the ESA to evaluate the potential effects of the fishery on sea turtles and would also have to complete a new biological opinion (BiOp) to determine if the fishery posed jeopardy for any listed species. The BiOp would enumerate reasonable and prudent actions that would allow the fishery to continue without jeopardizing the turtles.

Before that consultation was completed, the Center for Marine Conservation and the Turtle Island Restoration Network filed a federal lawsuit in February 1999 against NMFS through the Sierra Legal Defense Fund (later Earthjustice) in Hawai'i. Earthjustice alleged NMFS had violated the ESA by allowing the then 114 Hawai'i-based longliners to catch hundreds of turtles without filing a BiOp. The suit also alleged that NMFS violated the National Environmental Policy Act by failing to prepare a proper environmental impact statement to address the possibility of fatal interactions between turtles and Hawai'i-based longliners.

On November 23, 1999, Judge David Ezra, chief judge of the U.S. District Court for the District of Hawaiʻi, granted an injunction that closed the swordfish longline fishing grounds within the area north of 28 degrees north latitude and between 168 and 150 degrees west longitude while NMFS developed an environmental impact statement for the fishery. The closure was later modified in August 2000 with additional closed areas and effort limits, as well as requiring 100% observer coverage in the swordfish longline fishery.

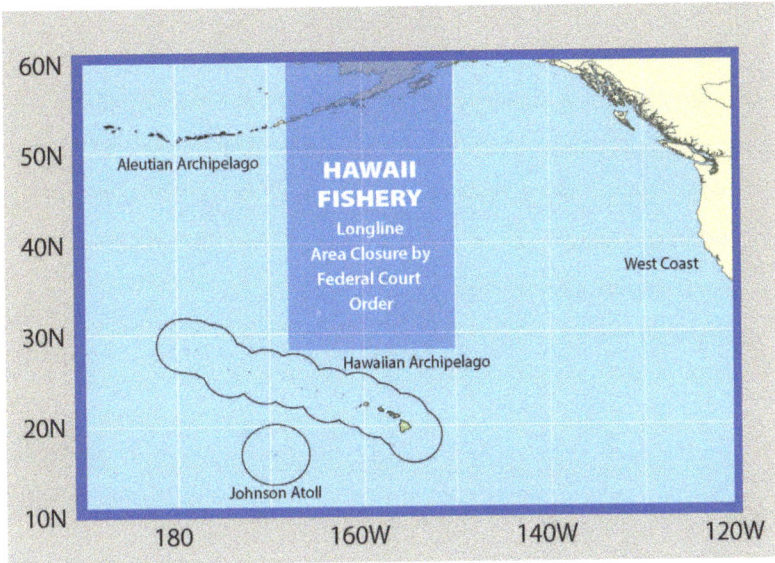

In November 1999, Judge David Ezra of the U.S. District Court for the District of Hawaiʻi closed the swordfish longline fishing grounds within the area north of 28 degrees north latitude and between 168 and 150 degrees west longitude while NMFS developed an environmental impact statement for the fishery. *WPRFMC illustration.*

On March 29 and 30, 2001, NMFS issued a new BiOp and a final environmental impact statement for the Hawaiʻi-based longline fishery. The BiOp concluded that the fishery was likely to jeopardize the continued existence of sea turtle species and included reasonable and prudent measures, such as prohibition of shallow-set longline fishing targeting swordfish north of the equator; prohibition of deep-set longline fishing targeting tuna from the equator to 15 degrees north latitude and from 145 degrees west to 180 degrees longitude between April 1 and May 31; and attendance of an annual protected species workshop by all Hawai'i longline vessel operators. The Hawaii Longline Association (HLA) filed a lawsuit in the District Court for the District of Columbia challenging the process and substance of the 2001 BiOp. The Council's executive director, SSC and others submitted comments to NMFS that were highly critical of the 2001 BiOp. Nonetheless, on June 9, 2002, the Council passed Regulatory Amendment 1, which incorporated the reasonable and prudent alternatives of the 2001 BiOp, and NMFS issued the final rule implementing the BiOp's reasonable and prudent measures on June 12, 2002 (NMFS 2002a).

During this time, the swordfish vessel owners converted to deep-set gear to target tuna, became inactive or left the region, including homeporting in California. The impacts of the swordfish longline fishery closure on turtle bycatch in Hawai'i's fishery were immediate. The catch of Hawai'i swordfish dropped by 93% (WPRFMC 2004b).

On November 15, 2002, NMFS issued a replacement BiOp, which showed that the average annual number of turtles caught from mid-2001 to mid-July 2002 dropped to an estimated eight green sea turtles, eight leatherbacks, 14 loggerheads and 26 olive ridleys (NMFS 2002b). HLA challenged the 2002 BiOp in its ongoing lawsuit.

TABLE 9: HAWAI'I BILLFISH CATCH (1,000 pounds), 1987–2004
(Note: Catches reflect the development and regulation of the fisheries.)
Source: Western Pacific Region pelagic fisheries 2003 annual report
(WPRFMC 2004b).

YEAR	SWORDFISH	BLUE MARLIN	STRIPED MARLIN	OTHER MARLINS	TOTAL
1987	60	686	667	144	1,557
1988	65	812	1,231	194	2,301
1989	635	1,502	1,403	340	3,880
1990	5,383	1,485	1,247	164	8,279
1991	9,953	1,418	1,551	208	13,129
1992	12,569	1,339	1,097	349	15,354
1993	13,036	1,434	1,191	266	15,927
1994	7,010	1,454	796	267	9,526
1995	5,994	1,952	1,313	464	9,724
1996	5,529	1,931	1,044	292	8,797
1997	6,368	1,908	861	354	9,491
1998	7,208	1,403	891	421	9,924
1999	6,856	1,432	866	605	9,758
2000	6,520	1,121	472	371	8,482
2001	500	1,494	873	352	3,219
2002	725	1,045	618	387	2,774
2003	323	1,163	1,371	581	3,438
2004	578	996	956	505	3,035
Average	4,961.8	1,365.2	1,024.9	347.9	7,699.8
Standard deviation	4,284.7	354.5	296.9	133.2	4,494.8

The closure of the Hawai'i-based swordfish fishery likely contributed to more turtles harmed than saved. After the closure of Hawai'i's fishery, swordfish from other nations continuously supplied the U.S. market. Much of the swordfish substituted for Hawai'i product was imported from Mexico, Costa Rica, Panama and South Africa (Bartram and Kaneko 2004). These foreign fisheries did not have conservation measures in place to protect sea turtles and other vulnerable bycatch species. Consequently, it is estimated that the transferred effects from the closure of the Hawai'i fishery resulted in an additional 2,882 sea turtle interactions in foreign fisheries (Rausser et al. 2009).

In an effort to deal with turtle bycatch as an international problem driven by market forces, the Council began to study turtle lifecycle and migration patterns. Working with NMFS researchers, the Council found that turtle catches took place in what scientists describe as a "turtle layer" near the surface. This is why turtles are rarely caught by vessels fishing for tuna using deep sets, but they are often caught by swordfish vessels using shallow sets. The tuna longlining and swordfish fishermen operate in different areas of the Pacific. Tuna longlining used to take place near the edge of a mid-ocean gyre system to the south of the Hawaiian Islands. However, more recently the fishery began operating to the north and east of the Hawaiian Islands and increasingly in the EPO (Howell et al. 2008). Shallow-set swordfish fishing takes place in the mid-ocean subtropical convergence frontal system to the north-northwest of the Hawaiian Islands (Seki et al. 2002).

Because shallow-set fishing took 10 times more turtles than the deep sets of tuna fishermen, the Council's priority became finding new methods or gear changes to implement. From a holistic viewpoint, Council members believed that technological changes in fishing gear that reduced incidental catches of turtles would have a greater impact on the survival of sea turtles in the long term than forced temporary closures in a limited area. Former Council Chair Jim Cook recalled that "after the lawsuits we had to rethink completely how it was that we took the turtles and seabirds and what we could do to prevent it." [61]

In 2002, Council members were made aware of sea turtle bycatch experiments that had been conducted on longline vessels in the western North Atlantic Ocean by NOAA researchers from the NMFS Pascagoula Laboratory in Mississippi (Watson et al. 2005). The researchers had found that turtles often were seriously injured by swallowing conventional J hooks used in swordfish fishing. While these hooks could potentially move through the digestive system, they were easily swallowed and could then embed in the esophagus or deeper in the digestive system, impairing feeding or causing internal bleeding and death. However, when fishermen used large circle hooks with steel points that had been offset by 10 degrees or less, swordfish could still be caught but the turtles were not easily able to swallow the hooks. Hawai'i researchers began to think that circle hooks might present an opportunity to protect the turtles while still catching swordfish. This information was presented to the Council.

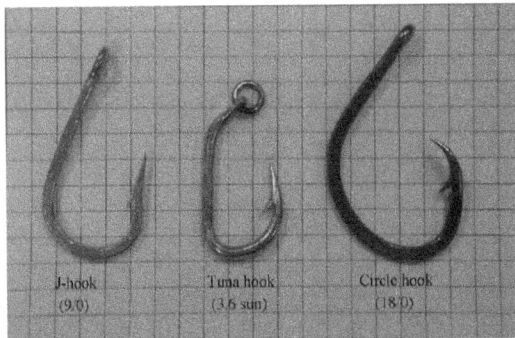

Circle hooks offset by 10 degrees or less can catch swordfish but are difficult for turtles to swallow. This research finding would lead to management measures requiring fishermen to switch from J hooks (left) to circle hooks (right) in the Hawai'i swordfish fishery. *NOAA illustration.*

---

[61] Jim Cook in discussion with the Michael Markrich, January 27, 2021.

Another important lesson learned from the Atlantic fishery had to do with the bait used by longliners. Research showed that loggerheads are less attracted to mackerel than to squid, so a change in bait might also mitigate interactions.

On August 31, 2003, U.S. District Judge Colleen Kollar-Kotelly responded to HLA's claim for relief by remanding the June 2002 measures as they were based on the 2001 BiOp, which NMFS had withdrawn while it prepared the 2002 BiOp. The court also vacated the 2002 BiOp as it was based on the assumption that the June 2002 regulations were in place. The court's decision had the unintended effect of leaving the fishery with no incidental take statement and no protection from ESA liability for the incidental bycatch of sea turtles. The Council stepped into action, holding an emergency meeting on September 23, 2003, to take final action on emergency interim rules and initial action on long-term rules.

Manuel Cruz of Guam chaired the Council in late 2003 when it held a series of monthly meetings to address the federal district court's decision regarding NMFS's ESA-related actions to mitigate Hawai'i longline interactions with sea turtles.

The Council proposed to amend the Pelagic FMP for the Western Pacific Region to reopen the Hawaiʻi swordfish fishery as a model fishery, demonstrating the use of circle hooks and mackerel-type bait to reduce interactions with sea turtles. Under Regulatory Amendment 3, the newly opened longline fishery was limited to 2,120 shallow sets per year, half of the historical effort level. The effort was controlled using single-set certificates that could be transferred between permit holders. In addition to the circle hooks and mackerel-type bait, fishermen were required to use special de-hooking devices to release hooked turtles with minimal trauma and to attend annual protected species workshops. As a further incentive to reduce turtle interactions, a "hard cap" was placed on the number of allowed interactions per year. If reached, the hard cap would trigger closure of the fishery for the remainder of the year. At the time the fishery reopened, the hard caps were set at 16 leatherbacks or 17 loggerheads, whichever came first. The fishery reopened with 100% observer coverage, which allowed real-time monitoring of the sea turtle hard-caps.

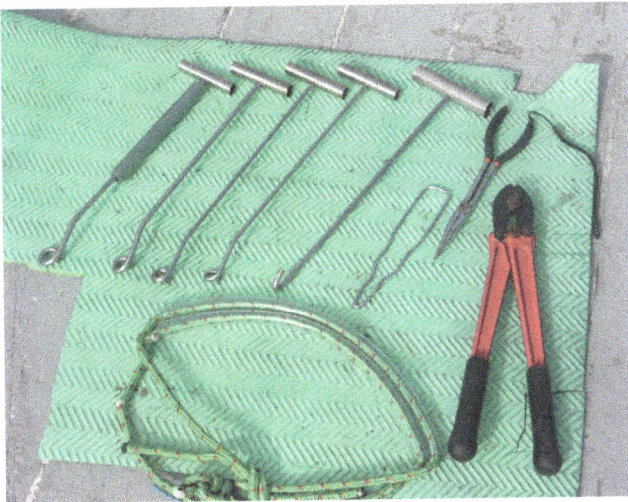

Turtle de-hooking devices, like those pictured, are required gear aboard Hawaiʻi longline vessels and are part of the sea turtle handling requirements initially implemented in June 2002 (Regulatory Amendment 1) and reimplemented in April 2004 (Regulatory Amendment 3). *NMFS photo.*

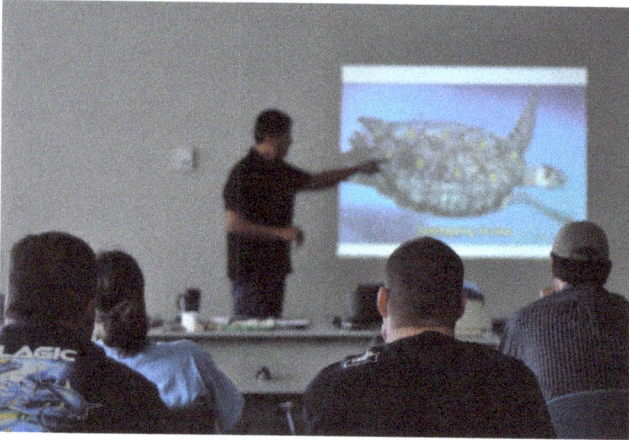

Hawai'i longline vessel owners and operators attend an annual NMFS protected species workshop as required under Framework Measure 2, implemented in June 2002. *NMFS photo.*

An NMFS observer monitors an olive ridley sea turtle interaction (circa 2008). Amendment 3 of the Pelagic FMP (1991) implemented provisions for a mandatory observer program to collect information on interactions between longline fishing and sea turtles. The Hawai'i shallow-set fishery coverage has been 100%, while the deep-set fishery coverage has been 20%. *NOAA Pacific Islands Regional Office photo.*

The new rules went into effect in April 2004, and the Council was soon able to see the impact of the new gear and bait regulations. Their success was strikingly clear. Between 1994 and 1999, NMFS had reported annual takes (interactions, not necessarily deaths) of 112 leatherbacks and 418 loggerheads. In the period after the fishery reopened from 2004 through 2009, observers (100% coverage) recorded a total of 27 leatherback and 48 loggerhead takes, all of which were released alive. The fishery reached the turtle hard cap twice—in 2006 for loggerhead turtles and in 2011 for leatherback turtles. The rate of takes has been reduced by approximately 90% for both loggerhead and leatherback turtles, and the nature of those takes is far less likely to result in serious injury or mortality (Gilman et al. 2007b). This is a great success story in the conservation of sea turtles.

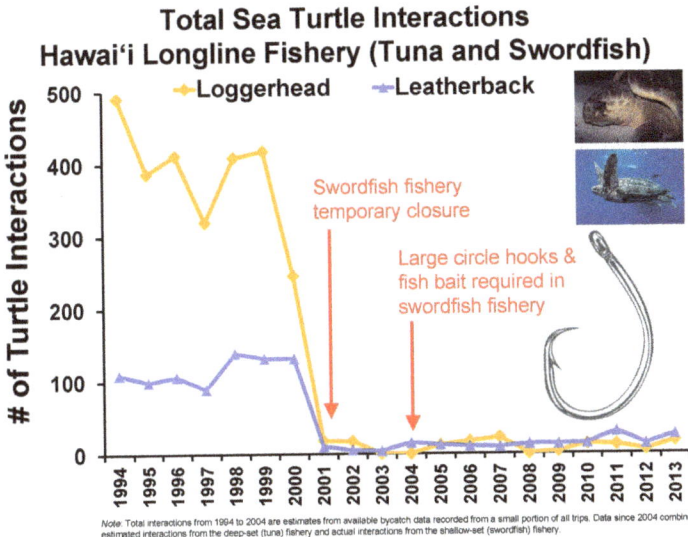

Total Sea Turtle Interactions
Hawai'i Longline Fishery (Tuna and Swordfish)

Note: Total interactions from 1994 to 2004 are estimates from available bycatch data recorded from a small portion of all trips. Data since 2004 combine estimated interactions from the deep-set (tuna) fishery and actual interactions from the shallow-set (swordfish) fishery.

An approximately 90% drop in Hawai'i longline interactions with sea turtles is the result of gear restrictions implemented in 2004. *WPRFMC illustration.*

Based on the success of mitigation efforts, the Council proposed amending the FMP to eliminate the annual limit on the number of sets and raise the cap on sea turtle interactions to 46 loggerheads and 16 leatherbacks, since existing release requirements had proven to be non-lethal with minimal injury to the animals. These were based on the determinations in the 2008 BiOp that the mortality associated with these levels of turtle take using the new gear would not endanger the continued existence of loggerhead or leatherback turtles.

This amendment was initially approved in December 2009, and a new BiOp was issued concluding that the new limits would not pose jeopardy to any species of sea turtles. Subsequently, the Sea Turtle Restoration Project filed a suit challenging the new regulations as being inconsistent with the ESA and other law. Surprisingly, and without opportunity for the Council or fishermen to participate or provide input, U.S. attorneys reached an out-of-court settlement that resulted in reversal of the new regulations and reinstatement of the 2004 regulations for turtle hard caps, pending completion of a status review of loggerhead turtles stocks and the possible designation of the North Pacific stock as a distinct population segment under the ESA. Following the settlement, NMFS issued a new BiOp in January 2012 and a new final rule in October 2011 that raised the hard caps to 34 loggerhead and 26 leatherback turtles.

Meanwhile, the Council continued to contribute significantly to a better scientific understanding of sea turtles and their protection worldwide.

Knowledge of loggerhead lifecycle has vastly improved, although there are still unknowns. Pacific loggerhead turtles hatch in Southern Japan and other areas of the Western Pacific. Current information indicates that the North Central Pacific and the Pacific Coast of Baja

California Sur, Mexico, serve as developmental habitat for loggerheads. Those congregating to forage in waters off Baja California were caught and killed in large numbers by Mexican fishermen using nylon mesh nets; thus fewer were able to mature and reproduce.

Less is known about at which stage they return to coastal Japan after growing up in their developmental habitat. The central North Pacific, north of Hawai'i, is one of the juvenile loggerhead foraging grounds, where they are encountered by Hawai'i swordfish longline fishermen. They eventually return to the western Pacific as they approach adulthood, where they stay for the rest of their lifecycle. The number of nesting loggerhead turtles in Japan declined by 50% to 90% between the 1950s and 1990, in large part due to egg harvest that continued until Japan prohibited it in 1973 and the long time lag between birth and initial reproductive maturity. Following the prohibition of egg harvest and with strong efforts by Japanese NGOs to protect nursing sites, nest counts trended upwards from the late 1990s, peaking at more than 15,000 nests in 2013. Loggerhead nesting is known to follow decadal variation patterns, but the North Pacific loggerhead population overall is showing a long-term increasing trend, growing at an average annual rate of 2.3%. The Council supported the Japanese NGOs for their nesting beach monitoring and conservation activities from 2004 to 2013 as part of the Council's Sea Turtle Conservation Program (Ishizaki 2015).

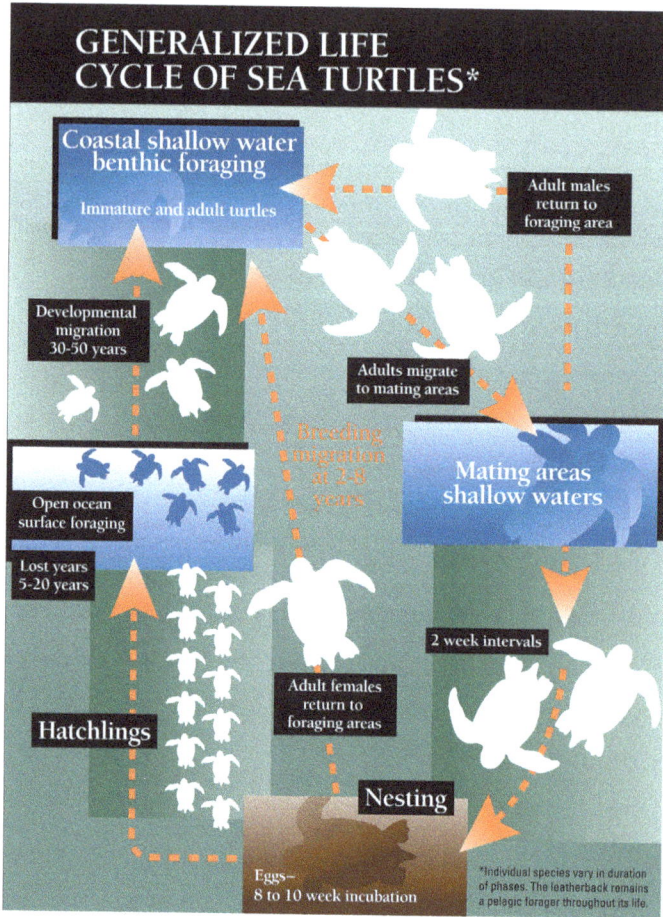

The Life Cycle of Sea Turtles poster shows the different habitats in which sea turtles are vulnerable to human impact. Regulatory Amendment 3 of the Pelagic FMP (2004) included conservation projects to protect sea turtles in their nesting and coastal habitats. *WPRFMC illustration.*

Initially, it was thought that the leatherback turtles that transited through Hawaiʻi waters were from nesting beaches in Mexico or Southeast Asia, though now it is fairly certain that most leatherbacks in the Council's area originate in the Western Pacific. In past decades, leatherback populations have declined due to predation and harvesting, interactions with fishing gear or other human activities, and habitat destruction. Scientists looking for specific causes of the decline in leatherback populations came to the preliminary conclusion that the turtle nesting areas in developing countries were being destroyed or damaged by human development and that people were killing nesting females and/or taking their eggs for food.

In order to protect the endangered turtle populations, nesting areas had to be identified and protected and incidental capture in fisheries would have to be reduced or mitigated. Congress provided funds to NMFS, some of which were made available to the Council, to support conservation projects at important loggerhead and leatherback turtle nesting and foraging areas in a number of countries (e.g., Papua New Guinea, Indonesia, Mexico, Ecuador, Japan, Vanuatu, Malaysia and the Solomon Islands).

The "Nesting Beach Conservation: A Community-Based Approach to Sea Turtle Recovery in the Pacific" brochure produced by the Council promotes management of sea turtles at nesting sites in foreign countries. Similar brochures were produced to assist community-based conservation at sea turtle foraging sites in foreign countries. *WPRFMC illustration.*

The Council also championed the improvement of the existing South Pacific turtle database that was developed by the Secretariat of Pacific Regional Environmental Program based in Apia, Samoa. This vast amount of turtle data, collected for decades, is now being standardized into a new database known as the Turtle Research Database System. The system was officially launched in February 2009 at the 29th Annual Symposium on Sea Turtle Biology and Conservation in Brisbane, Australia, and is now widely used by the Secretariat's member countries and territories. The database is a joint initiative of the Secretariat of the Pacific Regional Environmental Program, the Council, the Secretariat of the Pacific Community, Queensland Parks and Wildlife, Southeast Asia Fisheries Development Center, the Pacific Islands Fisheries Science Center and the Marine Research Foundation in Malaysia.

In addition, the Council, along with NMFS and the Rockefeller Foundation, sponsored an international meeting in Bellagio, Italy, in 2003 to address sea turtle conservation issues. The multi-disciplinary group of 25 experts drafted an action plan to restore Pacific turtle populations (Bellagio 2004). The *Bellagio Blueprint for Action on Pacific Sea Turtles* included protection of all nesting beaches, reduction of turtle take in at-sea and coastal fisheries, stimulation of pan-Pacific policy actions and encouragement of sustainable traditional use of sea turtles. The key issues presented and discussed at the meeting were published in *Conservation of Pacific Sea Turtles* by the University of Hawai'i Press in 2011 with funding provided by the Council (Dutton et al. 2011).

A follow-up meeting, the Bellagio Sea Turtle Conservation Initiative workshop was held In Malaysia in 2007 to further address the conservation of Pacific leatherback turtles and to develop a financial scheme to ensure continued U.S. funds for turtle conservation (Bellagio 2008).

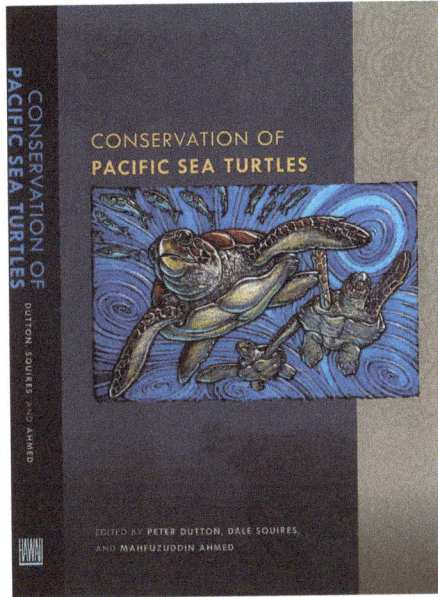

Key issues discussed at the Bellagio sea turtle conservation meeting are presented in *Conservation of Pacific Sea Turtles*, published with funding provided by the Council. *University of Hawai'i Press illustration.*

Participants of the Bellagio Sea Turtle Conservation Initiative pose with the leatherback mascot of their second meeting, held in Terengganu, Malaysia, in 2007. The participants developed a strategy for long-term funding of Pacific leatherback conservation. *WPRFMC photo.*

Because of its belief in the importance of engaging fishermen in conservation efforts, the Council has helped sponsor the Tri-national Exchange, a gathering of fishermen, turtle scientists and community leaders from Japan, Mexico and the United States. The group has met in Japan, Mexico and Hawai'i, key areas in the lifecycle for North Pacific loggerheads. One outcome of the 2007 Hawai'i meeting was a commitment by conference attendee Efrain de la Paz Regaldo, the owner of a large Mexican fishing fleet in Baja California, to no longer set bottom longlines. This is significant because de la Paz Regaldo estimated that he could save 700 loggerhead turtles per year by doing this (WPRFMC 2007a). "Prior to this program we didn't realize the global impacts of our local bycatch," de la Paz Regaldo said. "Once we understood, we saw that our bycatch was both an embarrassment and a grave problem. So we are fixing the problem, with input from our Hawaiian and Japanese partners" (ibid.).

Despite virtually eliminating turtle interactions by Hawai'i-based fishing vessels in Western and Central Pacific waters, the issue continues to be a focal point for environmental campaigners. Unfortunately, unilateral closure of the Hawai'i longline fleet cannot diminish the threat to turtles throughout the region. Only the adoption of strong habitat and other protective measures—including limiting sea turtle interactions and subsequent mortality by all nations, applicable to all fleets—can save turtle populations. In this context, it should be noted that further stresses on the turtle populations are coming from other distant-water longline fleets (principally Taiwanese and Chinese), with non-U.S. fleets setting a conservative estimate of 77 million shallow-water hooks annually in the Central and Western Pacific (Bartram and Kaneko 2004). This is why the Council continues to support the spread of information about turtle bycatch-reducing gear at industry forums throughout the world and fosters technology exchange, including shipboard training, whenever possible.

Creating a Seabird Protection Program

Black-footed albatross (*Phoebustria albetus*) and Laysan albatross (*P. immutabilis*) are two of the most numerous albatross species in the North Pacific. More than 90% of their populations breed in the NWHI during the winter months. Both species feed primarily on the surface of the ocean, searching for fish, fish eggs, squid and crustaceans.

Where the albatrosses' foraging area and fishing effort overlap, the seabirds may become attracted to fishing vessels due to availability of food in the form of bait and fish scraps. Some of these birds become accidentally hooked as they attempt to forage on bait as the hooks are being set; the birds are dragged beneath the surface and drown. Some are also captured during gear retrieval. In the Council's area of jurisdiction, the focus is on the longline fisheries for tuna and swordfish; however, the same problem arises in other longline fisheries for bottomfish (e.g., for sablefish off Alaska and the West Coast). It is estimated that thousands of seabirds were killed in the 1989–2000 period. For 1997 alone, it was estimated that Hawai'i longliners killed approximately 2,900 black-footed albatross and 1,645 Laysan albatross (WPRFMC 1998). For fishermen, the bird bycatch means not only the death of birds but also high expenses as the fishermen lose bait, hooks and valuable fishing time to tediously clear their lines. As fishermen reported this bycatch, it added to concerns of the Council, which was trying to reduce interactions with protected species. The SSC recommended that the Council require longline captains to be trained to minimize seabird interactions as part of their requirement for obtaining and keeping their fishery licenses.

The Council staff realized that, to effectively avoid catching seabirds, fishermen needed not only training but also a strong understanding of seabird lifecycles and habits. To gather this information, the Council hired a marine biologist to review records at the USFWS, Bishop

Museum and other facilities where albatross bird tag records were kept. An investigation of thousands of records going back to 1910 revealed 255 records of birds whose ages could be identified at the time of tagging at sea. From these data, it was determined that 44% were a year old or less and only 20.4% were five years or older. The fact that the greatest percentages of the birds tagged at sea were newly born or inexperienced young birds indicated that the older birds were more wary and less likely to get caught. This suggested that the answer to the seabird bycatch issue was to somehow discourage the young birds from foraging on the bait of the swordfish longlines (Cousins and Cooper 2000). As part of this effort, the Council organized a workshop in 1998 so the available data could be presented and discussed.

THE POPULATION BIOLOGY OF THE BLACK-FOOTED
ALBATROSS IN RELATION TO MORTALITY CAUSED
BY LONGLINE FISHING

At the Council-organized Black-Footed Albatross Workshop in 1998, participants reviewed an analysis of hundreds of tagged albatross and explored measures to reduce albatross interactions with the Hawai'i longline fleet. *WPRFMC illustration.*

Based on the results of the workshop, the Council then contracted with consultants Garcia and Associates to conduct experiments on reducing seabird bycatch. They found that employing blue-dyed bait (meant to blend in with the ocean) reduced bird hookings by between 63% and 95%. The young birds were not attracted to bait that they apparently could not see.

Other studies provided additional possible solutions. One simple technique that proved effective was discarding offal on one side of the boat while setting the hooks from the other side. The albatrosses foraged on the free offal food and ignored the side with the hooks. Another solution was side-setting—that is, setting longline gear from the side of the vessel rather than from the conventional position at the stern. By setting baited hooks close to the side of the vessel hull, it became possible to keep the birds from the hooks until they had sunk deeply enough to avoid bird interaction. Simplest of all was the idea that all fishing be conducted at night when the seabirds' ability to see was diminished and they were less actively foraging. When these techniques were applied together, seabird mortality dropped still further.

Armed with this information, in 1999 the Council recommended a suite of seabird mitigation measures for the Hawai'i-based longline fishery. These measures were implemented in 2001, with the side-setting provision added in 2006. These efforts resulted in a more than 90% reduction in total seabird captures in the Hawai'i-based longline fishery (Van Fossen 2007) and a 67% reduction in seabird bycatch rate in the deep-set component of the longline fishery (Gilman et al. 2008). These techniques, like those developed to reduce turtle interactions, have now been adopted by fishing fleets all over the world.

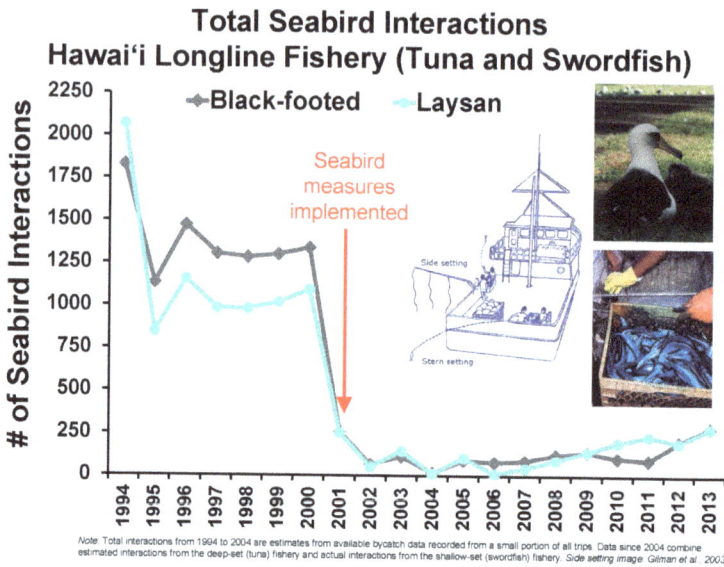

Total Seabird Interactions
Hawai'i Longline Fishery (Tuna and Swordfish)

Note: Total interactions from 1994 to 2004 are estimates from available bycatch data recorded from a small portion of all trips. Data since 2004 combine estimated interactions from the deep-set (tuna) fishery and actual interactions from the shallow-set (swordfish) fishery. Side setting image: Gilman et al., 2003

Seabird interactions in the Hawai'i longline fleet dropped by more than 90% as a direct result of mitigation measures recommended by the Council in 1999. *WPRFMC illustration.*

As in other instances of Council action to address protected species interaction problems, this success in reducing seabird interactions did not bring respite from criticism of commercial fishing. In 2006, an environmental NGO group, the Turtle Island Restoration Network, brought suit claiming that NMFS had violated the Migratory Bird Treaty Act. However, this case was ultimately dismissed as an administrative challenge to the MSA.

There were two positive lessons for the Council that came from this experience. First, with the right strategies and new technologies, nearly all the interaction problems between seabirds and commercial fishermen can be solved. Second, fishermen are usually eager to resolve bycatch problems and, when taught about their importance, will readily adopt effective strategies that involved them in the development and testing.

Unfortunately, a third lesson was also confirmed, which is that some anti-commercial fishing environmental organizations will continue to litigate bycatch issues in ongoing efforts to curtail or eliminate the longline fisheries. It does not matter to them that sea turtle and seabird interactions and mortalities have been reduced by 90% or more. Rather than accept or even commend the Council and industry for these achievements, such groups wish to eliminate these U.S. fisheries—even if the result is more turtle and seabird interactions and mortalities as foreign fisheries try to meet the U.S. market demand for tuna, swordfish and associated resources.

In 1988, NOAA attorney Martin Hochman warned the Council that the Sierra Club Legal Defense Fund had opened an office in Honolulu and, consequently, there would be more lawsuit and public monies spent. He added that the recently amended ESA provided for the recovery of legal fees in cases in which plaintiffs successfully challenged the federal government (WPRFMC 1988b, p. 6). An economic incentive had been created that would drive environmental litigation.[62]

Fortunately, other NGOs collaborate with industry and governments to address bycatch issues in a positive manner, and the Council works with these NGOs and the governments of fishing nations in the Pacific to address difficult bycatch issues.

In November 2000, the Council participated in the First International Fishers Forum—Solving the Incidental Capture of Seabirds in Longline Fisheries, held in Auckland, New Zealand. Fishermen, gear

---

[62] In 2014, the House Natural Resources Committee reported that attorneys were receiving as much as $700 or more per hour in taxpayer-funded fees from successful ESA litigants under the ESA citizen suit provision, and so it sought through HR 4318, Endangered Species Litigation Reasonableness Act, to limit the compensation to that afforded under the Equal Access to Justice Act, which has a $125/hour cap (U.S. House of Representatives 2014).

technologists, researchers and government officials were in attendance to exchange ideas and information on measures being used or developed around the world to reduce capture of seabirds on longlines. This information exchange aimed to coordinate and accelerate progress in solving this global problem (Baird 2001).

Two years later, the Council hosted the Second International Fishers Forum in Honolulu to address the bycatch of sea turtles as well as seabirds by longline gear. More than 200 representative from fishing industries, government agencies, NGOs and other interested parties from 28 countries in the Atlantic and throughout the Pacific participated in this second forum.

Inspired by the international embrace of this effort to solve protected species bycatch issues in the longline fisheries, the Council proceeded to support, along with the Japan Fisheries Agency, a third forum, held in Yokohama, Japan, in 2005 and hosted by the Organization for the Promotion of Responsible Tuna Fisheries. In addition to incidental seabird and sea turtle bycatch, the third forum addressed sustainable tuna and shark fisheries; fishing capacity; production; marketing; consumption monitoring; illegal, unregulated and unreported fisheries; cetacean depredation; and the employment of market-based mechanisms such as eco-labeling to influence marine capture fisheries' practices and management. The Council co-hosted the fourth International Fishers Forum held in Puntarenas, Costa Rica, in 2007, with the Costa Rica Fisheries and Aquaculture Institute (i.e., Instituto Costaricense de Pesca y Acuicultura) as well as the fifth forum (on marine spatial planning and managing fisheries bycatch) in Taipei in 2010, with the Taiwan Fisheries Agency (part of the Taiwan Council of Agriculture). The proceedings of these forums can be found on the Council's website.

Representing the hosts and sponsors of the third International Fishers Forum held in 2005 in Japan are (from left) Yuchiro Harada, managing director, Organization for the Promotion of Responsible Tuna Fisheries; Akira Nakamae, Fisheries Agency of Japan; Isao Nakasu, president, Organization for the Promotion of Responsible Tuna Fisheries and Japan Fisheries Association; Rebecca Lent, director, NMFS Office of International Affairs; and Kitty M. Simonds, executive director, Western Pacific Regional Fishery Management Council. Not pictured is Sean Martin, president, Hawaii Longline Association. *WPRFMC photo.*

Participants attend a plenary session at the fifth International Fishers Forum held in Taipei in 2010. The forum adopted the Taipei Declaration, which calls for the fishing industry to have an equal voice in marine spatial planning and ecosystem-based management. *Overseas Fisheries Development Council photo.*

Marine Monument Impacts on the Hawai'i Pelagic Fishery

The establishment of the NWHI CRER and subsequently the Papahanaumokuakea MNM, designated through executive order under the authority of the Antiquities Act of 1906, had little initial, direct impact on the Hawai'i longline fishery as the area covered by these designations mimicked the Protected Species Zone created by the Council in 1991, which banned longline fishing within it. However, the NWHI monument designation in June 2006 did have a negative impact on fishing at Midway in the NWHI, shutting down the sportfishing charters that had been offered there.

A sportfishing vessel cruises the waters of Midway Atoll, prior to the establishment of the NWHI Marine National Monument in 2006. *Cavan Images/Alamy Stock Photo.*

In 2009, President Bush's establishment of the Pacific Remote Islands MNM would impact the Hawai'i longline fishery, which had historically fished around Palmyra, Kingman Reef and Johnston. The monument designation would ban commercial fishing within 50 nm of these and other islands in the Pacific Remote Island Areas. The Bush monument would also negatively impact commercial troll and handline vessels from Hawai'i, which had made sporadic trips to these remote islands. Future expansion of the Pacific Remote Islands MNM and the Papahanaumokuakea MNM would, however, harm the fishery (see Chapter 11).

International Cooperation and Development of the Western and Central Pacific Fisheries Commission

The Western Pacific Council has been described as the most international of the regional fishery management councils by virtue of its jurisdiction over a widespread area of the WCPO, its shared borders with the EEZs of other nations and the necessity to work cooperatively to manage highly migratory species such as tuna, billfish, sea turtles, seabirds and marine mammals. Shortly after the culmination of UNCLOS in 1982, the fishing of straddling stocks (highly migratory transboundary stocks) on the high seas by distant-water fishing nations began to garner significant attention. The early 1980s saw a rapid development of the tropical tuna purse-seine fishery in the WCPO. As purse seiners from Japan, the United States and later from Taiwan and Korea began to be more active in the western Pacific, the need for greater understanding and international cooperation on the management of highly migratory fish stocks became apparent.

In the late 1979, the Forum Fisheries Agency was established by Pacific Island countries as a sub-regional organization to assist in fisheries development and fisheries management initiatives and to provide a common voice for the countries in negotiations to ensure consistent minimum terms and conditions for fishing access to their EEZs by distant-water fishing nations The national waters of several Pacific Islands are among the richest tuna fishing grounds. In the late 1980s and 1990s, however, when tuna catches from the WCPO began to rapidly increase, there was no international management organization to set limits or rules applicable to the high seas or to ensure that tuna stocks were managed effectively across their range. Distant-water fishing nations could negotiate bilateral agreements with Pacific Island countries to gain access to their economic zones at relatively low cost, normally paying a percentage fee per ton based on landed price. Such negotiations were not transparent to neighboring states, and Pacific Island countries had little negotiating leverage. Catches were not limited; there was no basis for setting limits based on sustainable yield or other science-based criteria.

Once tuna was included as "fish" under the Magnuson Act in 1992, an enormous effort was put into upgrading research on tuna stocks in the Council's area of jurisdiction. The 1992 MSA reauthorization established the Pelagic Fisheries Research Program based at the University of Hawai'i. With the establishment of this new program, whose steering committee was initially chaired by the Council's executive director, new efforts were made to gather information on tuna catches and biological and ecological aspects of the stocks, including migratory patterns and stock productivity estimates. This scientific information would complement the information gathered by the Western Pacific Fisheries Information Network—whose steering committee included NMFS, the Council and representatives of the U.S. flag entities in the Pacific—on local landings of fish, including tuna and associated pelagic species within the waters of the U.S. EEZ and high seas (i.e., international waters) surrounding the islands. The Pelagic Fisheries Research Program could also foster cooperation with scientists in other organizations (such as the South Pacific Commission, now called the

Secretariat of the Pacific Community) to begin review of fishery data and research results with a focus on determining potential yields and stock problems that needed international attention.

By the late 1980s/early 1990s, it was described that a global crisis was occurring on the high seas with regards to transboundary stocks and that UNCLOS did not provide a comprehensive framework for addressing high seas fishing of shared stocks. The international community responded by establishing the U.N.-endorsed Conference on Straddling Fish Stocks and Highly Migratory Fish Stocks, held between 1993 and 1995 in New York. Council Executive Director Simonds participated in several of the conferences as a member of the U.S. delegation and provided critical perspectives on the need for international management of tuna stocks in the Western Pacific Region. The conference was chaired by Fijian Ambassador Satya N. Nandan, with whom Simonds formed a long-lasting friendship based on mutual respect and collaboration that continued well past the conference.

The outcome of the conference was the 1995 United Nations Agreement for the Implementation of the Provisions of the United Nations Convention on the Law of the Sea of 10 December 1982 Relating to the Conservation and Management of Straddling Fish Stocks and Highly Migratory Fish Stocks, commonly known as the United Nations Fish Stocks Agreement.

By the early 1990s, international tuna fisheries in the Western Pacific Region were a growing concern in terms of economic development, stock sustainability and food security. This growth offered both prospective benefits as well as potential risks for the Council's island constituencies. Distant-water foreign fleets were active in the ports of the CNMI, Hawai'i, Guam and American Samoa. The Council did not want to unbalance those arrangements; however, foreign vessels could conceivably fish hard enough to adversely impact local fish abundance as well as Pacific-wide stocks. As early as 1992, Samuel Pooley of the NMFS Honolulu Laboratory pointed out that the Pelagic Plan Team, in its review of the Pelagic FMP, found two significant long-term implications for Hawai'i. First, Pooley warned (as Wadsworth Yee had insisted 15 years previously) that the harvest of small pelagic fish (skipjack, bigeye and yellowfin) in the equatorial waters by purse seiners was posing a potential problem for the health of fisheries in the Council's jurisdiction (WPRFMC 1992b). Second, and more alarming, was the emergence of drifting FAD technology in the tropical purse-seine fishery. These FADs made purse-seine fishing more efficient but also exerted differential impacts on the stocks involved. Because several species of tuna aggregate simultaneously around FADs, purse-seine sets capture juvenile yellowfin and bigeye as well as the target skipjack species. If such captures were not controlled, this bycatch of juveniles could negatively impact the stocks of mature yellowfin and bigeye available to other fisheries, in particular, longline fleets. Simply put, the more small fish that are caught, the fewer fish will survive to reproductive maturity and the fewer adult fish will be available to longline gear. Nonetheless, the use of FADs had steadily grown, and drifting FADs were equipped with GPS buoys and fish finding sonar that significantly improved purse-seine efficiency. From the purse seiners' standpoint, the ability to catch huge amounts of skipjack tuna outweighed the potential loss in yield of yellowfin and bigeye tuna, which were more valuable in longline fisheries.

Total tuna landings grew steadily during the 1970s and 1980s. With the growing use of FADs beginning in the early 1990s, scientists and managers began to express concern. The same problems that countries faced in the Atlantic in trying to manage a limited number of migratory tuna stocks among competing nation states were occurring in the Pacific. As the international market for tuna continued to grow, fishing pressures increased on tuna stocks, highlighting the need to set more stringent rules at the international level in order to balance the interests of competing fishermen. At the time, most Pacific island nations had tuna in their waters but did not have the vessels to fish for them, so the islanders gained income by selling licenses allowing foreign fleets to access their EEZs. This could happen through bilateral agreements between individual nations or through multinational treaties, as was the case for the U.S. purse-seine fleet through the South Pacific Tuna Treaty. Fishing was relatively unconstrained, limited only by the number of active vessels and available licenses. In accordance with the Palau Agreement, there was a limit of 205 large-scale purse-seine vessels. However, vessels could be stretched to increase capacity, and technological improvements continued that increased the efficiency of fishing effort.[63]

The United Nations Fish Stocks Agreement, which guides countries on how to cooperate on the conservation and management of highly migratory stocks, provides for the international establishment of regional fishery management organizations. Soon after the Agreement negotiations, the Forum Fisheries Agency convened a Multilateral High Level Conference (MHLC) in 1994. The second MHLC was held in 1997 in Majuro, Republic of the Marshall Islands, where it was agreed that participating countries would work for three years to establish a formal international agreement on the conservation and management of highly migratory stocks in the WCPO. The agreement is known as the Majuro Declaration. The next MHLC was held in Tokyo in 1998. Between February 1999 and September 2000, four MHLCs were held in Honolulu, culminating in the Honolulu Convention. The Council played a critical role in these final MHLCs, with staff assisting with the organization and administration of the meetings. These meetings were chaired by Satya N. Nandan.

---

[63]   David Itano in discussion with Michael Markrich, August 31, 2016.

Participants gather at the Hawai'i Convention Center prior to the conclusion of MHLC7 on September 5, 2000. This final MHLC meeting culminated in the Honolulu Convention and establishment of the Western and Central Pacific Fisheries Commission. *WPRFMC photo.*

U.S. Senator Daniel K. Inouye, MHLC chairman Satya N. Nandan and Ambassador Mary Beth West, deputy assistant secretary for oceans, U.S. Department of State, stand in front of Council posters showing the Pacific-wide distribution of tuna and tuna-like species, displayed during MHLC7, August 30 to September 5, 2000, in Honolulu. *WPRFMC photo.*

Kitty M. Simonds, WPRFMC executive director, and Satya N. Nandan, MHLC chairman, worked closely together to run MHLC7 held August 30 to September 5, 2000, in Honolulu. The Council helped the United States to organize the final four MHLC meetings, held in 1999 and 2000 in Honolulu. *WPRFMC photo.*

The purpose of the MHLC meetings was to find a collective means to manage and control the fisheries for migratory tuna stocks in the WCPO so as to encourage the long-term sustainability of the resources. From a management perspective, the Council was an early supporter and participant in this process, recognizing that without international agreement, the stocks were likely to be fished down to unsustainable harvests, which in turn could severely harm U.S. fisheries, including island-based fisheries in the U.S. territories. The Council also was aware that the U.S. longline fisheries based in Hawaiʻi were restricted by measures such as area closures, gear and technique requirements to protect sea turtles and seabirds, and observer requirements. This put the U.S. fisheries at a competitive disadvantage with foreign longline fleets. Therefore, the Council actively participated and supported U.S. participation in efforts to help resolve the issue of how much fish should be taken in the

Pacific, by whom and under what kinds of controls. In this context, the Council's experience in development and administration of its Pelagic FMP would be very useful. In addition, the Council encouraged participation of the U.S.-flagged island territories in the MHLC meetings and subsequent negotiations. Because of this, the territories were given a voice in the development of U.S. positions at the international level.

The MHLC series was successful in its goal to develop a Convention on the Conservation and Management of the Highly Migratory Fish Stocks of the WCPO. As the agreement was signed in Honolulu on September 5, 2000, it became known as the Honolulu Convention. It went into effect on June 19, 2004, and established the Western and Central Pacific Fishery Commission (WCPFC, or Commission). The Commission is the regional fishery management organization that co-manages the highly migratory fish stocks of the Pacific Ocean with the Inter-American Tropical Tuna Commission, which covers fisheries for tuna and other highly migratory species in the eastern Pacific (Fougner 2010).

The principal goal of the Honolulu Convention is to ensure that WCPO tuna stocks are maintained in perpetuity at productive and sustainable levels. The Honolulu Convention also recognizes the need to ensure that other living marine resources are not harmed in the process of fishing for tuna and associated species. Among the Commission's concerns are whether and how to limit the number or capacity of vessels fishing in the WCPO and how to balance the interests of the competing fishing fleets. Members of the Commission are aware that there will be growing demand for fish to feed growing world populations, especially in Asia, and this demand will place ever increasing demands on fish stocks in the WCPO, with the potential for adversely affecting other marine species, such as turtles and seabirds.

The Council is a member of the Permanent Advisory Committee to Advise the U.S. Commissioners to the WCPFC. It also helps by providing technical expertise, leadership and experience to the Commission members. The WCPFC has adopted several measures based at least in part on the success of many Council innovations in fishery management and incidental catch control.

As noted, the FAD fishery for skipjack tuna has impacts on yellowfin and bigeye tuna stocks due to bycatch of juveniles. The WCPFC has initiated some controls, such as a seasonal prohibition of FAD fishing, in order to protect bigeye tuna. The WCPFC has also established bigeye tuna catch limits for longline fleets in the region.

WCPFC limits continue to impact fisheries managed by the Council. For example, WCPFC Conservation and Management Measure 2008-01 developed an annual U.S. longline bigeye catch limit of 3,763 mt for each of the calendar years 2009, 2010 and 2011. Based on these limits, the Hawai'i longline fishery was closed before the end of the year in 2009 and 2010 as NMFS estimated that the quota would have been reached by the closure date. (Note that in both years, the total landed catch ended up slightly below the limit, meaning the Hawai'i fleet was not allowed to catch its full quota.)

To the extent necessary, the United States has promulgated regulations to implement WCPFC measures that apply to U.S. fisheries. NMFS implements WCPFC decisions through the Western and Central Pacific Fisheries Commission Implementation Act, which was part of the MSA reauthorization of 2006. The WCPFC Implementation Act includes provisions that require the selection of U.S. Commissioners to the WCPFC, whereby one of the five designated commissioner seats is obligated to the chair or member of the Council.[64] This is indicative of the important role that the Council plays in the management of tuna and other highly migratory stocks in the region. To further recognize the role of the Council in the development of U.S. positions on the international management of highly migratory species, the WCPFC Implementation Act requires the State Department, NOAA and the Pacific, North Pacific and Western Pacific Fishery Management Councils to establish a memorandum of understanding. The memorandum was agreed to in December 2009 and details the following:

1. Participation in U.S. delegations to international fishery organizations in the Pacific Ocean, including government-to-government consultations;

2. Provision of formal recommendations to the Departments of Commerce and State regarding necessary measures for both domestic and foreign vessels fishing highly migratory stocks;

3. Coordination of positions within the U.S. delegation for presentation to the appropriate international fishery organization; and

4. Recommendations for domestic fishing regulations consistent with the actions of the international fishery organization for approval and implementation under the MSA.

---

[64] The Western Pacific Council recognized early in its history, following inclusion of tuna in the MSA, that there should be a lead council for Pacific pelagic issues, especially as the international tuna management realm increased in complexity. The Council wanted to avoid the debacle of the East Coast where NMFS removed jurisdiction of tunas, billfish and pelagic sharks from the fishery management councils due to the inefficiency of multi-council jurisdiction and cooperation. The Western Pacific Council worked on the lead council issue in the early and mid-1990s, but the request for lead council status was denied by then NMFS Assistant Administrator Rolland Schmitten (1996). Hence, both the Pacific and Western Pacific Councils are jointly involved in WCPFC delegations.

On the local level, as indicated previously, the Council agreed to close federal waters around American Samoa, Guam and CNMI to purse-seine fishing in order to minimize the risk of localized disruption of tuna in waters used by local small-scale commercial and artisanal fishers (WPRFMC 2008a). On a broader level, the Council helped create The Sustainable Tuna Roundtable, which applies pressure at the consumer level to protect Pacific tuna stocks by providing information through "sustainable fisheries" certification, eco-labeling and stock assessments. This encourages consumers to demand that wild-caught marine fish are harvested using sustainable methods and places new and unprecedented pressure on the tuna industry to change their future fishing practices (WPRFMC 2008b).

Since its inception, the Council has worked to make the point that effectively managed fisheries—in which fishermen participate and provide accurate data—are likely to be healthy fisheries. It has also stressed that denying the impacts of fishing or simply transferring those impacts to areas outside the United States does not necessarily lead to healthy world marine environments. Unnecessary curtailment of American fisheries causes import substitution that frequently transfers unresolved environmental problems elsewhere—often to less developed communities with less ability to create or enforce effective management measures.

In order to find common ground and attempt compromise with anti-commercial fishing interests, in 2009 the Council supported the work of the National Center for Ecological Analysis and Synthesis. The Center established a working group to find common agreement among marine ecologists, fishery scientists and fishery managers regarding methods and scientific criteria. In contrast to the prevailing popular media view that all commercial fisheries throughout the world were on the point of collapse, two members of the Center, fishery scientist Ray Hilborn of the University of Washington and fishery ecologist

Boris Worm of Dalhousie University of Canada, presented a more balanced view. "Rebuilding Global Fisheries" (Worm et al. 2009) does not discount the fact that 63% of the world's fisheries are in need of better management or that others are in trouble, but it points out that the take is below the recommended maximum sustainable yield in seven out of 10 well-documented commercial fisheries and that catch rates are declining in five of the 10. The paper also makes the point that fisheries management must be comprehensive in scope and recognize the interdependence of living and non-living resources and the impacts of resource use on all aspects of an ecological complex.

In summary, the battle for tuna sustainability has not ended; it has simply entered a more challenging and complex phase at the international level. Given the Council's geographic jurisdiction, international cooperation will continue to form a critical activity in the Council's future.

Supporting Traditional Fisheries

Under the SFA, the U.S. secretaries of Commerce and the Interior are authorized to make direct grants through the Western Pacific CDPP of $500,000 annually for not less than three or more than five community demonstration projects in eligible Western Pacific communities, as recommended by the Council. The program aims to increase the involvement of Western Pacific indigenous communities in fisheries through the application and/or adaptation of methods and concepts derived from traditional indigenous practices in resource management, conservation and utilization. The CDPP grants fund fishery demonstration projects that address the following:

a)   Demonstrate the applicability and feasibility of traditional indigenous marine conservation and fishing practices;

b) Develop or enhance community-based opportunities to participate in fisheries; and/or

c) Involve research, community education, or the acquisition of materials and equipment necessary to carry out such demonstration projects.

The Act's language allows broad latitude in the kinds of projects that can be funded and may include projects with the following objectives:

a) Further the goals of the indigenous community;

b) Promote the development of social, cultural and commercial initiatives and enterprises to enhance opportunities for Western Pacific communities to participate in fisheries, fishery management or conservation; and/or

c) Enhance culture, support traditional and customary fishing practices, or seek new methods and activities.

Projects must involve the aboriginal community and the marine resources of the U.S. Western Pacific.

A CDPP Advisory Panel was established by the Council in 1997. In addition to evaluating and ranking grant applications, the panel also made recommendations and advised the Council in relation to CDPP preference programs. Eight members were selected, two from each of the island areas represented by the Council. Their terms of service are governed by the Council's Statement of Organization, Practices and Procedures.

In 1998, the Council finalized membership of the CDPP Advisory Panel and, in April of that year, approved eligibility criteria to apply for Western Pacific Community Development Program and CDPP, transmitting criteria to NMFS for review. The proposed rule and definitions were published for public comment in the Federal Register on July 27, 2001, and the final rule was implemented and published in the Federal Register on April 16, 2002. To be eligible to participate in either of these programs, a community shall meet the following criteria:

1. Be located in American Samoa, the CNMI, Guam or Hawai'i (Western Pacific Area);

2. Consist of community residents descended from aboriginal people indigenous to the western Pacific area who conducted commercial or subsistence fishing using traditional fishing practices in the waters of the western Pacific;

3. Consist of community residents who reside in their ancestral homeland;

4. Have knowledge of customary practices relevant to fisheries of the western Pacific;

5. Have a traditional dependence on fisheries of the Western Pacific;

6. Experience economic or other barriers that have prevented full participation in the western Pacific fisheries and, in recent years, have not had harvesting, processing or marketing capability sufficient to support substantial participation in fisheries in the area; and

7. Develop and submit a Community Development Plan to the Western Pacific Council and NMFS.

An initial list of funding priorities was developed by the CDPP Advisory Panel in 1998 and approved by the Council. These were the priorities:

1. Promote fishery resource stewardship by indigenous communities.

2. Promote economic growth and stability in indigenous communities.

3. Promote self-determination in indigenous communities.

4. Promote solidarity in indigenous communities.

For second and subsequent solicitations, the Council specified more specific priorities to include the following:

1. Community education;

2. Processing of fishery products and byproducts;

3. Feasibility studies for participation in fishery and fishery related activities;

4. Increased opportunities for participation in the WPRFMC activities and process; and

5. Demonstration of traditional, cultural fishing practices.

Fourteen community projects were approved by the Council (of which NMFS approved 13) in three solicitations: four in 2002, five in 2004 and five in 2005. The projects funded through the CDPP for Hawai'i included the following:

a) Training of young fishermen on the island of Moloka'i and on the Wai'anae coast of O'ahu to catch aku using traditional pole-and-line methods from small boats;

b) Grant writing workshops that help Pacific Islanders in Hawai'i more successfully apply for financial assistance related to projects for marine conservation;

c) Restoration and operation of an ancient Hawaiian fishpond used for educational purposes; and

d) Training in the propagation of *limu* (edible seaweed).

Unfortunately, after 2005, NMFS would state that funding the CDPP was discretionary and that it did not have the funds to do so.

To this day, Congress has not appropriated money to the CDPP section of the MSA. On February 21, 2020, at the listening session hosted in Honolulu by Congressmen Ed Case and Jared Huffman about the MSA and fisheries management in the Western Pacific Region, Council Advisory Panel Vice Chair Gil Kuali'i asked Congress to consider funding the CDPP.

Council advisor Keli'i Mawae (the honorary mayor of Moloka'i) sits at the rear of the vessel funded by the CDPP to train Hawaiians on the island of Moloka'i to fish for aku (skipjack tuna). *WPRFMC photo.*

8.3 American Samoa Pelagic Fishery

Between 1976 and 1982, the Council would find itself having to respond to enormous changes in the American Samoa industrial fishery. Prior to 1982, most of the fish brought to the canneries were yellowfin and albacore tuna from longliners and carrier vessels from Taiwan and Japan. However, this changed as U.S. purse seiners moved westwards across the Pacific in the late 1970s to avoid laws that forbade U.S. vessels from using purse-seine gear in the Eastern Pacific to catch tuna in a manner that also took dolphins. In the Western Pacific, dolphins did not associate closely with tuna as they did in the East. After 1982, most of the fish brought into American Samoa canneries came from U.S. tuna purse seiners. This was accelerated by the South Pacific Tuna Treaty, under which U.S. purse-seine vessels could fish in most of the EEZs of south and western Pacific states. As the presence of fishing vessels from the U.S. mainland grew in the Territory, so too did the desire of American Samoans to participate in the fishery.

Developing a Local Commercial Fishery

During the 1980s, fishermen in American Samoa increasingly expressed to the Council their concerns and desires for assistance in developing their domestic fisheries. Specifically the fishermen needed marketing assistance, help in transporting fish to and from Pago Pago, directions on what to do when sighting foreign vessels in their area, search and rescue service, protection for the Rose Island National Wildlife Refuge and help controlling the crown-of-thorns starfish that were damaging their reefs. At the same time, the size and importance of the canneries in American Samoa increased with the influx of the purse seiners, as did the need to upgrade the fishery data used for decision-making. The Council, in conjunction with the Western Pacific Fisheries Information Network, helped the American Samoa Department of Marine and Wildlife Resources (DMWR) upgrade its database resources. Help was also forthcoming for fishery projects, such as the creation of a pilot giant clam farm and the purchase of anchored FADs for local fishermen.

Other development projects, funded through the Western Pacific CDPP, helped local American Samoa captains use their alia (motorized double-hull canoes) to catch fish for the canneries. This was at first very successful and provided new economic opportunity for local fishermen who set their longline gear manually, using between 200 and 300 hooks per set. Eventually the American Samoa longline fishery would grow to encompass both large conventional longline vessels greater than 50 feet (15.25 meters), mostly U.S.-based, and a locally owned alia fleet (Kaneko and Bartram 2005). Other CDPP projects in American Samoa included construction of a four-room cold storage restoration system; the reduction of fisheries bycatch through the creation of new value-added fishing products, such as canned wahoo; and grant writing workshops to help Pacific Islanders in American Samoa more successfully apply for financial assistance related to projects for marine conservation.

The cold storage refrigeration system helped fishermen on smaller alia vessels to comply with new U.S. Food and Drug Administration regulations that were mandated in December 1997, i.e., the Hazard Analysis and Critical Control Points for Seafood Safety. To address this issue, the Council also sent technical experts, one being John Kaneko of the Hawai'i Seafood Council, to hold training classes on HAACP with local fishermen.

The Council continued to support efforts to increase local participation in and benefit from the conservation and management of fishery resources, always with an eye toward fostering local community input and control. In 2007, it commissioned a study to explore three different scenarios that might lead to economic growth. These scenarios included fresh fish export to Hawai'i, value-added processing of longline fish and development of a domestic albacore longline marketing cooperative.

The fragility of the American Samoan economy has led to calls for closer cooperation with its neighbor Western Samoa. In the March 2009 meeting in American Samoa, the Council voted to explore options to support collaboration between American Samoa and the neighboring independent Government of Samoa to address management of albacore tuna, bottomfish and reef fish. By working together the two Samoas can look for mutual means to improve their economies.

Addressing Gear Conflicts

Recognizing the potential conflict between the growing fleets of small and large longline vessels, in 1998 the Council recommended that the EEZ waters within 50 nm of shore around American Samoa be closed to longline vessels greater than 50 feet (15.25 meters) in length. Underlying this action was the concern that American Samoa fishermen were being displaced by outsiders who had not traditionally fished in these waters as a result of early closures of the fishery in Hawai'i waters. A sociocultural poll sponsored by the Council and conducted by Craig Severance of the University of Hawai'i at Hilo indicated that the most frequently reported concern among American Samoa fishermen was "overfishing" and in particular "fear that there might be overfishing by outsiders and foreign vessels" (WPRFMC 1998).

Council members decided that a 50-nm limit was needed as a precautionary policy for the following reasons:

1) To avoid gear conflict;

2) To keep local catch rates of albacore at viable levels for local alia fishermen; and

3) To prevent a possible "boom and bust" cycle of development that might result in significant financial losses for American Samoans who might otherwise rush into the fishery.

In 1998, the Council took the additional step of recommending that EEZ waters within 50 nm around American Samoa be closed to large-scale pelagic fishing vessels. This effectively protected the local alia fishery by providing them with 37,902 square nm (130,000 square km) in which to fish without direct competition from larger vessels.

NMFS was reluctant to give approval to such an exclusionary action that would benefit American Samoa fishermen over other U.S. fishermen as MSA standards allow all U.S. citizens the equal right to fish in U.S. waters.[65] Consequently, NMFS disapproved the initial draft amendment to implement the area closure (Hogarth 1999). After concerted Council effort and insistence, the Large Vessel Prohibited Area (LVPA) went into effect on March 1, 2002, closing the area to all pelagic fishing vessels greater than 50 feet (15.25 meters) in length. This responded to the twin concerns of American Samoa fishermen about competition between small (less than 50 feet) and large (greater than 50 feet) longliners and the occasional nearshore fishing by the U.S. purse-seine fleet. In the meantime, the number of large vessels working in the fishery increased from three in 2000 to 30 by March 2002.

The interests of the alia fishermen were protected, but soon what was at issue was the ability of the small number of large vessels owned by American Samoans to compete. The owners and operators of these larger locally owned longliners would face additional hardships with increasing expenses and decreasing catches (attributed to a large offshore Chinese fleet).

---

[65] Unlike citizens of other U.S. territories who are U.S. citizens, American Samoans are U.S. nationals. They are entitled to all of the legal protection of a U.S. citizen but do not have the complete political rights of a U.S. citizen.

Local alia longline vessels fill Pago Pago harbor as the American Samoa fleet rapidly increased tenfold between 2000 and 2002.

The crews of large, locally owned and operated American Samoa longline vessels provided tours to visitors during Fishers Forum Day at the American Samoa Port Administration main dock, Fagatogo Village, on October 17, 2015. The event, co-sponsored by the Council and the American Samoa DMWR, aimed to educate the public about the Territory's different local fisheries. *WPRFMC photo/Sylvia Spalding.*

Echoing problems that had already arisen in Hawai'i, Ufagafa Ray Tulafono (then director of the American Samoa DMWR) found himself caught between his fellow American Samoans and his role as a Council member as the longliners came in. "The American Samoan fishermen would come to me and say, 'Why do you let them fish here?' I have to tell them that it's U.S. law that we have to let any U.S. vessel in. Sometimes they complain and they say, 'Why are you siding with them?'"[66]

Aware that allowing large, conventional, mono-hulled longliners to enter the American Samoa fishery unchecked would lead to increased competition and possibly localized depletion of albacore, the Council established a limited-entry program for longliners that capped fishing effort. A special provision was then made for American Samoans so that they might have an advantage when applying for limited-entry permits based on the length of their past experience in the American Samoa longline fishery.

Frank McCoy of American Samoa chaired the Council in 2005 and 2006 when the American Samoa limited-entry program was established.

The tuna canneries in Pago Pago were very important to the economy of American Samoa and to the fisheries of the region. A 2006 South Pacific albacore longline conference convened by the Council in Honolulu indicated that virtually all the individual fisheries in

---

[66]   Ufagafa Ray Tulafono in discussion with Michael Markrich, January 26, 2021.

the region brought their fish to Pago Pago for sale and processing (WPRFMC 2006). The canneries were the primary private-sector employer, and together with federal aid represented nearly 92% of the Territory's economy (Levine and Allen 2009). However, even as the American Samoa longline fishery grew in size during the late 1990s, the two canneries operating in American Samoa—Chicken of the Sea and StarKist—struggled to survive financially. Federal minimum wage requirements made canning operations in American Samoa significantly more costly than in Thailand and other areas of Asia where comparable packing plants had been built. Tuna canned in American Samoa enters the U.S. market duty free, and efforts were made through Congress to subsidize the canneries through federal budget earmarks. Nonetheless, the American Samoa Chicken of the Sea cannery closed for economic reasons on September 27, 2009.

Two days later, on September 29, 2009, the island was struck by a devastating tsunami that destroyed a newly constructed Star-Kist processing facility as well as other equipment and severely damaged the longline and local native alia fishing vessels moored in the harbor. The Council's response to the disaster was immediate. Its staff worked with government agencies in American Samoa to compile information on the tsunami's effects on the fisheries, canneries, harbors and related infrastructure, and it facilitated the filing of claims to the Federal Emergency Management Agency.

Mitigating Green Turtle Interactions in the Longline Fishery

The American Samoa longline fishery began in 1995 with one vessel targeting albacore tuna to be delivered to the local canneries. The fishery soon expanded from approximately 21 mostly small vessels in 1997 to 75 vessels of a variety of sizes in 2002. In response to the developing fishery, the Council began management under the Pelagic

FMP, establishing a longline prohibited area for large vessels and a limited-entry system for pelagic longline vessels fishing in the U.S. EEZ waters around American Samoa. Monitoring mechanisms were also put in place, requiring American Samoa longline vessel operators to complete federal logbooks, use VMS and carry federal observers if requested by NMFS.

When the federal observer program for the American Samoa longline fishery started in 2006, it was discovered that the fishery had occasional green turtle interactions. During the first few years, the fishery had one to three observed interactions with green turtles annually, leading to an estimate of approximately 30 per year for the entire fishery. The Council considered this interaction level low compared to other foreign fisheries operating in the Pacific. Nonetheless, it began developing mitigation measures upon receiving advice from NMFS in 2008.

The American Samoa longline fishery was already using circle hooks and fish bait, which were the main measures that resulted in significant sea turtle bycatch reductions in the Hawai'i-based swordfish longline fishery. The focus of mitigation measures was directed to the depth of the fishing gear because available observer data showed that most of the green turtle interactions occurred on the shallowest hooks.

In 2009, the Council recommended a measure to mitigate green turtle interactions by requiring longline gear configuration in such a way that all hooks are set deeper than 100 meters. Since the measure became effective in 2011, the estimated total interactions have been about seven green turtles per year, a substantial reduction compared to the estimated 30 per year prior to implementation of the gear configuration rules.

Tuna Purse Seiners

Adding tuna to the list of managed species in 1992 further increased the scope of the Pelagic FMP and the range of issues that the Council would have to address, including the purse-seine industry. Analysts that year had alerted the Council that purse seiners could become a significant issue for local fisheries in the U.S. EEZ of the Western Pacific Region.

Between 2006 and 2008, the U.S. purse-seine fleet more than doubled from 12 vessels to 26, and then it expanded to 39 by 2015. The new vessels were built in Taiwan but were 51% U.S.-owned and under the U.S. flag. They could legally claim American Samoa as their home port, yet work out of other ports of the region. The Council took proactive steps to restrict their access to the U.S. EEZ after U.S. Congressman Eni Faleomavaega took steps to legislate that these purse-seine vessels would be eligible to fish within the U.S. EEZ. In 2008, the Council recommended prohibiting purse seining in all EEZ waters surrounding Guam, CNMI and American Samoa.[67] It was concerned because, while the South Pacific Tuna Treaty limits the number of U.S. purse seiners permitted to operate in the waters of other treaty nations to 40, it does not cap the number of U.S. purse seiners fishing on the high seas or within the U.S. EEZ. The Council's recommendation for a purse-seine fishing spatial closure in the U.S. EEZ in the Western Pacific Region is the only Council recommendation in 40 years to be denied by the Secretary of Commerce for not meeting National Standard 2, use of the best scientific information available.

---

[67] Purse-seine fishing is not an issue in Hawai'i waters where it is too windy for purse-seine gear.

Sean Martin of Hawai'i chaired the Council from 2007 to 2009 as it addressed purse-seine and international tuna management issues and closure of the Pacific Remote Island Area waters from 0 to 50 nm offshore through presidential proclamations.

The Council's efforts—which followed the MSA and National Environmental Policy Act processes of national standards, environmental evaluation and public comment—were not successful in banning purse-seining in the EEZ near the inhabited islands of the Western Pacific Region. However, in 2009, President Bush through executive order under authority of the Antiquities Act of 1906 unilaterally banned all commercial fishing in EEZ waters within 50 nm of the Pacific Remote Islands of Baker, Howland, Jarvis, Johnston, Kingman Reef and Wake. These waters had historically been used by the U.S. purse-seine fishery that landed fish in American Samoa for the canneries. The presidential proclamation established these waters as the Pacific Remote Islands MNM.

Meanwhile, at the international level, the difficult task of managing U.S. and foreign purse seiners catching large amounts of not only skipjack but also juvenile bigeye and yellowfin tuna would continue to be a challenge.

8.4 Guam Pelagic Fishery

Guam is a small island territory of approximately 210 square miles (544 square kilometers) of land with an ancient fishing tradition. The indigenous Chamorro make up approximately 37% of the population. The population of Guam has grown from approximately 96,000 people in 1976 to nearly 170,000 in 2020, with the prospect of additional population growth due to a transfer of U.S. military personnel from other U.S. Pacific military bases. The strategic importance of Guam is due to its proximity to Asian countries. Guam is intermittently affected by large typhoons that destroy property, damage infrastructure and adversely affect natural resources.

Vessels at the Gregorio D. Perez Marina in Agana were thrown ashore by Super Typhoon Pongsona, which hit Guam on December 8, 2002. It was one of the most intense typhoons to ever strike the island (NWS 2003). *WPRFMC photo/John Calvo.*

In 1977, the Guam Fisherman's Cooperative Association (GFCA) was formed. A contract with the Guam Department of Education to provide frozen fish for the student lunch program provided sufficient regular cash flow to stabilize the fledgling organization. Soon after the formation of the GFCA, the government of Guam funded construction of a new fish market. Manny Duenas, the longtime president of the association, would later serve on the Council.

Customers line up at the GFCA facility on April 11, 2005. The co-op specializes in the sale of seafood items caught daily by its members and processed in the store. *Paul Bartram photo.*

Manny Duenas is the long-time president of the GFCA. He served as a Council member from Guam from 2003 to 2012 and as the Council chair during the last two years. *WPRFMC photo.*

The first Guamanian members on the Council included Paul Bordallo and Paul Callaghan. From the outset, Guam leaders voiced concern to the Council about the many potential impacts of foreign nations on the Territory's fisheries and the many factors that impeded the local fisheries from developing. For example, Guam representatives wanted it known that they were very worried about potential agreements between the United States and Japan regarding nuclear waste dumping and nuclear testing in the Pacific.

Also from the outset, Guamanians wanted to ensure that billfish, which were important to their growing tourism charter fishery, were not targeted by foreign longliners near Guam. For this reason, in 1977 they asked that the Billfish PMP include a measure prohibiting foreign vessels from fishing for billfish within at least 50 nm of Guam (WPRFMC 1977b). Later they would ask for additional regulation that would exceed the Council's recommendation—asking for a 200-nm no-take zone around Guam. Guam representatives were also concerned that foreign or out-of-state vessels could come to the islands and, without Guam's permission, take their baitfish. Guam residents and government worried that this would have long-term negative consequences to their main fishery. For these and other reasons, the same 1992 amendment that established the longline area closure around Hawai'i (Amendment 5 to the Pelagic FMP) also established a 50-nm longline area closure around Guam and its offshore banks to prevent gear conflicts and for safety vessel issues.

As a U.S. territory, Guam was exempted from the maritime law known as the Nicholson Act of 1950, which prohibits foreign-flag vessels from landing in a port of the United States its catch of fish taken on board the vessel on the high seas, or fish products processed or fish or fish products taken on board the vessel on the high seas. This exemption allowed Guam to develop into a tuna transshipment hub.

Significant numbers of distant-water fishing vessels, both foreign and domestic, came to Guam not only to transship but also for fuel, crew transfer, repair and provisions. To accommodate them, a new fisheries facility was built by the YTK Company. In addition, an air transshipment service was initiated to send top-quality sashimi to Japan and Europe. Guam transformed from an isolated port to a major transshipment point for Japanese and Taiwanese fishing vessels and, for a short time during the 1980s, as a base for a dozen U.S. purse seiners. The Department of Commerce (Division of Economic Planning and Development) reported that 328 Japanese and Taiwanese longliners participated in the industry, unloading 12,729 mt in 1990 (Maeda 1992). The estimated value of the fish transshipped from Guam in 1996 was more than $94 million, and fleet expenditures to resupply and for repair services earned Guam an additional $68 million in 1998 (WPRFMC 1999a).

Tuna are offloaded a Guam Commercial Port on August 10, 2003, to be processed for transshipment The industry expanded dramatically in the late 1980s to include not only Japanese and Taiwanese vessels but also a few Korean longliners and American longliners and purse seiners. *WPRFMC photo/Sylvia Spalding.*

"Guam was a transshipment hub," explained Kitty M. Simonds, Council executive director. "However, this industry declined when countries in Micronesia began requiring vessels to offload in their ports in order to fish in their waters."[68] Consequently, the last transshipment activity on Guam occurred on December 28, 2020, and the last transshipment agent would leave the territory in the weeks following (O'Connor 2021).

Guam had specific marine and fisheries development issues during the early 1970s that were difficult to address. Chief among these were the lack of U.S. and Territorial government spending on infrastructure

---

[68] Simonds 2021 op. cit.

in the form of docks, boat ramps and small boat harbors that could facilitate the development of small boat fisheries and charter fishing.

By 1990, Guamanian fishermen were more active in the fishery and new facilities were built, including boat ramps at Umatac Village and at the Agat Village Marina. Later, funds were obtained from the USFWS Sportfish Restoration Fund to build facilities at Merizo Pier in Southern Guam. Additionally, 40 vessel mooring buoys were installed to protect coral reefs from indiscriminate anchoring damage and 10 anchored FADs were purchased and put in service at various points around the island. The Council would augment these initiatives with the purchase and outfitting of a longline vessel, the *Galaide*, for use by Guam fishermen through the CDPP in 2002.

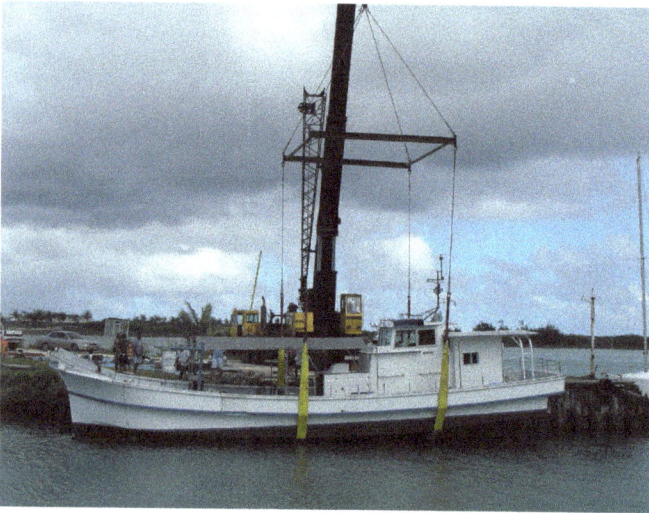

After months of preparation, the *Galaide* is launched at Agana Marina on September 30, 2005, to serve as a longline fishing platform for Guam. The purchase and renovation of the vessel was part of a 2002 CDPP project that included a training program and marketing research. *WPRFMC photo.*

Solomon Monteverde lands a pelagic fish caught on a small trolling vessel, typical of the craft used for the fishery in Guam. *Felix Reyes photo.*

In 1993, Council Executive Director Kitty M. Simonds, Nancy Foster (NMFS acting assistant administrator for fisheries) and Al Stayman, Department of the Interior deputy assistant secretary for territorial and international affairs, held meetings in the territories to review laws that were creating obstacles to the local governments' desires to develop their fisheries.[69]

Eventually, U.S. law would allow qualifying vessels not built or rebuilt in the United States that are eligible for documentation to be issued a fishery endorsement so they could engage in fishing in the territorial sea and fishery conservation zone adjacent to Guam, American Samoa and the Northern Mariana Islands (U.S. Code 2001). As the distance from Guam to the continental United States is 6,000 miles (9,656 kilometers), this helped keep the cost that fishermen in the territory had to pay for a fishing vessel from increasing significantly.

---

[69]   Simonds ibid.

Management of Guam's fisheries also faced difficulties, particularly due to inadequate fisheries data, especially in the early years of the Council. For this reason, the Council devoted many years of staff time to develop and support the Western Pacific Fisheries Information Network in an effort to upgrade fishery data collection in Guam and throughout the region. The Council also worked closely with the Guam Division of Aquatic and Wildlife Resources to make substantial progress in quantifying marine catch and coral reef data in Guam. One example is the Voluntary Guam Fishery Data Collection Program. The Western Pacific Community Development Program gives the Council authority to create preferential administrative programs in the fisheries it manages. Guam fishermen proposed to the Council that, rather than have new federal rules forcing them to provide data, they would create a voluntary data collection program that they themselves would administer. The Council then provided training to the GFCA in data collection, computer use and data analysis in conjunction with the Guam Division of Aquatic and Wildlife Resources, which would analyze the collected data. The data are provided to the Pacific Islands Fisheries Science Center for use in managing Guam ocean areas.

Still many data quality issues continued to exist.

8.5 Northern Mariana Islands Pelagic Fishery

The CNMI consists of 14 islands (three inhabited full-time) that stretch some 300 miles (483 kilometers) northward from Guam in the North Pacific. The total land area encompasses 184 square miles (477 square kilometers), and there are approximately 56,608 residents (SPC 2020). CNMI has a special commonwealth status with the United States governed by a Covenant, few resources and no "name recognition" on the U.S. mainland. As such, it has come to depend on the Council for help in getting its ocean-related needs addressed by the federal government.

The CNMI government has aggressively pursued development of its fishing industry. Around the early 1980s, tuna was transshipped for about three years from Guam to Japan through Saipan, where containers were put on larger planes.[70] Also around this time a fish market facility was built in Garapan on Saipan, which actively bought and sold various fish to local and off-island customers.[71] In the late 1980s and early 1990s, the island of Tinian had a major skipjack tuna transshipment facility, which was built primarily to service the vessels of the Zuanich company fleet and Taiwanese purse-seine vessels. However, the facility closed in the mid-1990s and the purse-seine vessels dispersed elsewhere, particularly to Papua New Guinea where tuna canneries were opening operations (WPRFMC 2016).

Over the next 20 years, CNMI established a FAD fishing buoy system for its fishermen and had as many as 11 active FADs at one time. There was also a significant investment in harbor infrastructure on Tinian, Lower Base, Rota West Harbor and Tanapag. In addition, the launching ramp was expanded at Smiling Cove Marina on Saipan.

In 2004, the Council facilitated a scoping meeting in CNMI that explored the size and accessibility of the pelagic fisheries resource, arranged for bathymetric mapping of the reef around Farallon de Medinilla and encouraged funding for the deployment of 12 new FADs. It also helped the CNMI Department of Fisheries and Wildlife with planning for creel surveys and arranged for John Kaneko to conduct workshops on Hazard Analysis and Critical Control Points so that CNMI fish could be exported to the U.S. mainland. In 2011, the Council established a 30-nm longline closure around the islands of the CNMI.

---

[70]   Jack Ogumoro in discussion with Sylvia Spalding, September 22, 2016.

[71]   Ogumoro ibid.

Through the CDPP, the Council in 2004 supported the CNMI with the organization of the Saipan Fishermen's Association and cooperative. This work would evolve into SFF projects that would develop public market places on Saipan where fishermen could sell their catch.

As it had elsewhere in the Pacific, the Council played a key role in facilitating data collection. With the support of the Western Pacific Fisheries Information Network, an up-to-date data base was created for CNMI fisheries that included commercial fish purchases, boat and shore creel surveys, and the monitoring of importation of seafood products.

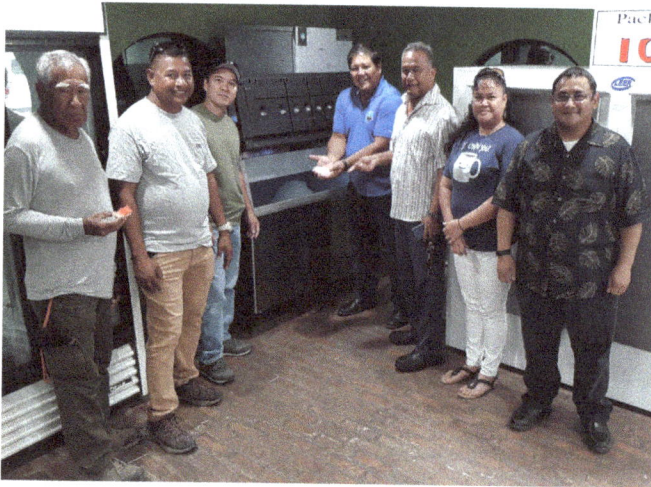

CNMI Department of Lands and Natural Resources Secretary Anthony Benavente (center) at the installation of the ice machine at the Garapan Public Market, Saipan, on June 17, 2019, is joined by Council Advisory Panel members Lino Olopai and Cecilio Raiukiulipiy, market manager Leroy Pangelinan, Advisory Panel member Raymond Tebuteb, market administrative assistant Paula Teregeyo and the Council's CNMI Island Coordinator Floyd Masga. The development of seafood markets was among the CDPP projects supported by the Council. *WPRFMC photo.*

8.6 Pacific Remote Islands Pelagic Fishery

The Pacific Remote Islands MNM proclamation prohibited commercial fishing in waters within 50 nm of Baker, Jarvis, Johnston, Howland, Kingman Reef and Palmyra but allowed for the possibility of recreational and noncommercial fishing. On Palmyra, visitors can sportfish offshore for tuna, wahoo, mahimahi and other pelagic species.

TABLE 10: PELAGIC FMP: SUMMARY OF COUNCIL ACTIONS		
DATE	ACTION	MEASURES
1987 Mar 23	FMP came into effect	Managed species included billfish, wahoo, mahimahi and oceanic sharks. Measures prohibited drift gillnet fishing within the region's EEZ waters and foreign longline fishing within certain areas of the EEZ.
1991 Mar 1	Amendment 1	Defined recruitment overfishing and optimum yield for each federally managed pelagic species.
1991 May 26	Amendment 2	Required domestic pelagic longline fishing and transshipment vessel operators to have federal permits, maintain federal logbooks, mark longline gear with official number of the permitted vessel and be subject to observer placement on board if directed by NMFS to fish within 50 nm of the NWHI. Incorporated EEZ waters around the CNMI into the FMP managed area.

1991 Oct 14	Amendment 3	Created a 50-nm longline exclusion zone around the NWHI to protect endangered Hawaiian monk seals. Implemented framework provisions for establishing a mandatory observer program to collect information on interactions between longline fishing and sea turtles.
1991 Oct 10; 1994 Dec 15	Amendment 4	Established a three-year moratorium on new entry into the Hawai'i-based domestic longline fishery. Provided for the establishment of a mandatory VMS for domestic longline vessels fishing in the Western Pacific Region. Required Hawai'i-based longline vessels to carry and use a NMFS-owned VMS transmitter.

| 1992 Mar 4 | Amendment 5 | Created a domestic longline vessel exclusion zone around the main Hawaiian Islands ranging from 50 to 75 nm and a similar 50-nm exclusion zone around Guam and its offshore banks to prevent gear conflicts and vessel safety issues. The main Hawaiian Islands zone included an October–January seasonal reduction in size, prohibiting longline fishing within 25 nm of the windward shores of all main Hawaiian Islands except Oʻahu, where it is prohibited within 50 nm from the shore. |
| 1992 Nov 27 | Amendment 6 | Included tunas and related species as federally managed species. Applied the longline exclusion zones of 50 nm around the island of Guam and the 25- to 75-nm zone around the main Hawaiian Islands to foreign vessels. |

1994 Jun 24	Amendment 7	Replaced Amendment 4's moratorium with a limited-entry program for the Hawai'i-based domestic longline fishery with transferable permits, a limit of 164 vessel, a 101-foot length overall maximum vessel size and a framework procedure for future implementation of certain types of new regulations.
1999 Feb 3	Amendment 8	Addressed new requirements under the 1996 SFA regarding designations of EFH and descriptions of some fishing communities. On April 19, 1999, NMFS approved portions that address EFH, overfishing definitions, bycatch and some fishing communities. On August 5, 2003, NMFS approved provisions regarding Hawai'i fishing communities.
2002 Mar 1	Framework Measure 1	Prohibited fishing for pelagic species by vessels greater than 50 feet (15.25 meter) in length overall within EEZ waters 0–50 nm around the islands of American Samoa. An exception was made for vessels that landed pelagic MUS in American Samoa under a federal longline general permit prior to November 13, 1997.

2002 Jun 9	Regulatory Amendment 1	Incorporated the reasonable and prudent alternative of NMFS's March 2001 BiOp. To mitigate interactions with sea turtles, shallow-set pelagic longlining was prohibited north of the equator by vessels managed under the FMP and in waters between 0 and 15 degrees north latitude from April through May of each year. Instituted sea turtle handling requirements for all vessels using hooks to target pelagic species in the region's EEZ waters. Extended the protected species workshop requirement to include the operators of vessels registered to longline general permits.
2002 Jun 13	Framework Measure 2	Required Hawai'i-based pelagic longline vessel operators to use blue-dyed bait, strategic offal discards and line shooters with weighted branch lines to mitigate seabird interactions when fishing north of 23 degrees north latitude. Required all Hawai'i-based longline vessel owners and operators to annually attend a NMFS protected species workshop.

2002 Jun 14	Amendment 10	Prohibited the harvest of managed species in the no-take MPAs established by the Council in areas around Rose Atoll in American Samoa, Kingman Reef, Jarvis, Howland and Baker. NMFS final rule implemented the amendment on February 24, 2004.
2002 Oct 4	Regulatory Amendment 2	Established federal permit and reporting requirements for any vessel using troll or handline gear to catch pelagic MUS in EEZ waters around Kingman Reef, Howland, Baker, Jarvis, Johnston, Palmyra, Midway and Wake.

2004 Apr 2	Regulatory Amendment 3	Required Hawai'i-based shallow-set swordfish fishery to use circle hooks with mackerel bait. Limited fishing effort to 50% of the average annual number of sets between 1994 and 1999 (just over 2,100 sets) allocated between fishermen applying to participate in the fishery. Implemented a hard limit on leatherback (16) and loggerhead (17) turtle interactions that could occur in the fishery. Re-implemented earlier sea turtle handling and resuscitation requirements. Included a number of conservation projects to protect sea turtles in their nesting and coastal habitats. Required night setting on Hawai'i-based longline vessels targeting swordfish north of 23 degrees north latitude.

2005 Aug 1	Amendment 11	Established a limited access system for pelagic longlining in EEZ waters around American Samoa based on historical participation in the fishery. Allowed limited vessel upgrade. Required longline vessel to obtain federal permits; complete federal logbooks; carry and use VMS installed, owned and operated by NFMS on vessels greater than 40 feet (12 meter) in length; carry federal observers if requested by NMFS; and follow sea turtle handling and resuscitation requirements.
2005 Dec 15	Regulatory Amendment 4	Required vessels with longline general permits making shallow sets north of the equator to use 18/0 circle hooks with mackerel-type bait and dehookers to release any accidentally caught turtles. Required both operators and owners of these vessels to annually attend protected species training workshops. Required operators of these vessels to carry and use specific mitigation gear to aid in the release of sea turtles accidentally hooked or entangled by longlines. Required operators of non-longline pelagic vessels (e.g. trollers and handliners) to follow handling guidelines and remove trailing gear.

2006 Jan 18	Regulatory Amendment 5	Allowed operators of Hawai'i-based longline vessels fishing north of 23 degrees north latitude and those targeting swordfish south of 23 degrees north latitude to utilize side-setting to reduce seabird interactions in lieu of the seabird mitigation measures required by Framework Measure 1.
2007 Mar 28	Regulatory Amendment 6	Removed the delay in effectiveness for closing the Hawai'i-based longline shallow-set swordfish fishery as a result of it having reached one of its turtle interaction limits.
2007 May 16	Amendment 14	NMFS approved the portion regarding international management action to end overfishing of bigeye and yellowfin tuna stocks.
2007 May 17	Regulatory Amendment 7	Provided pelagic fishery participants the option of using NMFS approved electronic logbooks in lieu of paper logbooks.

2008 Nov 21	Amendment 15	Added pelagic squid species *Ommastrephes bartramii*, *Thysanoteuthis rhombus* and *Sthenoteuthis oualaniensis* as MUS. Required owners of U.S. vessels greater than 50 feet (15.25 meters) in length overall that fish for pelagic squid in U.S. EEZ of the Western Pacific Region to obtain federal permits, carry federal observers if requested by NMFS and report any Pacific pelagic squid catch and effort either in federal logbooks or via existing local reporting systems.
2009 Dec 10	Amendment 18	Removed the 2,120 set limit for the Hawai'i-based shallow-set longline fishery. Implemented a new loggerhead sea turtle hard cap of 46 annual interactions. Maintained existing requirements including 100% observer coverage, use of circle hooks and mackerel-type bait and onboard handling and release techniques

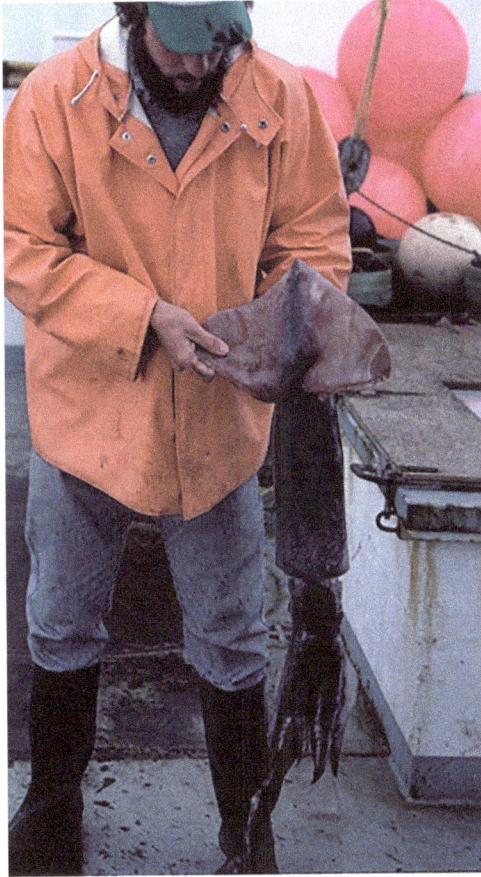

A fisherman displays a pelagic squid caught in the Pacific (circa 2001). Three species of pelagic squid, which are targeted for food and bait, became federally managed species under Amendment 15 of the Council's Pelagic FMP on November 21, 2008. *NOAA photo.*

Chapter 9

Coral Reef Ecosystem Fisheries— Developing the Nation's First Ecosystem-Based Fishery Management Plan (2001–2009)

9.1 Overview of the Coral Reef Ecosystem Plan

Planning for ecosystem-based fishery management has an extended history in the Western Pacific Region. Council Member Rufo Lujan of Guam and Executive Director Kitty M. Simonds were active participants in NOAA's first workshop on the ecosystem approach to fisheries management in 1986. The NMFS Program Development Plan for Ecosystems Monitoring and Management, published the following year, called for reorienting monitoring and management of the nation's fisheries from a single species/species complex perspective a multi-species ecosystem approach (Evans et al. 1987).

In October 1987, the Council announced at the North Pacific Rim Fisherman's Conference on Marine Debris in Kona that it would change the way it managed fisheries to the ecosystem approach.

In 1993, a symposium entitled Global Aspects of Coral Reefs: Health, Hazards and History held in Miami, Florida, attracted 125 coral reef scientists. Subsequently the United Nations declared 1997 as the International Year of the Reef. The Council responded to this worldwide movement and contracted Cynthia Hunter (1995) and Alison Green (1997) to assess the status, use and value of coral reef resources in the Western Pacific Region.[72]

In June 1998, as a response to bio-prospecting activities (the practice of hunting the sea for sources of potential pharmaceutical value), the Council created a CRE Plan Team and began developing the CRE FMP for the region. It would become the nation's first ecosystem-based FMP.

Prior to 1998, there were no regulations to protect coral reef fishery ecosystems in the U.S. EEZ waters in the Western Pacific Region. Because the commercial exploitation of reef areas could lead to overharvesting, the Council took the precautionary step of creating the CRE FMP. In 1998, U.S. Senator Daniel Inouye announced that $15 million had been appropriated for Hawaiian ocean research initiatives, with $1.23 million earmarked for coral reef research and $230,000 to be used by the Council to develop its CRE FMP.

As work began on the CRE FMP, the Council found itself beset by conflicting mandates, namely providing for the national benefit through harvesting at maximum sustainable yield/optimum yield and at the same time protecting the EFH as required under the 1996 SFA. This was further made difficult because there was little scientific knowledge regarding the life history and habitat requirements of the thousands

---

[72] As noted in Chapter 5, Section 5.1, an unexpected consequence of Hunter's report would be its controversial use to estimate that the coral reefs in the NWHI were the greatest in the nation in order to politically justify establishment of the NWHI CRER and Papahanaumokuakea MNM.

of coral reef fish species to be conserved and managed under the CRE FMP. Once again the Council would be pioneering uncharted waters.

When the CRE FMP and accompanying draft environmental impact statement were presented to the Council at its 101st meeting in October 1999, controversy arose over how the plan was to be implemented. At the meeting, it was explained that the ecosystem approach was needed because coral reefs were intricate and unique. The conventional FMP approach manages fish species or species complexes for sustainable use, whereas the ecosystem approach considers the impact of catches on both the targeted fish stocks and other species as well. This was taking management to a far more elaborate level. What made the task even more difficult was that coral reefs encompassed vast, largely unexploited areas of the Western Pacific, including the NWHI.

The Council's experience with establishing the CRE FMP was difficult and adversarial. Besides opposition from academics, conservation NGOs and USFWS to the Council's management of coral reef fisheries, the CRE Plan Team, composed mainly of coral reef biologists, had an animus toward fishing on coral reefs that greatly slowed progress on the plan.

Members of the Plan Team complained that, as written, the draft CRE FMP excluded species and species complexes already managed by existing FMPs. For example, lobsters were managed under the Crustacean FMP in a manner that favored a species-based approach over an ecosystem-based approach. USFWS coral reef ecologist James Maragos singled out the decline of the lobster fishery in the NWHI as an example, in his opinion, of all that was bad about individual species management under species specific FMPs (WPRFMC 1999b).

Underlying the different viewpoints was an intense philosophical debate about how the coral reefs of the ocean were to be properly managed. Were they to be fished for the benefit of people as they had in the past, or were they to be set aside forever in the long-term interest of the coral reefs and associated resources? For the Council members it was not acceptable to close all coral reef areas to fishing and thereby discard the work accomplished over 25 years to develop the NWHI fishery. They were confident that the new CRE FMP could coexist with the existing FMPs, even if the coral reef scientists were not.

Setting aside designated areas as no-fishing zones had been a fisheries management tool for many years. During the 1980s, MPAs became popularized as a marine planning tool. MPAs could act as buffers between fisheries and could also be used to protect spawning grounds, nursery areas and vulnerable habitats. Most controversial were MPAs that allowed nothing to be taken and that set aside the area indefinitely. The Council had used the concept of MPAs since 1983 when it created an MPA near Necker Island as part of its Precious Coral FMP and around Laysan as part of the Crustacean FMP.

The philosophy behind no-take MPAs is to set aside certain ocean areas so that marine life inhabiting them can reproduce and spawn without being impacted by fishing pressure. The argument made in favor was that establishing no fishing areas adjacent to fishing areas would allow the fish in the no-fishing areas to reproduce in large numbers and their offspring would spill over into the open fishing areas. This recruitment overflow would compensate for the loss of biomass in the areas being fished. Known as the "spill-over effect," this idea was to become the theoretical underpinning for the establishment of MPAs all over the world.

Driving the no-take MPA movement was the belief that all existing methods of government regulated fisheries management were badly flawed and that only by completely eliminating use of an area is it possible to protect diverse species. It is a similar theory to the "hot spot" approach, popular with many NGOs, which came originally from British ecologist Norman Meyers (Dowie 2009). The hotspot approach argues for the saving of the most species at the least cost as a means to set conservation priorities (Meyers 2003). One popular champion of this idea for ocean protection was Callum Roberts, a professor at England's University of York, who advocated the establishment of large-scale MPAs in the open sea to protect up to one-third of the stock from fishing. These kinds of beliefs would have future repercussions on fishing in the U.S. EEZ.

Critics of both hot spot and no-take MPAs concede that fish biomass does grow in areas that are not fished relative to areas that are fished. Unfortunately, there is no evidence that the spill-over effect works in all environments or for all species. Fish that grow and swim out of the MPAs are often caught by waiting vessels clustered around the borders. More controversially, when large portions of fish stocks are off limits to fishing, the remaining portions are targeted more intensely and the impacts of fishing are focused instead of being distributed over a wider area. For this reason, hot spot MPAs are not seen by most fishery managers as a panacea for fishery problems but rather as one of many tools that fisheries managers can use in appropriate habitats to reserve certain fish stocks.

The SSC recommended that the Council's new CRE FMP include MPAs in most areas of the NWHI. They were not to be no-take, but the intended rules were to be highly restrictive. All fishermen operating in the area would require special CRE permits. There would be no taking of coral or live rock without a special permit. Fisheries covered by existing FMPs would continue to be covered by their respective permit and reporting mechanisms. These rules for the NWHI would include all substrate from the shoreline to a depth of 50 fathoms. Later this would be broadened to include a no-take MPA zone from 0 to 10 fathoms and a special permit for any fishing from 10 to 50 fathoms.

From the Council's perspective, the reefs in the NWHI were so extensive and healthy that a level of restriction that permitted some fishing was sufficient. But some biologists on the CRE Plan Team disagreed. They saw the Council position as unacceptable. As Maragos said during a Council meeting in October 1999, "The regulations of the MSA and the [SFA] were geared towards species-specific plans; this ecosystem plan is a new area where there is no precedent. The other FMPs need to be incorporated, particularly how they interact with the ecosystem" (WPRFMC 1999b).

The position of the coral reef scientists in this argument would later be countered by commercial fishermen such as Timm Timoney, a full-time NWHI fisherman. She pointed to the absurdity of the presumption that specific areas in the NWHI were pristine when in fact a number of the areas had been used for industrial and military purposes and had become home to alien populations of ants, rats, dogs, rabbits and other harmful invasive species and were covered with large amounts of marine debris.

There was no middle ground between the two positions. It pitted those who saw themselves protecting the ocean and reef areas from commercial fishing—while allowing nonconsumptive economic gains associated with marine tourism—against those who saw accessing the reefs and ocean waters in the NWHI as essential for acquiring food and sustaining livelihoods. As it had in the past, the Council sought balance. After lengthy discussion, the Council voted unanimously to accept the SSC's recommendation regarding the CRE FMP, which included both coral reef protection and the coexistence of the existing

FMPs. Despite Council Chair Jim Cook's pleas for everyone to work together, advocates were polarized.

Ultimately, the members of the Council voted unanimously to keep all their existing FMPs while allowing for the adoption of the newly created CRE FMP, which would complement rather than supersede the existing FMPs. Dialogue between the groups became heated. The coral reef scientists on the Plan Team and environmentalists at the meeting would not settle for anything less than dismantling the existing FMPs. The custom of the Council in regard to solving difficult questions was give and take; but, on the issue of closing down fishing in the NWHI, many coral reef biologists wouldn't compromise. This schism would eventually lead to the creation of the NWHI CRE Reserve and subsequently the NWHI MNM, later renamed Papahanaumokuakea MNM.

The FMP included specific management measures to promote sustainable fisheries while providing for substantial protection of CRE resources and habitats throughout the Council's jurisdiction.

1. A network of MPAs in the Pacific Remote Island Areas (both no-take and low-use MPAs were proposed for the NWHI in the FMP but were disapproved by NMFS).

   (a) Fishing prohibitions in all no-take MPAs designated in this section.

   (b) Anchoring prohibition of all fishing vessels measuring more than 50 feet (15.25 meters) length overall in the U.S. EEZ seaward of the Territory of Guam west of 144 degrees 30 minutes east longitude, except in the event of an emergency caused by ocean conditions or by a vessel malfunction that could be documented.

(c) No-take MPAs in the U.S. EEZ waters landward of the 50-fathom curve at Jarvis, Howland, Baker, Kingman Reef and Rose Atoll.

(d) Low-use MPAs where fishing would be allowed under special fishing permits in all U.S. EEZ waters between the shoreline and the 50-fathom curve around Johnston, Palmyra and Wake.

2. A special permit and federal reporting system for controlling and monitoring the harvest of certain CRE MUS for which there was little or no information. Requirement for special permits to fish in all areas designated as low-use MPAs. Use of data collected under existing local reporting systems to monitor the harvest of currently fished CRE management unit species.

3. Prohibition on the use of destructive and non-selective fishing gears.

4. Prohibition on the harvesting of coral and live rock, but allowance of a limited take under the special permit system for collection of seed stock by aquaculture operations, and religious/cultural use by indigenous peoples.

5. An adaptive management approach using a framework process for rapid regulatory modifications in the event of major changes within coral reef ecosystems or coral reef fisheries.

6. Incorporation of the historical and cultural dependence of coral reef resources by indigenous people in management.

7. Identification and prioritization of coral reef related research needs for each island area, including socioeconomic and cultural research for future potential allocation of resources.

The Council's CRE FMP was completed in October 2001 and partially approved by NMFS in 2002. NMFS disapproved the portion of the plan that governed fishing in the NWHI west of 160 degrees 50 minutes west longitude because it was inconsistent with or duplicated certain provisions of the NWHI CRE Reserve. A final rule implementing the CRE FMP was published on February 24, 2004.

This achievement was groundbreaking in the MSA system. Among other things, the plan established a process to assess and control ecosystem effects on coral reef environments from bottomfish, precious coral, crustacean and pelagic fisheries operating in federal waters under then-existing FMPs. The CRE FMP would be the precursor to further actions to solidify the ecosystem-based approach to marine resources management. The provisions of the CRE FMP were subsequently incorporated into a series of island-based fishery ecosystem plans (FEPs).

9.2 Hawai'i Coral Reef Fishery

Archaeological evidence reveals that seafood, particularly coral reef fish species, was part of the customary diet of the earliest human inhabitants of the Hawaiian Islands. Fish were also imbued with cultural significance, and certain species were venerated as *aunukua* (personal, family or professional gods or guardian spirits).

In contemporary times, the majority of the commercial catch of inshore fishes, invertebrates and seaweed in Hawai'i comes from the nearshore reef areas in state waters around the main Hawaiian Islands. However, harvests of some coral reef fish species also occur in federal waters, such as around Penguin Bank. Most reef fish are caught for subsistence and recreation, and the catch often goes unreported. It is estimated that the value of reef-associated commercial fisheries in Hawai'i is $1.8 million/year, out of a total value of $385 million/year for Hawaiʻi's coral reef ecosystem in general, which includes the value of tourism and other services (Markrich and Hawkins 2016).

In 1999, the Council co-organized and co-sponsored a special Fisheries Forum 2000: Reclaiming Our Ocean Resources at the University of Hawaiʻi's William S. Richardson School of Law. Initiated by the NGO Malama Na ʻIa, this forum had the support of administrative and legislative branches of the State of Hawaiʻi and attracted the attendance of numerous fishermen, fishery scientists and managers, Native Hawaiians and environmental organizations. The main topics covered were akule, moi, Native Hawaiian fishery issues and community-based fishery management. Besides airing 45 times on ʻOlelo Community Television, the Forum was featured in the *Honolulu Star-Bulletin* in the article "Fisheries forum 'a big step' toward cooperation."

Fishermen pull in a net of akule (bigeye scad) in waters off Hawaiʻi (circa 2009). Akule became a managed species under the Coral Reef Ecosystem FMP in 2001. *Leo Ohai photo.*

Through the CDPP, the Council has supported projects in Hawai'i to restore and revitalize the He'eia Kea fishpond and to provide training in the propagation of limu (edible seaweed).

The Council has also supported—through the Ke Ola Ke 'Aha Moku program—community participation in fishpond restoration and maintenance of fishponds at Kalauha'eha'e in Maunalua, O'ahu, and at Haleolono in Keaukaha, on the island of Hawai'i. Fishponds are important cultural sites, foster community cohesion and provide important habitat for fisheries, the nature of which has yet to be fully understood.

The Council has also supported the development of a community resource monitoring protocol, which has been tested at Mokauea Island in Honolulu and at Wailua on Kaua'i. If the program can be developed with a high confidence level, the data may be usable to inform fisheries management. The program enhances community cohesion and cooperation and contributes to community resilience.

Visitors are shown the *makaha* (fish gate) at He'eia Kea fishpond. Through the CDPP and other sources, the Council has supported the renovations of several fishponds in Hawai'i. *WPRFMC photo.*

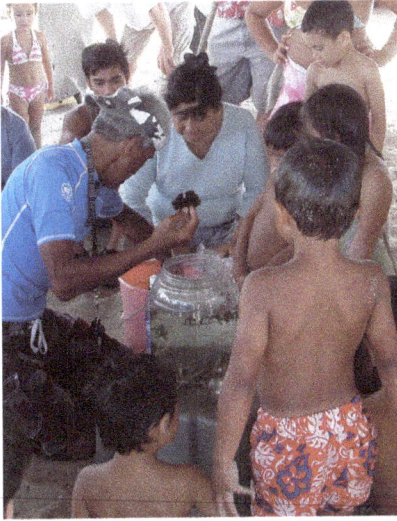

Council advisor and honored *kupuna* (elder) Henry Chang Wo Jr. teaches *keiki* (youth) from an elementary school class about limu (edible seaweed) at ʻEwa Beach, Oʻahu. The Ewa Beach limu project was supported by the Council through the CDPP. *WPRFMC photo/Charles Kaʻaiʻai.*

9.3 American Samoa Coral Reef Fishery

The American Samoa coral reef fishery was historically a subsistence and cultural fishery, and some fishermen still sell their catches to small markets or from the side of the road. Traditional fisheries continue to be associated with seasonal runs of certain species, such as atule (bigeye scad), juvenile surgeonfish (Acanthuridae), goatfish (Mullidae) and, notably, palolo worms (*Palolo viridis*).

Over time, the Council became an increasingly important resource used by American Samoans as a neutral party finding solutions to sensitive local fishing matters. One such issue was the use of fishing methods that damaged fish habitat.

"I remember how it used to be when the Council started 30 years ago," said Ufagafa Ray Tulafono, a former Council member and director

363

of the American Samoa DMWR. "When you went out on the reef there were holes 20 feet (6 meters) wide where the coral was shattered into tiny pieces. These were the places where people used dynamite. It was terrible. When the Council came, we were able to educate people to find another way to fish."[73]

In 1985, American Samoan fishermen asked members of the Council for support in ending the practice of fishing with explosives and poisons in American Samoa (WPRFMC 1985b). The Council sent Advisory Panel member Frank Farm Jr., a Native Hawaiian fisherman and fishing expert who would become a Council member, to American Samoa. In a series of village meetings and interpersonal discussions, he explained to fishermen that it would be harmful to continue dynamiting. Farm was able to influence the American Samoa village leaders to stop the use of dynamite on their reefs. His efforts caused little friction or discord and were a great success.

Ufagafa Ray Tulafono directed the American Samoa DMWR for more than 20 years and in that capacity served on the Council during three periods between 1985 and 2012 as the Territory's designated state official. *WPRFMC photo.*

---

[73] Tulafono op. cit.

An avid diver and fisherman, Frank Farm Jr. served as vice chair of the Council's Advisory Panel from 1986 to 1989, as a Council member from Hawai'i appointed by the Secretary for four three-year terms between 1988 and 2005 and as the Council chair in 2002. *WPRFMC photo.*

While a majority of the coral reef fishery catch is believed to come from waters under the jurisdiction of the Territory of American Samoa, some reef areas in federal waters are also fished. However, reporting systems have not allowed managers to parse out catch by federal versus local waters.

9.4 Guam Coral Reef Fishery

The second International Year of the Reef in 2008 was a muted celebration on Guam as new regulations designed to conserve coral reefs and promote tourism were displacing long-time fishermen. In a protest against their displacement, two Guam fishermen were arrested on December 13, 2007, after challenging the law by openly fishing illegally in the waters of the Tumon Bay Marine Preserve. The men claimed that it was their cultural right as Chamorro to fish in the area.

The mayor of the nearby village of Tamuning-Tumon-Harmon had informed the Council in 2007 that the island's growing tourism industry was causing community access to beaches to decline. The mayor made the statement during the meeting on Guam of the Council's Mariana Archipelago Regional Ecosystem Advisory Committee (REAC). The Council had established a REAC on each of the island areas under its jurisdiction in 2007 to address land-based issues that might be affecting fisheries. The Tumon Bay Marine Preserve was created in 1997, the year of the first International Year of the Reef. Fishermen complained that the regulations not only denied them access to traditional fishing grounds but also deliberately promoted a tourism industry that, fishermen believed, was harming the near-shore environment by employing commercial beach rakers to systematically remove sand, rock and marine organisms.

Additional restrictions preventing access to shore fishing grounds came from the U.S. military bases, which occupied nearly one-third of the island. This presented a significant and growing problem because Guamanians traditionally depend on reef fishing for food and recreation.

Juan Benavente releases a traditional *talaya* (throw net) at Tumon Bay (circa 2007). Leana Peters, of George Washington High School, Guam, won the Council's 2007 high school photo essay contest with this photo of her grandfather and an essay, in which she exalted the patience and endurance of *talaya* fishermen. *Leana Peters photo.*

9.5 Northern Mariana Islands Coral Reef Fishery

The coral reef fishery in the CNMI is important for subsistence and to maintain social and cultural cohesion. The local government allows the use of the traditional *talaya* (throw net), but other traditional net fishing practices were banned for many years. A representative from Tanapag village in Saipan requested assistance from the Council during a REAC meeting to revive their traditional *chenchulun umesugon*, a fishing practice that is similar to the *pai-pai* (pushing fish into the center) netting method used in Hawai'i. The Council facilitated a meeting between villagers and the CNMI Department of Lands and Natural Resources and its Division of Fish and Wildlife that enabled the villagers to get an exemption permit for use on special feast days. On March 9, 2008, a demonstration was arranged for the Council of this traditional technique. The fishing method was videotaped as part of a documentary on traditional practices being produced by the Northern Marianas College.

Through the CDPP, the Council assisted the CNMI with three nearshore fishery projects. Working with the Northern Islands Mayor's Office, it assisted in the establishment of a fishing station on the remote island of Alamagan in 2003; the first harvest was sent to Saipan in 2005. Working with the Northern Marianas College Cooperative Research, Education and Extension Service, the Council helped to revive and restore the traditional fishing practices of the Chamorro and Carolinian communities in the Commonwealth in a project that involved cultural experts from the island of Rota. Working with the governor's Carolinian Affairs Office, the Council helped to partner community and indigenous groups to train children on Saipan about traditional fishing, such as *chenchulu*.

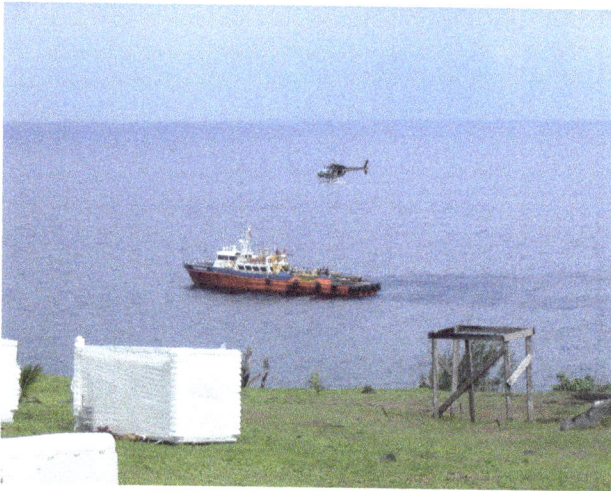

A helicopter and ship deliver goods and materials to Alamagan in the Northern Islands of the CNMI. The activity was part of a CDPP project that established a fishing station on the remote island in 2003 and began sending fish to Saipan for sale in 2005. *CNMI Northern Islands Mayor's Office photo.*

Stan Taisacan carves a traditional canoe and paddle on the island of Rota, CNMI. The revival and restoration of traditional fishing practices of the Chamorro and Carolinian communities in the CNMI was the goal of a 2005 CDPP project. *WPRFMC photo/Charles Kaʻaiʻai.*

Fishermen demonstrate traditional *chenchulu* (net) fishing at Tanapag, Saipan, as part of a 2004 CDPP project undertaken by the Carolinian Affairs Office in the CNMI. *WPRFMC photo/Sylvia Spalding.*

9.6 Pacific Remote Islands Coral Reef Fishery

The Pacific Remote Islands MNM proclamation in 2009 prohibited commercial fishing but allowed for the possibility of recreational fishing.

On Palmyra, recreational bonefishing is conducted on a catch-and-release basis with artificial flies and barbless hooks. No sportfish are allowed to be shipped off the atoll. Bottomfish and other reef fish are not allowed to be targeted. Those caught accidently are to be released immediately. Jacks can be fishing on a catch-and-release basis, but none can be consumed or retained.

On Johnston, once a possession of Hawai'i known as Kalama Atoll, a cooperative agreement between the USFWS and the military stopped the once common practice of shipping coolers of fish and live coral back to Hawai'i. In the late part of the 20th century, an average of 1,300 military personnel and civilian contractors were stationed on the atoll, which is comprised of four islands. In 2003, after 70 years of military control, the USFWS took possession of the atoll as a national wildlife refuge and prohibited fishing out to 12 nm.

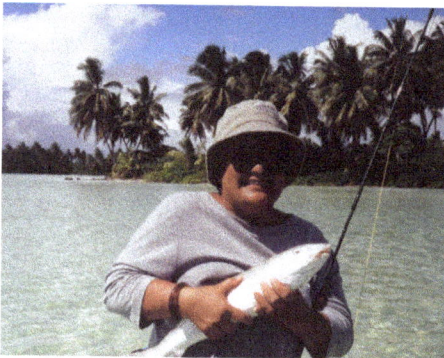

Council Executive Director Kitty M. Simonds grips a bonefish she caught at Palmyra Atoll, one of eight U.S. Pacific remote islands whose fisheries are managed by the Council under the MSA. The presidential proclamation establishing the Pacific Remote Islands MNM placed further restrictions on the fisheries in these remote islands. *WPRFMC photo.*

TABLE 11: CORAL REEF ECOSYSTEM FMP: SUMMARY OF COUNCIL ACTIONS		
DATE	ACTION	MEASURES
2001 Nov 28	FMP completion	The Council completes plan and requests transmittal to the Secretary of Commerce.
2002 Jun 14	NMFS Record of Decision	The CRE FMP is partially approved including provisions that prohibited the harvest of federally managed crustacean, precious coral and pelagic species in the no-take MPAs established under the plan in areas around Rose Atoll in American Samoa, Baker, Kingman Reef, Jarvis and Howland. NMFS disapproved provisions that would have governed fishing in the NWHI (including Midway) west of 160 degrees 50 minutes west longitude because they would be inconsistent with or duplicate certain provisions of executive orders that established the NWHI CRE Reserve.
2004 Mar 24	Final rule to implement the FMP	Established a CRE regulatory area, MPAs, permitting and reporting requirements, no-anchoring zone, gear restrictions and a framework regulatory process.

PART THREE

EVOLVING BEYOND SPECIES-BASED MANAGEMENT BY RETURNING TO AN ECOSYSTEM APPROACH

(2010–2020)

Chapter 10

Ecosystem-Based Monitoring and Management

10.1 Overview of the Fishery Ecosystem Plans

The ecosystem approach to fisheries management is, in a sense, a return to the traditional resource management methods that have been practiced in the U.S. Pacific Islands for millennia prior to Western contact. The indigenous Chamorro, Carolinian, Samoan and Hawaiian communities sustained themselves on land-limited islands and atolls for thousands of years through holistic management of the natural resources on which they depended (Kirch 2000). These methods are adaptive and consider elements from the ocean to the mountain tops and beyond into the atmosphere, including the recognition and respect for the spirit world and the needs of the local community.

Considering that 12 of the 13 voting members of the Council are local marine users and local agency representatives (the exception being the NMFS Pacific Islands regional administrator), it is not surprising that the ecosystem mindset was reflected in the FMPs developed by the Council. While the first four FMPS were species-based, they included measures to safeguard the ecosystem, including several expansive MPAs and the prohibitions against the use of numerous potentially destructive fishing gear and methods. The fifth FMP was the nation's first ecosystem-based FMP.

The Council had experienced firsthand the importance of being able to predict change in the ecosystem in order to manage fisheries when the unanticipated North Pacific oceanic regime shift in the 1980s coincided with the development and collapse of the lucrative NWHI lobster fishery.

Following the reauthorization of the MSA in 1996 (i.e., the SFA), the FMPs further incorporated ecosystem principles with the inclusion of EFH descriptions. The SFA also called for an Ecosystem Principles Advisory Committee (EPAP) to develop recommendations to implement ecosystem principles in fisheries management (Wilkinson and Abrams 2015). The EPAP's Ecosystem-Based Fishery Management report to Congress was published in 1999. Its primary recommendation was for each council (including NMFS in the case of the Atlantic highly migratory species) to develop FEPs as the mechanism to meaningfully integrate ecosystem principles, goals and policies into species-/species-complex-based FMPs.

In 2005, the Council recommended restructuring its five FMPs into FEPs for the pelagic, archipelagic and remote island areas in the Western Pacific Region. To help inform this reorganization, Council staff conducted three ecosystem-based fishery management workshops covering the biophysical sciences in 2005, social sciences in 2006 and management and policy in 2007. The proceedings of the workshops were made available in 2008 and were published as a book by Wiley-Blackwell in 2011 with assistance from the Council (Glazier 2011).

Participants of the Biophysical Workshop pose outside the Council office in Honolulu. The workshop held in 2005 would be the first of three workshops hosted by the Council to help inform the restructuring of its FMPs into FEPs. *WPRFMC photo.*

The Council's FEPs for the Pacific Pelagic, Hawaiʻi Archipelago, American Samoa Archipelago, Mariana Archipelago (Guam and CNMI) and Pacific Remote Island Area fisheries were approved by the Secretary of Commerce in September 2009 and codified in 2010. They incorporated and replaced the Council's existing Pelagic, Bottomfish and Seamount Groundfish, Crustaceans, Precious Corals and Coral Reef Ecosystems FMPs and reorganized their associated regulations into a place-based structure.

The objectives of the FEPs were as follows:

1. To maintain biologically diverse and productive marine ecosystems and foster the long-term sustainable use of marine resources in an ecologically and culturally sensitive manner

377

through the use of a science-based ecosystem approach to resource management.

2. To provide flexible and adaptive management systems that can rapidly address new scientific information and changes in environmental conditions or human use patterns.

3. To improve public and government awareness and understanding of the marine environment in order to reduce unsustainable human impacts and foster support for responsible stewardship.

4. To encourage and provide for the sustained and substantive participation of local communities in the exploration, development, conservation and management of marine resources.

5. To minimize fishery bycatch and waste to the extent practicable.

6. To manage and co-manage protected species, habitats and MPAs.

7. To promote the safety of human life at sea.

8. To encourage and support appropriate compliance and enforcement with all applicable local and federal fishery regulations.

9) To increase collaboration with domestic and foreign regional fishery management organizations, other governmental organizations, NGOs, communities and the public at large to successfully manage marine ecosystems.

10. To improve the quantity and quality of available information to support marine ecosystem management.

The organizational structure for developing and implementing the FEPs explicitly incorporated community input and local knowledge into the management process. To enhance this process, the Council established not only the REACs in each of the island areas under its jurisdiction but also a Marine Planning and Climate Change (MPCC) Committee. It continued to convene meetings of its Protected Species Advisory Committee and Social Science Planning Committee. It also reorganized its Standing Committees, Advisory Panels and Plan Teams to align with the place-based nature of the new FEPs; increased communication and collaboration efforts with fishing, indigenous and international communities; and strengthened its outreach and education efforts, including the establishment of an Education Committee.

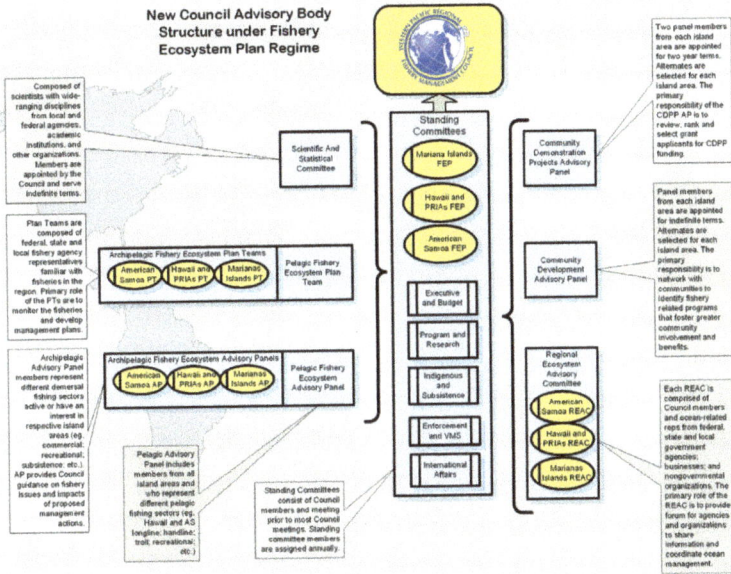

This chart outlines how the Council in the early 2000s restructured its advisory bodies to support the move from species-based FMPs to place-based FEPs. *WPRFMC illustration.*

The Council's Advisory Panel members gathered in Honolulu in 2003 to be introduced to the restructured Council organization that groups them into place-based subpanels rather than species-based subpanels. *WPRFMC photo.*

In 2014, the Council began its five-year review of the FEPs, which included a series of community meetings. The Council has been working with NMFS to finalize the updated FEPs, which contain refined objectives and enhanced ecosystem sections. However, to this day the updates have not been completed because NMFS has not made it in a priority.[74]

A key responsibility of the Pelagic and Archipelagic Plan Teams is the development of annual Stock Assessment and Fishery Evaluation (SAFE) reports for each of the FEPs. The annual reports for each FEP since 2015 have included a chapter that monitors ecosystem considerations such as protected species, climate change, marine planning and socioeconomic factors. The archipelagic annual reports also monitor coral reef ecosystems (i.e., reef fish biomass and habitat condition), life history and length-derived parameters. Each annual

[74]   Simonds 2021 op. cit

SAFE report also includes a chapter that integrates fishery and ecosystem data into efforts to better understand and predict changes in the fishery. The public has online access to both the annual reports and the data used to produce them. Additional information about the Council's restructuring its management into FEPs can be found in the Council's *Pacific Islands Fishery Monographs* issue 12 (Martell and Spalding 2020).

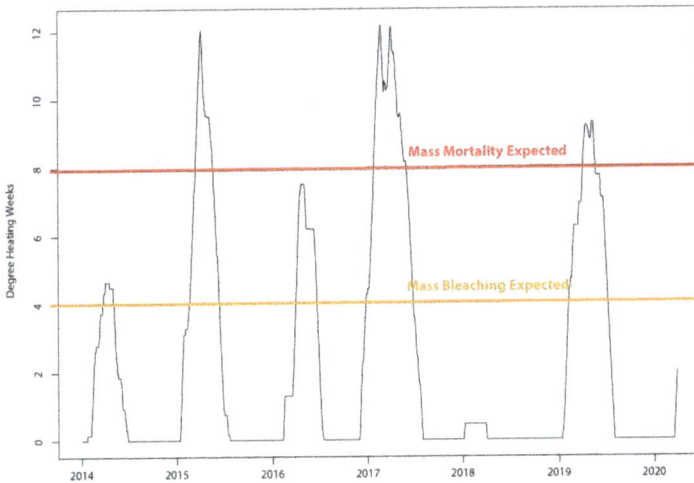

The Council's annual SAFE reports monitor a variety of climate factors. The graph from the 2019 report for the American Samoa Archipelago FEP indicates that prolonged high temperatures associated with mass coral bleaching occurred five times between 2014 and 2019. *WPRFMC 2020.*

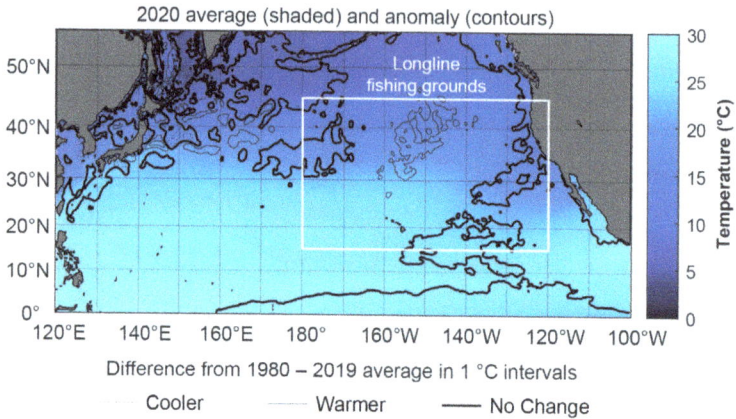

Among the climate factors monitored in the annual SAFE report for the Pelagic FEP is the monthly average sea surface temperature over the fishing grounds targeted by the Hawai'i longline fleet (white square). This graph from the 2019 report shows pockets of water northeast of the Hawaiian Islands that have been warming relative to average temperatures from 1985 through 2018. *WPRFMC 2020.*

In 2015, the NMFS Office of Sustainable Fisheries reviewed 10 FEPs that had been developed by four regional fishery management councils, comparing them against the eight recommendations by the EPAP in 1999. It noted that the FEPs for the Western Pacific Region were "unique from those created by the other councils because they are also full FMPs" (Wilkinson and Abrams 2015).

10.2 Developments with Regionwide Impacts on Fisheries

While the FEPs address fisheries at a place-based level, some issues during the first decade of their implementation have had an impact on multiple fisheries in the region.

Annual Catch Limits—Addressing Data-Poor Fisheries

The 2006 reauthorized MSA required that all federally managed species (with a few exceptions) be managed by ACLs and accountability measures. The exceptions are species managed internationally, ecosystem component species and species with a one-year or less lifespan. To establish realistic ACLs, fishery managers and scientists need information about fishing catch and effort, the status of the stock and the life history of the species. Nearly 800 species are taken in the Council's managed fisheries, and fundamental information was lacking for nearly all of them. In response, the Council established the Fisheries Data Collection and Research Committee to replace the Fisheries Data Coordinating Committee so as to better integrate fishery monitoring and research programs in the Western Pacific Region.

Beginning in 2010, the Council and its SSC worked to identify which species could be exempted from the ACL mandate and ranked the other species according to the quality of the data and stock assessment information available for them. The Council at first attempted to bundle species into complexes to reduce the number of ACLs that had to be generated each year through a detailed scientific methodology. Despite such bundling, the Council still had to generate 115 ACLs for the region. In 2017, the Council decided to move most of the species into the ecosystem component category where they would continue to be monitored but would not be required to have maximum sustainable yield, optimum yield, ACL and EFH specified. This action would allow the Council, NMFS and local fishery managers to focus resources on the most important fishery species.

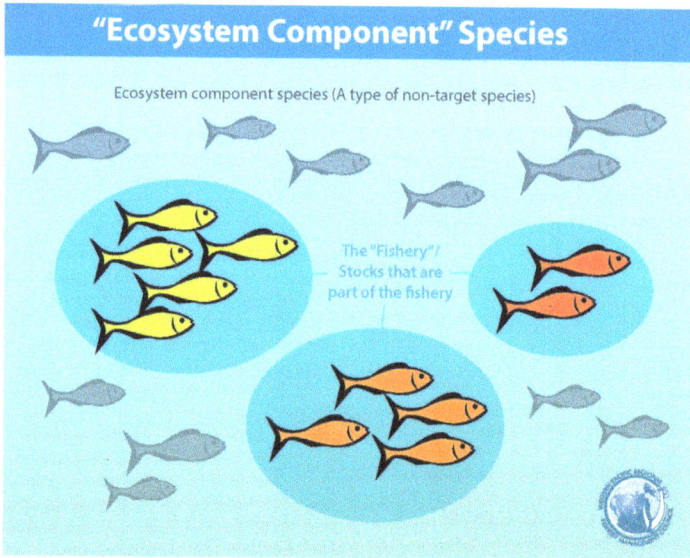

Fundamental information is lacking for most of the nearly 800 species taken in the Western Pacific Region, leading the Council in 2017 to move most of the species into the ecosystem component category where they continue to be monitored but are not be required to have maximum sustainable yield, optimum yield, ACL and EFH specified. *WPRFMC illustration.*

At the same time as these categorization efforts were taking place, the Council continued to improve fisheries data collection in the region. A series of workshops were convened in 2011, including the Workshop on Establishing ACLs for Coral Reef Fisheries, the Hawai'i Noncommercial Data Workshop and the Workshop on Improving Data Collection. The results of the latter workshop were presented to the Council in March 2012, along with an overview of the region's data-limited situation and the lack of funding to support fishery-dependent data collection. NMFS Deputy Assistant Administrator for Regulatory Programs Samuel Rauch III, the person in charge of proper implementation of ACLs, was in attendance at the meeting. As a direct result, NMFS in 2013 created the Territorial Science Initiative to increase locally based science, build local scientific and monitoring capabilities and enhance fisheries science capacity in the U.S. territories in the Pacific and Caribbean. Up to that point, funding sources for creel surveys and other data collection projects in the region had been through the SFF and NOAA Coral Reef Conservation Program. Later funds would also come from Marine Recreational Information Program, a state-regional-federal partnership. In 2014, a three-day workshop resulted in a Regional Strategic Plan to Improve Fishery Data Collection. The Council, NMFS and local fishery agencies subsequently supported projects to capture data from fisheries with seasonal runs and from vendors and fishermen cooperatives.

In spite of these many efforts, the 2019 Territorial Bottomfish MUS Benchmark Stock Assessment released by NMFS found the American Samoa fishery to be subject to overfishing and the American Samoa and Guam stocks to be overfished. It is generally accepted that these assessments reflect the lack of accurate and complete data for the fisheries. The Council and NMFS agreed that a review of the territorial data collection elements was paramount (Sabater 2021), and, in August 2019, they convened the Pacific Island Fisheries Monitoring and Assessment Planning Summit. Among the recommendations was the requirement of mandatory reporting for all fisheries, mandatory licensing for fishermen and vendors, the promotion of electronic reporting and linking catch and dealer reports on a real-time basis. To reach these goals and in collaboration with the Pacific Islands Fisheries Science Center and territory fishery agencies, the Council developed the Catchit Logit progressive web application, which allows fishermen and vendors to provide accurate, timely data tracked against applicable ACLs.

In August 2019, the Council and NMFS convened the Pacific Island Fisheries Monitoring and Assessment Planning Summit in Honolulu to review fishery data collection in the U.S. Pacific territories. *WPRFMC photo.*

In efforts to meet the recommended mandatory reporting for all fisheries, promotion of electronic reporting and linking catch and dealer reports on a real-time basis, the Council collaborated with the Pacific Islands Fisheries Science Center and territory fishery agencies to developed the Catchit Logit web application, which allows fishermen and vendors to provide accurate, timely data tracked against applicable ACLs. *WPRFMC illustration.*

The owner of MJ Fishing in Saipan transcribes daily purchase receipts into the Catchit Logit app. *WPRFMC photo.*

Safeguarding Traditional Fishing Practices— Customary Exchange, Green Sea Turtles

Presidential proclamations establishing marine national monuments in 2009 prohibited commercial fishing in U.S. waters out to 50 nm surrounding Rose Atoll in American Samoa and three Northern Islands (Uracas, Maug and Asuncion) in the CNMI. However, the proclamations allowed the possibility of noncommercial fishing. The distance of these monument waters from the populated islands where fishermen live meant only wealthy individuals would have the discretionary funds needed to fish in the monuments recreationally. Moreover, it would be culturally inappropriate for local fishermen to spend the time and money to travel hundreds of miles merely to eat fish for themselves and not bring some back to share with their family. By law, the Council is provided the opportunity to draft the fishing regulations for monument waters. In doing so, the Council and the SSC in 2010 defined "customary exchange" as a way by which local fishermen could continue to access the monument waters by allowing some reimbursement for fuel costs without a profit or "commercial" motive.

The concept and practice of customary exchange explains why fishermen in the Western Pacific Region traditionally fished and why giving and sharing fish is so important to them. It shows that the MSA's clear distinction between purely commercial fishing (defined as the sale, trade or barter of even one fish) and purely recreational fishing (without having a subsistence or cultural fishing category) creates a real problem for the Western Pacific Region, where many small boat fishermen fish in order to share and give fish at a variety of cultural and ceremonial occasions. Even if there are no ceremonial obligations, island fishermen share a significant portion of their catch with friends and neighbors with no real expectation of return. Those fish then flow through the community in ways that contribute to people's health and give people a sense of sharing and common identity. As a Samoan high chief said, "Fish is culture!"

The motivation to give fish relates to the role and reputation of fishermen in their community and to their social networks. There is no negotiation or expectation of return gifts. The value is in the perpetuation of social relationships in culturally valued ways. In Hawai'i, many people have social relationships with good fishermen and may turn to them when fish are needed for a first birthday *luau*, a graduation party, a wedding or other kind of cultural need. In American Samoa, fish is expected at a number of ceremonial occasions, and many high chiefs have boats and are served by a *tautai* (master fisherman) from their village who often fishes when there is ceremonial need. Atu (skipjack tuna) are still targeted and formally distributed in quarters at certain ceremonies including the one-year anniversary of a death and funeral. The largest fish landed on a small boat trip is given to the chief or high chief. In Guam and the CNMI, many important ceremonies provide cultural pride, solidarity and continuity for the indigenous Chamorro and Refaluwasch. Fish are a desired and expected contribution at the annual village fiestas, baptisms, confirmations, weddings and other special ceremonies. The giving of fish to one's clan members enhances the fisherman's reputation but gives no measurable or immediate economic return.

Fisheries anthropologists call this "generalized reciprocity," a form of nonmarket, noncommercial exchange. Fisheries economists with mainland mindsets often have difficulty grasping this concept. They think in terms of capital investment, economic rent and profit from fisheries. Their analysis tools measure these things but cannot easily evaluate a value for good social relations and cultural continuity.

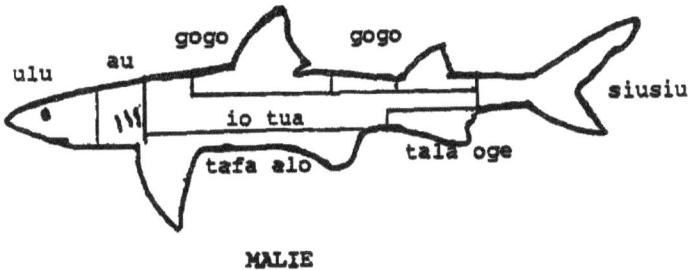

MALIE

Ceremonial cuts and distribution of fish in American Samoa differ depending on the species. For malie (sharks), the *gogo* and *io tua* portions go to the *ali'i sili* (minister); the *tafa alo* and *tala oge* to other ranking chiefs; the *siusiu* (tail) to the talking chief presiding over the ceremony; and the *ulu* (head) often to a young untitled man who distinguished himself on the *lepaga* (shark excursion) or in preparation of the feast and *'ava* ceremony. *Illustration from Severance and Franco 1989, with permission.*

Atulai (mackerel) are divided in equal piles for the traditional *patte* (division of sharing) at the Umatac Village mayor's office, Guam (circa 1970s). *Ramon Topsana photo.*

Other regionwide indigenous efforts undertaken by the Council have centered on the green sea turtle. The species has been of particular interest to the Council due to the cultural significance to island populations and the fact that there are breeding sites in areas in the Council's jurisdiction. The Council convened a Mariana Green Turtle Workshop in Saipan, CNMI, from January 25 to 27, 2011, during which 50 traditional leaders, cultural practitioners and federal and local sea turtle program partners from throughout the U.S.-affiliated Pacific Islands shared knowledge about the species. The meeting aimed to find a balance between cultural needs and modern-day recovery goals. Among the knowledge shared was the cultural importance of sea turtles as a symbol of peace, use in women's menstrual houses and use in navigation. Traditional practitioners know hundreds of turtle migratory paths, which modern technology is now confirming.

The Council also held a Fishers Forum on the Future of Honu [Hawaiian green sea turtle] Management in Honolulu in 2011, a video of which is available on the Council's website and on DVD.

Collaborating with Native Americans and Other
Indigenous Peoples

Other Council initiatives have aimed to support traditional fishing practices at not only the regional but also the national and international levels, which aligns with the ecosystem approach of collaborating with a wider array of partners.

On February 26, 2014, the Council hosted the American Indian and Indigenous U.S. Pacific Islander Roundtable on Fishing and Other Native Rights. Participants included the renowned environmental leader and treaty rights activists Billy Frank Jr. from the Nisqually tribe (State of Washington); Tapa'au Daniel F. Aga, dean and director of American Samoa Community College's Community and Natural Resources Division; Malia Akutagawa, assistant professor of law at the University of Hawai'i; Marie Alailima, attorney practicing in American Samoa; Peter Apo, Office of Hawaiian Affairs trustee; D. Ululani Beirne-Keawe, cultural resource consultant and planner; Mahina Paishon Duarte, president of the Paepae o He'eia Board; Michael Duenas, Guam fisherman; Michael Grayum, Northwest Indian Fisheries Commission executive director; Ed Johnstone, Northwest Indian Fisheries Commission treasurer; Ke'eaumoku Kapu, traditional fisherman and farmer; U'ilani Kapu, Ku 'Ikahi LLC president; Lorraine Loomis, Northwest Indian Fisheries Commission vice chair; Kepa Maly, Lana'i Culture and Heritage Center executive director; Felicidad T. Ogumoro, CNMI House of Representatives member; Kelson "Mac" Poepoe, Hui Malama o Mo'omomi founder; Begnino Sablan, Marianas Trench MNM Advisory Council chair and former Council member; Kitty M. Simonds, WPRFMC executive director; Ramsay R. M. Taum, cultural resource and Life Enhancement Institute of the Pacific founder; Mililani Trask, Native Hawaiian attorney; Richard Trudell, American Indian Lawyer Training Program founder; and Patricia Zell, a Navajo/

Arapaho and partner in Zell and Cox Law, specializing in laws affecting indigenous Indians, Alaska Natives and Native Hawaiians. Several Council staff members, Council SSC member Craig Severance and law student Shaelene Kamakaʻala attended as observers.

American Indians and indigenous U.S. Pacific Islanders met at the Council office on February 26, 2014, to discuss their mutual concerns about fishing and other native rights. Among the participants were (front from left) Felicidad Ogumoro, Lorraine Loomis, Kitty M. Simonds, D. Uluani Beirne-Keawe, Billy Frank Jr.; (standing from left) Mililani Trask, Ed Johnstone, Charles Kaʻaiʻai, Michael Duenas, Kelson "Mac" Poepoe, Kepa Maly, Uilani Kapu, Tapaʻau Daniel F. Aga (behind Kapu), Keʻeaumoku Kapu, Benigno Sablan, Mahina Paishon Duarte, Michael Grayum (behind Duarte), Patricia Zell, Ramsay Taum (behind Zell) and Marie Alailima. *WPRFMC photo/Sylvia Spalding.*

In 2019, the Council and the Ocean Networks Canada co-organized the indigenous delegation to OceanObs'19 in Honolulu. This decennial conference of 1,500 participants for the first time in its 30-year history included nearly 100 indigenous members. The conference sets the stage for ocean observation for the next 10 years and was particularly important in 2019 as preparations were underway for the United

Nations Decade of Ocean Science for Sustainable Development (2021–2030). The indigenous delegation presented a declaration entitled 'Aha Honua, asking the ocean observing community to establish meaningful relationships with indigenous communities; to learn and respect each other's ways of knowing; to negotiate paths to design, develop and carry out ocean observing initiations; and to share ocean observing responsibilities and resources. It was noted that scientists in the past would often appear without notice in the waters of an indigenous community, conduct their research and then leave without sharing the results to the community or asking the community for their input.

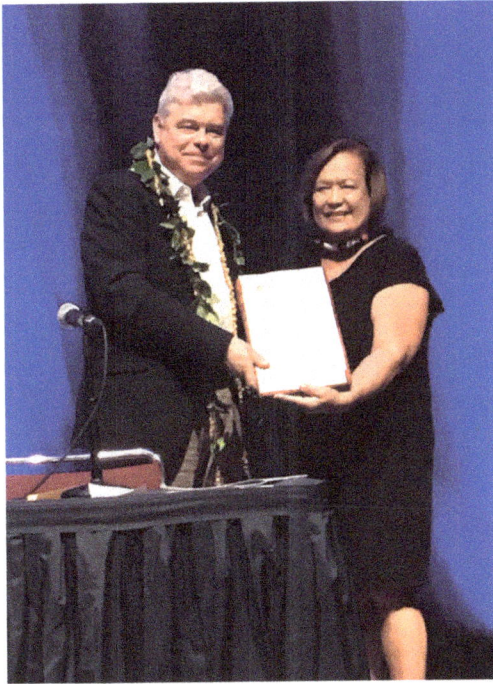

Council Executive Director Kitty M. Simonds presents the 'Aha Honua indigenous declaration to Vladimir Ryabinin, executive secretary of the United Nations. Intergovernmental Oceanographic Commission, at the OceanObs'19 conference. *WPRFMC photo/Sylvia Spalding.*

Applicable Laws—Listing of Corals under the Endangered Species Act

About the time that the FEPs were implemented, the Council faced a new set of challenges with significant potential to affect federal fisheries management. Several environmental NGOs petitioned NMFS to list new marine species as threatened or endangered under the ESA, including the Hawai'i insular population of false killer whales, bumphead parrotfish, humphead wrasse, several species of hammerhead and thresher sharks, eight species of pomacentrid reef fish (clownfish and damselfish) and 83 species of reef-building corals.

When NMFS receives a petition to list new species, it goes through a process set out under the ESA to review and determine whether the petitioned action is warranted. Within 90 days of receiving a petition, NMFS must make an initial determination of whether the petitioned action (e.g., listing a species) *may* be warranted (called "90-day finding"), based on the information presented in the petition and other readily available information. If the petition makes it through this initial stage, NMFS conducts a full status review and is required to make a finding within one year of the date the petition was received. If NMFS finds that the petition is not warranted, the petition is denied and the process concludes. If NMFS finds that a listing is warranted, a proposed rule will be issued and public comments are solicited. NMFS must then issue a final decision within the next 12 months.

With each new petition process, the Council turned to its extensive group of advisors to assist with reviewing information contained in the petition as well as in the associated status reviews produced by NMFS. The Council often found that the petition contained limited scientific information in regard to abundance, distribution and population trends. Furthermore, the timeline associated with the status reviews meant that

NMFS relied on available information rather than being able to conduct new studies. As a result, NMFS's decisions were often based on less than complete scientific data.

One of the most controversial petitions was the one requesting the listing of 83 species of reef building corals, most of which were found in the Pacific Ocean. Submitted in October 2009, the petition focused on potential climate change impacts to corals but contained very little data on the abundance, distribution and status trends for each species. Yet, NMFS announced a 90-day finding in February 2010 and proceeded to conduct a status review for 82 of the species.

Acropora retus is a species of coral found in waters surrounding Guam, American Samoa and the U.S. Pacific remote islands. It is one of the 83 species of coral that NGOs petitioned in 2009 to be listed under the ESA. *NOAA photo.*

In an unprecedented step, however, NMFS released the status review on the 82 species in April 2012, in advance of its proposed decision and held listening sessions and scientific workshops to gather more information from the public and outside experts. The Council's analysis of the status review noted that substantial scientific information had not been considered and raised questions about the extinction risk assessment that relied heavily on expert opinions due to the lack of species-specific information.

In December 2013, NMFS issued a proposed rule to list 66 of the 82 coral species as threatened or endangered under the ESA. After conducting a thorough review of the proposed rule package, the Council identified further concerns. For example, the determination tool developed by NMFS for the purpose of this decision process was biased toward listing. Further, misidentification of some coral species had likely resulted in inaccurate distribution ranges, and there was a general lack of quantitative abundance data for Indo-Pacific coral species, suggesting that a decision to list these species was premature. The proposed rule was also met with a surprising amount of opposition from the scientific community, which held that the decision was based on poor information.

Following the proposed rule comment period, the Council learned that NMFS had not considered a large volume of scientific information being compiled by the world renowned coral researcher John "Charlie" Veron. The Council requested that NMFS extend the deadline for final rule publication to consider the new information and subsequently worked in partnership with Veron and the Pet Industry Joint Advisory Council to make available substantial scientific information previously not considered by NMFS, submitting the information package in February 2014.

This new information was incorporated into NMFS's decision-making process. Subsequently, NMFS's final rule, published on September 10, 2014 (79 FR 53851–54123), listed 20 species under the ESA as threatened and no species were listed as endangered.

The Council continues to provide scientific review of ESA-related actions and to submit comments to ensure the best available information is considered by NMFS. The Council's contribution to the ESA process is essential, given that the MSA requires the Council to comply with other applicable laws, including ESA.

National Ocean Policy—Preparing Communities for Marine Planning

President Barack Obama on June 19, 2010, issued Executive Order 13547, *Stewardship of the Ocean, Our Coasts, and the Great Lakes.* The order adopted the *Final Recommendations of the Interagency Ocean Policy Task Force* and directed executive agencies to implement those recommendations as the National Ocean Policy. A third of the Task Force document addressed coastal and marine spatial planning (CMSP), a tool with roots in terrestrial land use planning.

To prepare U.S. Pacific Island communities for CMSP, the Council sponsored a training workshop in Honolulu from July 31 to August 4, 2011, bringing together 125 members of the indigenous and fishing communities of Hawai'i, Guam, CNMI and American Samoa. The training curriculum was provided by planning and resource management professionals and was led by Anne Walton of the NOAA National Marine Sanctuaries International Program. Assisting in the training workshop were the Hawaiian Islands Humpback Whale National Marine Sanctuary and HDAR.

Participants were taught how to use the CMSP process by developing mock plans for selected areas based on a mixture of real and made-up information that included historic uses of the ocean (fisheries, transportation and recreation) and contemporary factors, such as sewage outfall, research activities, tourism, MPAs, military training, aquaculture and renewable energy facilities. The exercise helped participants to understand the complexities involved in conducting planning activities and to realize that the quality of planning depends on the information available, including intergenerational traditional knowledge. The workshop reinforced the vital importance for government planners to engage with local and indigenous communities and ocean user groups in their CMSP pursuits.

Following on the regionwide workshop, the Council held community workshops and teacher trainings on CMSP in American Samoa, Guam and the CNMI.

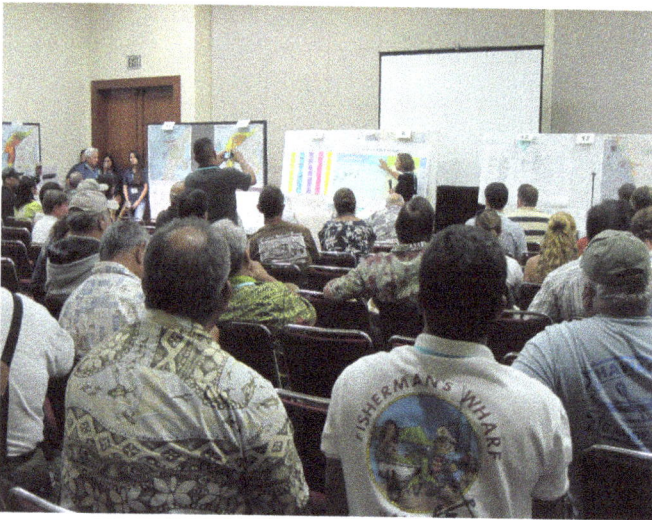

Anne Walton from the NOAA Office of National Marine Sanctuaries' international office leads the Council's training workshop to help prepare communities for the national CMSP initiative. More than 100 indigenous and, community members and fishermen from throughout the U.S. Pacific Islands attended the July 31 to August 4, 2011, workshop in Honolulu. *WPRFMC photo.*

The Community-Based Management Plan for the Malesso Coastal and Marine Resources is the outcome of the dedicated efforts of Ernest Chargualaf, mayor of Merizo (third from left), community members, contracted facilitators and Council staff Mark Mitsuyasu (far left) and Charles Kaʻaiʻai (third from right). Workshops to develop the plan were held in August and November 2013. *WPRFMC photo.*

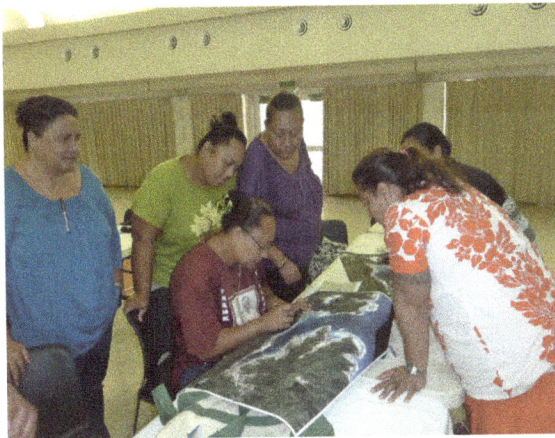

Classroom teachers in American Samoa map out where they will take water samples to monitor the health of Pago Pago bay. The exercise was part of The Sustainable Science Teachers Workshop: Coastal Water Quality Monitoring, held March 9, 2013, in Utulei and co-organized by the Council, NOAA OceanWatch and the American Samoa DMWR. *WPRFMC photo/Sylvia Spalding.*

To fulfill other recommendations of the National Ocean Policy, the Pacific Islands Regional Planning Body was formed in April 2013. The Council held a seat on the planning body, which was charged with developing a marine plan for the region. In 2015, the planning body decided to create one plan per jurisdiction, similar to the Council's FEPs. It envisioned these plans as roadmaps to facilitate coordination among all the local and federal agencies having mandates within the EEZ from 3 to 200 nm from shore, as well as those that manage shoreline to from 0 to 3 nm from the shore.

In 2015, the Council adopted its MPCC Policy, drafted by the Council's MPCC Committee, to help the Council coordinate development and amendment of its FEPs, programs and other relevant activities. The MPCC Policy recognizes a set of overarching and specific principles and policies for consideration by the Council, its advisory bodies and its staff when developing FEPs. The MPCC Policy recognizes marine planning as an appropriate approach to reconciling intersecting human uses, ocean resources and ecosystem health at multiple geographic scales. It also recognizes that traditional resource management systems, such as the *ahupua'a* system in Hawai'i and *fa'a Samoa* in American Samoa, can provide an appropriate context for marine planning. The policy states that MPAs can and should be used as a tool in marine planning for climate change reference and human use and impact research. The MPCC Committee disbanded in 2019 as key members became occupied in the restructured Plan Teams and other Council committees.

In 2012 and 2014, the Council was instrumental in the organization of two First Stewards symposia, Coastal Peoples Address Climate Change and United Indigenous Voices Address Sustainability: Climate Change & Traditional Places, respectively. The Council supported the participation of U.S. Pacific Islanders in these symposia and associated

Living Earth Festivals at the National Museum of the American Indian, Washington, D.C. For the 2012 events, the Council produced the video *Little Changes Have Big Impacts on Little Islands*. For both symposia, the Council held student essay contests on climate change related topics in the region and provided the winners from each island area an opportunity to present their winning essays at the nation's capital. Climate change was also the topic of student art contests held throughout the region, for which the Council developed lesson plans to encourage teachers to include the contest as part of their science, art and/or cultural studies curricula. The Council also engaged the American Samoa Community College's Samoan Studies Institute to do research on traditional knowledge about climate and weather in the Pacific and brought the student researchers and the director of the Samoan Studies Institute to Washington, D.C., to present their work.

Okenaisa Fauolo, director of the American Samoa Community College's Samoan Studies Institute, and her students present their Council-contracted research on traditional stories about weather and climate at the Living Earth Festival in Washington, D.C., July 18–20, 2014. *WPRFMC photo/Sylvia Spalding.*

Enhancing Financial Support for the Territories' Marine Conservation Plans

As noted previously, the 1996 MSA reauthorization (i.e., the SFA) addressed concerns by the U.S. Territories that, unlike other Pacific Island countries, they could not license foreign vessels to fish in the waters surrounding their islands because they were not independent of the United States. As the United States had not implemented any agreements for foreign fishing for federal waters surrounding the U.S. Pacific islands, the territories were not benefitting from this potential revenue source. The SFA attempted to address this concern. It required that the governor of each U.S. Pacific Island Territory develop a three-year MCP that identified how funds from any Pacific Insular Area Fishing Agreement would be utilized, and it created the SFF where funds from any such fishing agreement and other specific sources would be deposited to implement the MCPs through the Council. Sources could include, but were not limited to, fines for illegal foreign fishing in the U.S. EEZ around the Pacific Remote Island Areas and specified fishing agreements between a U.S. territorial government and U.S. fishing vessels that hold a valid permit authorized by the Pelagic FEP. Amendment 7 of the Pelagic FEP, implemented in 2014, established a management framework and process for specifying fishing catch and effort limits and accountability measures for pelagic fisheries in American Samoa, Guam and the CNMI and a framework to authorize the government of each territory to allocate a portion of its specified catch or effort limit to a U.S. fishing vessel or vessels through a specified fishing agreement. Amendment 7 also established criteria that a specified fishing agreement must satisfy.

Since implementation of Amendment 7, the funds available to the U.S. Pacific territories to fulfill their MCPs have increased significantly. In the early years, payments made by the U.S. vessels to the territories were calculated on an as-used basis, but in recent years they have been paid in advance, whether the amount agreed to is used or not. The Council can also use the funds in the SFF from foreign fishing fines in the Pacific Remote Island Areas for projects in Hawai'i.

Quota Management Inc. President Khang Dang (center) observes while Hawai'i Longline Association President Sean Martin and CNMI Governor Ralph Torres shake hands at the April 2016 signing of an agreement to transfer 1,000 mt of CNMI's annual quota of bigeye tuna for use by the Hawai'i longline fleet. *Edwin Ebisui Jr. photo.*

TABLE 12: TRANSFER OF TERRITORY LONGLINE BIGEYE TUNA QUOTA THROUGH SPECIFIED FISHING AGREEMENTS		
YEAR	TRANSFER AMOUNT (metric tons)	TERRITORY
2011	567	American Samoa
2012	815	American Samoa
2013	492	CNMI
2014	1,000	CNMI
2015	1,000	CNMI
	831	Guam
2016	884	CNMI
	939	Guam
2017	1,000	CNMI
	700	American Samoa
2018	1,000	CNMI
	100	American Samoa
2019	1,000	CNMI
	1,000	American Samoa
2020	1,000	CNMI
	1,000	American Samoa

COVID-19 Impacts to the Region's Fisheries

The coronavirus (COVID-19) pandemic had a devastating impact on the fisheries in the Western Pacific Region, where the economies are tied to tourism and pan-Pacific transportation, both of which were curtailed.

NOAA Fisheries released a National Snapshot, January–July 2020, that summarized the impacts of COVID-19 on U.S. fisheries. It noted that tourism in the U.S. Pacific Islands declined by 99% in April through July relative to 2019, which significantly impacted both the seafood sector and for-hire operations. In Hawai'i, commercial landings revenue declined 42% ($22 million) relative to the five-year baseline for the months of March through July, with the Hawai'i longline fishery incurring a 45% loss in landings revenue for this period. The Hawai'i charter industry was effectively closed since mid-March, with an estimated 99% decline of reported charter fishing trips for April through July relative to the baseline.

Longline vessels tie up at Honolulu harbor in March 2020 as restrictions related to COVID-19 cause the demand and prices for fresh fish to plunge. *WPRFMC photo.*

Similar to Hawai'i, travel restrictions and sharp declines in tourism had significant impacts on island economies and communities in American Samoa, Guam and the CNMI. The StarKist Samoa cannery operated at full capacity, but keeping up with demand was challenging due to COVID-19 restrictions.

If there was a bright side to the pandemic, it was the recognition of the importance of locally caught seafood not only as a commodity but also as essential food and protein for the islands and the nation. Historically, the importance of fisheries compared to agriculture has been overlooked, which was evidenced in the funding given to the two sectors through the Coronavirus Aid, Relief and Security Act. The revenue of the U.S. fishing industry in 2019 was $10.9 billion, and the Act appropriated $300 million in funds to assist affected fishery participants. By comparison, America's farms generated $132.8 billion in revenue and were provided $16 billion in direct payments to producers and more than $1.7 billion in commodity purchases for distribution to food banks and faith-based organizations. As Council Chair Taotasi Archie Soliai and Executive Director Simonds noted in the Spring 2020 issue of the Council's newsletter, "Put simply, while both fisheries and agriculture feed our nation, the fisheries sector is receiving COVID-19 relief equal to about 3% of its revenue while the agriculture is receiving relief greater than 16% of its revenue" (pages 1–2).

Taotasi Archie Soliai has served as the Council chair since 2018 and as a Council member since 2016. He resigned from his managerial position at StarKist Samoa in October 2020 and became director of the American Samoa DMWR in January 2021. *WPRFMC photo.*

As the pandemic continued and meat-packing plants on the U.S. mainland became COVID-19 hot spots, local governments in Hawaiʻi began to mention fisheries in their sustainability discussions. For example, Honolulu Mayor Kirk Caldwell announced a new "fish to dish" program, stating that "Hawaiʻi's longline fishermen provide a valuable source of food to our island, and fortifying this industry not only provides our community with some of the freshest fish in the world, but sets a sustainable network to solidify our food security ahead of future disasters" (Wu 2020). The city committed $2.6 million to help the industry with $1.7 million going to the HLA and $275,000 earmarked for the Honolulu fish auction, run by the United Fishing Agency, to upgrade the facility.

This was a long overdue recognition given that the State imports about 85% to 90% of the food consumed on island (Leung and Loke 2008, p. 2) and that fish tops the list of locally produced food in value, generating more than $100 million ex vessel annually, with the majority staying on island (see Table 13). Nonetheless, the State's "Increased Food Security and Food Self-Sufficiency Strategy," released in 2012 by the Hawai'i Department of Business Economic Development & Tourism, had focused entirely on agriculture, mentioning "fish" only once, in reference to food safety certification assistance.

TABLE 13: TOP 10 HAWAI'I FOOD CROPS, 2017
(in millions $ at farmgate or dockside)

Sources: USDA Annual Statistics Bulletin; FEP Annual SAFE report 2017.

Note: Hawai'i crop seed production valued at $120.8 million in 2017.

FOOD	$ MILLIONS	KEPT IN HAWAI'I
Hawai'i commercial fish landings	120	80%
Macadamia nuts	54.9	NA
Cattle	43.9	24%
Coffee	43.7	NA
Aquaculture	41.7	NA
Algae	35.2	0%
Papayas	9.4	50%
Milk	9.2	100%
Lettuce	8.7	100%
Bananas	6.1	100%

John Kaneko, Hawaii Seafood Council program manager (2nd from right), gives Christos Michalopoulos, NOAA Office of Education deputy director, a personal tour of Honolulu fish auction on October 7, 2010. Prior to the COVID-19 pandemic, more than $100 million of fish ex-vessel would be landed annually in Honolulu, consistently ranking it as one of the nation's top 10 ports based on the value of landed seafood. *WPRFMC photo.*

Enhancing Outreach and Education Efforts

With the call for a wider array of partnership collaborations under ecosystem-based management, the Council enhanced its outreach and education efforts. In 2010, it augmented its regular publications to include the *Pacific Islands Fishery Monographs* series; each issue provides in-depth coverage of a specific topic. The Council also began publishing an annual

Western Pacific Region Status of the Fisheries, a reader-friendly summary of the Council's annual SAFE reports, beginning with the 2016 reports.

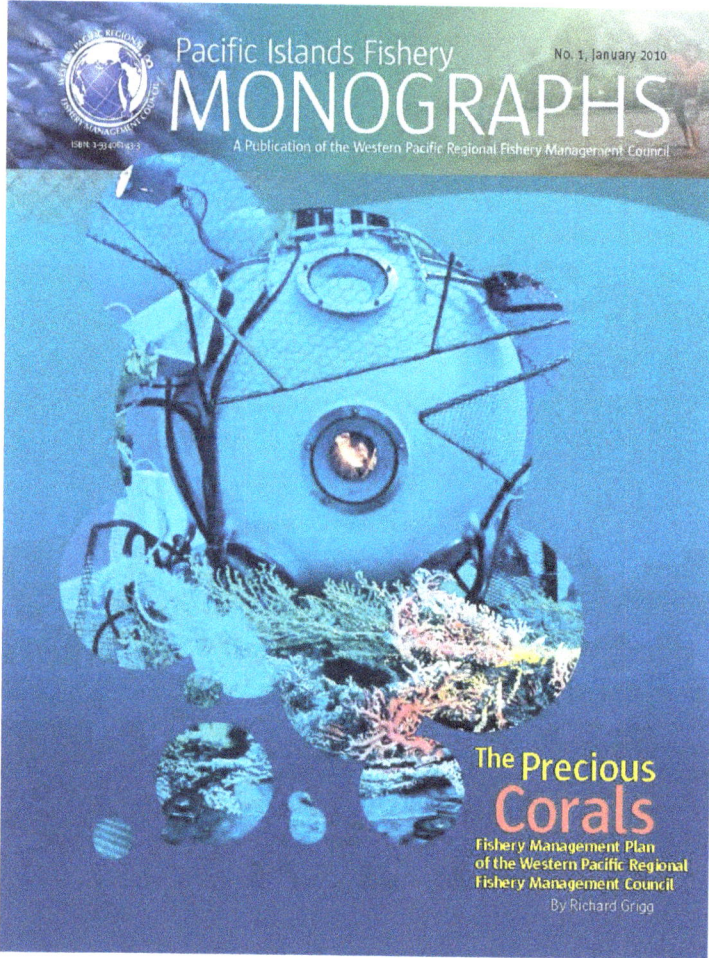

The inaugural issue of the *Pacific Islands Fishery Monographs* featured Council management of the precious coral fishery. The monographs are published regularly to provide an in-depth, historic overview of a specific regional fishery topic. *WPRFMC illustration/D. Doubilet photo.*

The Council continued its long history of broadcasting videos, with episodes airing on *Let's Go Fishing, Hawaii Goes Fishing* and 'Olelo community television in Hawai'i and on the government station in

American Samoa. *Mythbusting* (2015) and *Traditional Navigation and Seamanship* (2015) were among the video series produced, while longer, stand-alone videos aired in multiple venues included *Little Changes Have Big Impacts on Little Islands* (about climate change impacts on indigenous fishing communities in the Western Pacific Region) (2012); *'Ahi, the Yellowfin Tuna: Managing Our Fisheries* (2014); *Shutdown—The Effects of Arbitrary Bigeye Tuna Quotas on Hawai'i's Fishing Community and Consumers* (2015); *From Boat to Table* (2016); and *Kaulana Mahina (Hawaiian Lunar Calendar)* (2018), among others.

The Council co-sponsored the *Go Fish with Mike Buck* radio talk show on KHNR in Hawai'i and a local talk show on KKMP in the CNMI covering fishery issues in English, Chamorro and Refaluwasch.

Another highly successful, regional outreach effort was the Fishers Code of Conduct, developed by the Council from the wisdom shared by *kupuna* (elders) and traditional practitioners in Hawai'i. At the request of various entities throughout the region, the code has been translated in multiple languages and has been made available on posters, on signs and in videos. It has also served as a non-regulatory management strategy to address conflicts between longtime local fishermen and newcomers to the islands from the U.S. freely associated states, as well as to educate visitors to the islands about acceptable conduct and relationship with the ocean and local marine resources. In 2014, the Council met with the Consul Generals of the Federated States of Micronesia and Palau to enlist support and cooperation in these endeavors.

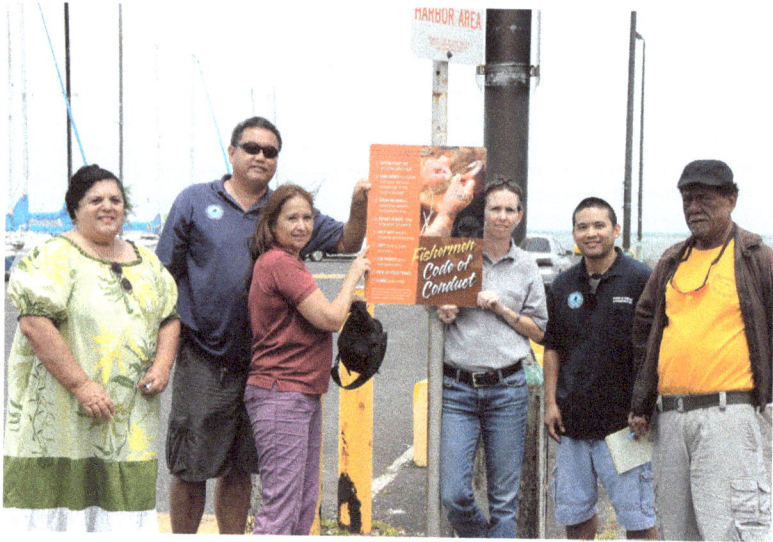

'Aha Moku representatives Leialoha "Rocky" and Jerry Kuluhiwa (far left and far right), harbormaster Ernie Choy, Council staff Sylvia Spalding, Hawai'i Division of Boating and Ocean Recreation's Meghan Statts and Mark Herron (harbor agent) post a Fishers Code of Conduct sign at He'eia Kea harbor in 2015. Participants of the Council-supported puwalu (conference series) leading to official recognition of the traditional 'Aha Moku system of natural resource management identified a code of conduct as a nonregulatory method of management. *WPRFMC photo.*

The Fishers Code of Conduct poster in Chamorro, one of the official languages spoken in Guam and the CNMI. The poster encourages nine non-regulatory behaviors and illustrates the final one—share your catch. *WPRFMC illustration.*

Additionally, the Council has held student symposiums on coral reefs for high school students, funded in part by the NOAA Coral Reef Conservation Program, and teacher workshops on a variety of fishery management related issues. The first teacher workshops were

413

held in Honolulu in partnership with the *Honolulu Advertiser* and its Newspaper in Education program, beginning in 2009. It focused on the Hawai'i seafood industry and federal management of the Hawai'i-based fisheries. Subsequent workshops held throughout the region included CMSP, water quality and climate change.

Lucas Moxey (right), operations manager of NOAA OceanWatch–Central Pacific, leads the Council's 2014 train-the-trainers workshop on geologic history, sea-level changes and water resources in Honolulu, with participants from throughout the region. *WPRFMC photo/Sylvia Spalding.*

In 2010, the Council partnered with the University of Hawai'i at Hilo and the American Samoa Coral Reef Advisory Group on an undergraduate marine science fellowship program that provided students from American Samoa the opportunity to pursue a marine science degree in Hilo. At the request of the Council, efforts were made in 2013 to expand the program to include students from Guam and the CNMI. Consequently, the Council formed an Education Committee, which resulted in an aspirational memorandum of understanding signed by the University of Hawai'i at Hilo, the Hawai'i Institute of Marine Biology, Hawaii Pacific University, University of Guam, American Samoa Community College, Northern Marianas College, the three territorial fishery agencies, the Council, the USFWS and the NMFS Pacific Islands Regional Office and Pacific Islands Fisheries Science Center to work cooperatively on a suite of measures to educate students from the territories about fisheries and marine science with the goal of having them be the future scientists and managers of their local marine resources. Among the objectives was the creation of the U.S. Pacific Territories Fishery Capacity-Building Scholarship Program. To date, the program has provided more than a dozen college students from the territories with funding for their upper-class and graduate education in fisheries-related marine science. The program requires that the students work for one year at their local fishery or marine-related agency for each year that they receive a scholarship. The program was so successful, that after five years, all of the parties signed an updated memorandum that also included the University of Hawai'i at Manoa and the Hawai'i Department of Land and Natural Resources. Added goals include vocational education and education about fisheries in the K–12 classrooms.

In 2011, select videos from the Council's *FishQuest* series were edited into a *Fish Forever* podcast series and augmented with additional videos with versions targeted to students and to fishermen. The videos are available on the Council's website, have aired on television and were distributed on DVDs to fishing clubs and teachers. Additionally, the lesson on fish habitat was converted into an interactive computer game, which has been used as part of the Council's informational booth at many community and educational events.

The Council's outreach efforts specific to Hawai'i include the following:

a) A series of workshops entitled "What's in Store for Hawaii Fisheries?" held throughout Hawaiian Islands in 2010 in ongoing efforts to engage communities in the management of fisheries through an ecosystem-based approach;

b) Development of the video *Fishing Management and Us* and support for Roy Morioka to present the video and lead a discussion on it with boat and fishing clubs throughout the state in efforts to engage more fishermen in the federal fishery management process;

c) Cultural tents at the annual Hawaii Seafood Festival;

d) Development and updating of a Hawai'i speakers bureau; and

e) Development of a highly successful internship and student research partnership with Hawaii Pacific University.

In American Samoa, the Council's outreach efforts have included a meeting on sustainable fisheries development and ecosystem management with the pastor, *pulenu'u* (village mayor), residents and young students in Matu'u and Faganeanea communities in 2011 and a similar meeting with more than 50 mayors in Utulei. The Council held a series of Fishers Forums, including one on Fishing: Food, Life, Future that featured a high school exhibit competition in 2017 and an outdoor Fishers Forum Day that included tours of longline, purse-seine, bottomfish and alia boats at the Port Authority docks.

As for education efforts, the Council hosted a teacher workshop in 2013 on water quality monitoring and co-sponsored a 2015 course for Manu'a students to learn about Muliava (Rose Atoll). Partners included the American Samoa DMWR, the Council and NOAA

OceanWatch–Central Pacific with cooperation from American Samoa Department of Education, American Samoa Community College's Samoan Studies Institute and the Manu'a District governor and paramount chief, Misaalefua Hudson. Funding for the Manu'a course came from a NOAA Marine National Monument grant.

Council outreach and education efforts specific to the Mariana Archipelago since implantation of the FEPs included Fishers Forums and student coral reef symposiums on Guam and Saipan in 2010; community meetings with villages in Saipan, Tinian and Rota; teacher workshops on monitoring local watersheds in Guam and the CNMI in 2012; and support for the annual *Gupot Fanha'aniyan Pulan CHamoru* (Chamorro Lunar Calendar Festival) and annual Mariana Islands Fishing and Seafood Festival.

Chapter 11

Pacific Pelagic Fishery Ecosystem

Highly migratory pelagic fisheries dominate the Western Pacific Region in terms of value and contributions to the nation. The Hawai'i pelagic fishery, for example, has an ex-vessel value of $120 million (2017) and continually places Honolulu as one of the nation's top 10 fishing ports in terms of the value of seafood landings. It is also the premier food producer in the State of Hawai'i, followed by macadamia nuts, cattle and coffee, each at less than half of the value of the fishery (see Table 13). The Hawai'i fishery produces 80% of the nation's domestically caught bigeye tuna, 55% of its swordfish and 65% of its yellowfin, and it accounts for 60% of the total U.S. bigeye market, 14% of the U.S. swordfish market and 4% of the U.S· yellowfin market (WPRFMC 2019, page 3). The Western Pacific Region accounts for approximately half of the entire 4.4 million square mile (11.4 million square kilometer) U.S. EEZ, which is the second largest EEZ globally after the French EEZ. The fish from the Western Pacific Region contributes to the overall U.S. seafood. However, this represents only a small fraction of the total as 70% to 85% of the seafood consumed by Americans by weight is imported (NOAA FishWatch).

Three key challenges have kept the U.S. pelagic fisheries in the Western Pacific Region from reaching their full potential: access to grounds, access to quota and access to markets. These challenges exist through actions initiated by U.S. presidents, international regional fishery management organizations and the U.S. Congress during the second decade of the 21st century. An additional challenge involves illegal, unreported and unregulated fishing by foreign fleets, which has a large impact on the region. Another issue is forced labor. An Associated Press article alleged some Hawai'i longline vessel owners/operators were involved in such activity (Mendoza and Mason 2016), but a Congressional investigation exonerated the U.S. fleet of such wrongdoing (U.S. DOJ 2021). The Council addressed each new challenge as it arrived and continued its efforts to help the U.S. Pacific Islands to develop and maintain their sustainable fisheries through fishery and community development efforts and outreach and education initiatives.

Marine National Monument Claim 52% of the U.S. Waters in the Region

Under the influence of what had become a continuous lobbying effort by NGOs, President Obama expanded the Pacific Remote Islands MNM boundaries around Wake, Jarvis and Johnston from 50 nm to the full extent of the 200 nm EEZ boundary on September 25, 2014. Obama's proclamation under the authority of the Antiquities Act enlarged the Pacific Remote Islands MNM, from 87,000 square miles (225,000 square kilometers) to more than 400,000 square miles (more than 1 million square kilometers), about three times the size of California. The proclamation prohibited commercial fishing in the Pacific Remote Islands MNM. The lost access to the EEZ around Johnston Atoll represented around 10% of historic fishing grounds for the Hawai'i longline fleet.

During the development of the Pacific Remote Islands MNM expansion in 2014, the Council voiced strong opposition on the grounds that available scientific information does not support the claim that large pelagic MPAs will protect populations of highly mobile species such as tuna and billfish, seabirds, sea turtles and marine mammals. Council representatives communicated its opposition all the way to the White House. President Obama's senior advisor, John Podesta, invited the Council to meet with him and other high-level federal officials in the Roosevelt Room of the West Wing, which ironically, is called The Fish Room. Council representatives presented several reasons why monument expansion, which initially was reported to include expansion out to the EEZ boundaries of all of the Pacific Remote Island Areas, was not in the best interest of U.S. fisheries in the region. They pointed out that monument expansion would eliminate U.S. vessels from U.S. fishing grounds, displacing them to fish on the highs seas, where international restrictions were already in place for the U.S. purse-seine fleet and might soon be for the U.S. longline fleet. However, it was clear that the administration had an environmental agenda and wanted to secure President Obama's "blue legacy."

Council members, staff and advisors (front row) Paul Dalzell, Claire Poumele, Arnold Palacios, Kitty M. Simonds, Sylvia Spalding, (back row) Sean Martin, Svein Fougner, Eric Kingma and Edwin Ebisui Jr. leave the White House West Wing on September 9, 2014, after discussing President Obama's plans to expand the Pacific Remote Islands MNM with John Podesta, the White House's counselor to the president. *WPRFMC photo.*

Not more than three months after President Obama expanded the Pacific Remote Islands MNM, the United States agreed in the WCPFC to further reductions on high-seas fishing by U.S. vessels, putting U.S. fisheries at a further disadvantage.

U.S. fisheries in the region suffered further limitations when Obama utilized the Antiquities Act again on August 26, 2016, to expand the Papahanaumokuakea MNM to the full extent of the U.S. EEZ surrounding the NWHI, enlarging that monument from nearly 140,000 square miles (362,000 square kilometers) to nearly 600,000 square miles (1.5 million square kilometers). The expansion was opposed by former Hawai'i governors, Native Hawaiian leaders and seafood representatives.

U.S. Senator Daniel Akaka speaks at a press conference on July 26, 2016, asking President Obama for adequate time and a transparent process to discuss the costs and benefits of the proposed expansion of marine national monument in the NWHI before a decision is made. Other speakers with the same message included former Hawai'i Governor George R. Ariyoshi, Office of Hawaiian Affairs Trustee Peter Apo, Association of Hawaiian Civic Clubs President Annelle Amaral and Leon Siu representing various Native Hawaiian organizations. *WPRFMC photo.*

The marine monuments in Western Pacific Region now claimed 52% of the U.S. EEZ in the Western Pacific Region. With other commercial fishing restrictions, 83% of the EEZ waters around Hawai'i were closed to the local longline fishery.

Five years after the expansion, two reports studied the impact of the Papahanaumokuakea MNM on the local longline fleet. A study funded by the Pew Ocean Legacy Project found minimal impacts, while a separate study by the University of Hawai'i and NMFS researchers found that, after the expansion, catch per unit effort fell by 7%, revenue per trip dropped 9% and vessels lost out of $3.5 million in revenue the first 16 months after the expansion (Orlowski 2020). At the same time, HLA reported that about 90% of the fleet effort was now on the high seas.

Marine national monuments established by presidential proclamations comprise 52% of the U.S. EEZ waters in the Western Pacific Region. Other large areas that restrict fishing in the region include various zones established by the Council under the MSA and the Southern Exclusion Zone established by NOAA under the Marine Mammals Protection Act. *WPRFMC illustration.*

Fishing Effort in the Pacific Ocean

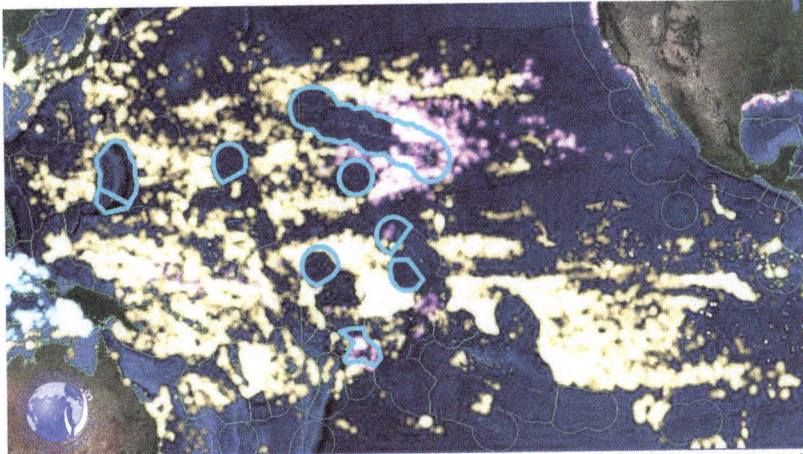

Three months fishing effort (November 9, 2018–February 9, 2019)   Data source: Global Fishing Watch
⬤ Foreign fishing vessels ⬤ U.S. fishing vessels ○ U.S. exclusive economic zone
Vessels are predominately purse seine, longline, and pole and line vessels targeting tuna and swordfish.

Satellite data from Global Fishing Watch for the period November 9, 2018, to February 9, 2019, reveals the extent of foreign fishing occurring immediately outside the boundaries of the marine national monuments. Monument regulations that closed traditional fishing grounds for U.S. fleets have forced them to compete against the foreign fisheries on the high seas. *WPRFMC illustration.*

The displacement of American fishing vessels as a result of marine monument proclamations has had widespread repercussions in the region. Diminishing access to tuna grounds along with supply and prices were cited by Tri Marine as reasons it closed its American Samoa canneries in December 2016, leaving hundreds of workers unemployed (Smith 2016). The Tri Marine facility had brought optimism to the Territory, when the future of American Samoa as a major canning center in the region looked uncertain following the closure of the Chicken of the Sea cannery in 2009 and the loss of at least 2,000 full-time equivalent jobs (AS DOC, n.d., page 9). Tri Marine was one of the largest tuna producers globally and had pumped a reported $70 million into its Samoa Tuna Processors facility, which opened in January 2015. It featured a fresh fish processing area, additional infrastructure to support fisheries development for fresh and value-added products and exporting of fresh tuna by air freight for the premium tuna markets in Japan and the mainland United States. A new blast freezer and two storage units capable of freezing and holding fish at minus 60 degrees centigrade were commissioned in support of the premium quality processing operation. In addition, a 2,000-ton capacity cold storage facility located at the Samoa Tuna Processors site would support of the American Samoa-based fishing fleets to offload and store their catch.

The American Samoa government, in its Comprehensive Economic Development Strategy 2018-2022, said the following about the shutdown:

> [The Samoa Tuna Processors] laid off as many as 600 employees, which led to hundreds of additional job losses in related and support industries in 2016 and 2017. StarKist, the final remaining tuna processor in the territory, also faced setbacks in 2017. Availability of landed fish, along with a number of federally mandated equipment upgrades, forced the company to temporarily halt operations for five weeks in the fourth quarter of 2017. The shutdown left thousands of employees without salaries to cover basic expenses and cost the local government more than half a million dollars in income tax revenues. The combined impact of the cannery closures contributed to a spike in the unemployment rate in the territory, rising from 10.5% in 2016 to 14.3% in 2017, and led to stagnant spending in both the private and public sectors. As a result, the economy contracted by 2.9% in 2017. Transportation

costs rose modestly that year, as well, causing real [gross domestic product] to have contracted by a clip of 5.9%.

On September 15, 2016, after yet another lobbying effort, Obama would proclaim his last marine national monument, the Northeast Canyons and Seamounts—the only marine national monument not in the Western Pacific Region—encompassing less than 5,000 square miles (less than 13,000 square kilometers). On June 5, 2020, President Donald Trump signed a proclamation to lift the ban on commercial fishing in this Atlantic monument but did not address similar pleas from the governments, fishing industry and the Council to lift the fishing restrictions in the marine monuments in the Western Pacific Region.

In his last days in office, Trump signed a bipartisan funding bill on December 27, 2020, which included a provision to restart the process to designate the NWHI as a national marine sanctuary. That process initially began with President Clinton's executive order establishing the NWHI CRE Reserve on December 4, 2000, but was abruptly halted in 2006 with Present Bush's proclamation of the NWHI as a marine national monument.

Ensuring access of U.S. longline and purse-seine vessels to tuna grounds in the tropical Pacific continues to be an issue that the Council monitors at the local, national and international levels.

In the international arena, a movement popularly known as Biodiversity beyond National Jurisdictions is developing an international legally binding instrument under UNCLOS that could potentially convert 30% of the high seas into no-take MPAs by 2030. Two organizational meetings were convened under the auspice of the United Nations in New York in 2018 and another two in 2019. Due to COVID-19, a meeting scheduled in 2020, which the Council was registered to attend, was postponed until 2021.

In the meanwhile, the Council from June 15 to 17, 2020, hosted a virtual international workshop on area-based management of blue water fisheries. Thirty-four science and management experts from around the globe participated in efforts to develop a roadmap to effective area-based management of blue water fisheries. The meeting was co-chaired by Ray Hilborn, an SSC member and professor at the University of Washington, and Vera Agostini of the United Nations Food and Agriculture Organization. The participants agree that closing large sections of the ocean is not a silver bullet for managing blue water fisheries and their ecosystems; that MPAs are merely a single element in the toolbox of area-based management; and that economic, cultural and social objectives need to be considered and alternative management explored. A roadmap and a science-policy paper are expected to emerge from the meeting.

Uneven Playing Field: Low Quotas, High Fees for U.S. Vessels in the Pacific

In 1999 and 2000, the Council helped the U.S. government to host the final four of seven sessions leading to the establishment of the WCPFC. A regional fishery management organization to develop an international scheme for highly migratory tuna and tuna-like species in the WCPO seemed to be the best avenue to ensure the sustainability of these highly migratory species. However, dominated by Pacific Island countries and operating by consensus, the Commission has come to be a stumbling block for the longline fishery based in Hawai'i, which operates under the U.S. quota, one of the lowest quotas for longline-caught bigeye tuna, despite its management being the international gold standard for pelagic longlining. The WPCFC set the national longline bigeye catch limits from 2017 to 2020 at nearly 18,000 mt for Japan;

nearly 14,000 mt for Korea; more than 10,000 mt for Chinese Taipei; nearly 9,000 mt for China; nearly 6,000 mt for Indonesia; and less than 4,000 mt for the United States. Meanwhile, no catch limit was set for small island developing states and territories (including the U.S. Pacific territories) so as to encourage development of their fisheries. Although the WCPFC has set no longline bigeye quotas for the U.S. Territories, the Council set a catch of 2,000 mt for each and, under Amendment 7 of the Pelagic FEP, allows for transfer of part of the quota to federally permitted Hawai'i-based U.S. longline vessels.

While adult bigeye tuna are targeted by the longline fleets for the sashimi market, the purse-seine fleets, which principally target skipjack for the canned tuna market, have increasingly caught juvenile bigeye tuna since they began utilizing FADs.

Between 2014 and 2015, the Council convened three international workshops to address the incidental catch of bigeye. The first was held at the Council office in Honolulu from April 22 to 24, 2014. The Workshop on Pacific Bigeye Movement and Distribution brought together experts on the biology, fisheries, population dynamics and management approaches relevant to Pacific bigeye tuna and helped to set the stage for further discussion. The movement, distribution and impact from fisheries were examined through tagging studies, otolith chemistry, genetics, and climate mediated impacts. Alternate stock assessment assumptions, modeling and management approaches were explored. The second workshop was held April 8 to 10, 2015, in Honolulu, and included fishing industry, scientists, WCPFC and Inter-American Tropical Tuna Commission staff, government officials and NGO representatives. The meeting was chaired by Drew Wright, former WCPFC executive director, who noted that the workshop was successful in its endeavor to create a noncommission environment in which the industry could be heard. The third conference was held from August 19 to 21, 2015, in Majuro, in coordination with the Republic of Marshall Islands Marine Resources Authority. Both of the latter conferences attempted to identify solutions to the purse-seine bigeye problem and to share such solutions with the WCPFC for use in conservation and management measures. The workshops were successful in bringing people together and identifying potential solutions that need further evaluation; however, action by the Commission has yet to produce effective purse-seine measures that cover the range of the stock.

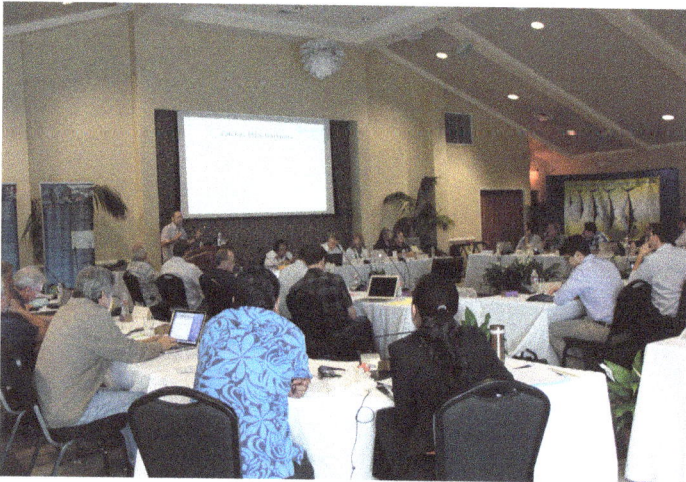

Participants listen to a presentation on the stock status and catch trends of bigeye tuna at the WCPO Purse-Seine Bigeye Management Workshop, held April 8 to 10, 2015, in Honolulu. Hosted by the Council, the workshop explored regulatory and nonregulatory measures to address the incidental take of juvenile bigeye by purse-seine fleets. *WPRFMC photo.*

On the surface, the Honolulu Convention gives authority to the Commission to establish measures that are applicable throughout the range of the stocks concerned in waters both on the high seas and in the members' respective EEZs. Under this view, the management measures of member nations within their EEZs are to be compatible with the measures of the WCPFC. However, some members expressed the view that the WCPFC's jurisdiction was limited to the high seas and that WCPFC should take measures applicable to fisheries on the high seas that are compatible with measures taken for domestic waters. This view was most strongly expressed by the group of nations referred to as the Parties to the Nauru Agreement, a separate agreement covering eight members of the WCPFC. These eight members represent the largest tuna purse-seine fisheries for skipjack tuna. They are the Federated States of Micronesia, Kiribati, Marshall Islands, Nauru, Palau, Papua New Guinea, Solomon Islands and Tuvalu. These countries have agreed among themselves to fishery management measures such as a "vessel day scheme," which ostensibly commits members to limit purse-seine fishing effort in their waters.

Access by U.S. purse-seine vessels to their waters are regulated by the South Pacific Tuna Treaty. Per day fishing fees for U.S. purse-seine vessels to fish in these waters have tripled and the number of fishing days is limited (WPRFMC 2019).

While nations have every right to limit the number of licenses or permits they issue and the number of fishing days for licensed vessels, if the cumulative effort allowed under separate national programs are too large, the impacts are felt by all. Actions by the Parties of the Nauru Agreement can potentially serve to undermine broader international agreement through WCPFC to implement controls necessary to conserve the stocks. This disagreement is likely to continue being a disruptive problem with the WCPFC.

At the same time, there is continuing concern that the U.S. purse-seine and longline fleets are disproportionately burdened by WCPFC actions despite the fact that the United States is apparently one of the few nations to effectively enforce measures adopted by the WCPFC. The playing field is not level.

Meanwhile, in the Eastern Pacific Ocean (EPO), the Inter-American Tropical Tuna Commission's bigeye conservation measures included a quota for U.S. longliners greater in 24 meters (~79 feet) in length. In 2015, the limit was set at 500 mt. Eventually, the bigeye tuna catch limit in the EPO would increase to 750 mt for U.S. longline vessels greater than 24 meters (~79 feet) in overall length to begin in calendar year 2021.

Fishing conditions in the Hawai'i longline fishery the first half of 2015 were excellent, with a 36% increase in catch per unit effort. As a result, Hawai'i vessels reached the U.S. quota for bigeye tuna early in the WCPO, and that area was closed to them on August 5, 2015. The fleet also reached the EPO quota for vessels greater than 24 meters (~79 feet), and the EPO was closed to those vessels on August 12. The only U.S. longline vessels able to operate in the WCPO were those with dual American Samoa longline permits (about 20 vessels), but they were able to operate only on the high seas. The WCPO opened to Hawai'i longline vessels on October 9, 2015, after NMFS eventually approved a fishing agreement between the Guam government and Hawai'i longline vessels for a transfer of 1,000 mt of bigeye quota.

The EPO, however, would remain closed to Hawai'i longline vessels over 24 meters in length until the end of the year. While smaller longline vessels could theoretically fish in the area, it would require traveling a distance of 1,100 nm, which could prove unsafe for the smaller vessels.

In early October, the Council visited the idle longline vessels owners and operators in Honolulu. They estimated that the loss of income for the two-month period between August and October amounted to about $1.4 million per week, or $11.4 million. Despite the loss of income, the vessel owners still had to contend with dock fees, crew wages, crew food, loan payments, insurance and vessel maintenance. Foreign crew members were not allowed to work on the vessels when they were tied up, as to do so was outside the visa conditions, according to the Department of Homeland Security.

The Council produced a nine-minute video (*Shutdown*) about the impacts of arbitrary quotas on the U.S. longline fleet based in Hawaiʻi (WPRFMC 2015).

The Council continues to argue for strong, effective negotiations to support the U.S. fishing industry at international regional fishery management organization negotiations and for timely processing by NMFS Pacific Islands Regional Office of the paperwork necessary for the legal transfer of longline bigeye quota in the WCPO between the U.S. Territories and the Hawaiʻi longline fleet.

Fishing Area of the Hawaii Longline Fleet

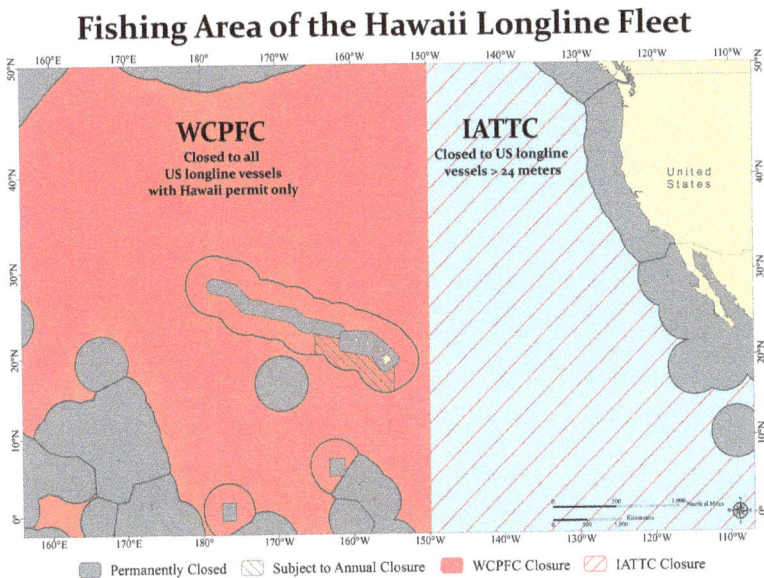

Excellent fishing conditions in early 2015 led to the Hawaiʻi longline fishery reaching its WCPO bigeye quota on August 5, 2015, and its EPO quota for vessels greater than 24 meters (~79 feet) on August 12. Only the 20 U.S. longline vessels with dual American Samoa permits were able to operate in the WCPO during the closure, but they could operate only on the high seas. *WPRFMC illustration.*

Billfish Conservation Act Amendment—Closing
U.S. Pacific Island Seafood Markets

In addition to fishing ground closures and limited allowable catch of longline-caught bigeye tuna, the U.S. pelagic fisheries in the Western Pacific Region have been hit with a closure of an important market. In 2018, President Trump signed into law a Congressional bill introduced by U.S. Representative Darren Soto (R-Florida) to amend the Billfish Conservation Act of 2012, such that U.S. caught billfish landed in the U.S. Pacific Islands by U.S. fishermen could no longer be sold in continental U.S. markets, including Alaska and Puerto Rico. The amended Act did not include swordfish in its definition of billfish, but it did include the several marlin, sailfish and spearfish species that the Council and NMFS had been monitoring and managing since the inception of the Council in 1976.

Prior to the bill's passage, the Council had received letters from both the NOAA Assistant Administrator for Fisheries Chris Oliver and the U.S. Secretary of Commerce Wilbur Ross stating that the bill was unnecessary and would not lead to improved billfish conservation.

Sales of foreign-caught billfish in the United State and commercial harvest and sales of U.S.-caught billfish in the Atlantic, where several billfish species are overfished or experiencing overfishing, had been prohibited since 1988. During the ensuing decades, the NMFS-administered Billfish Certificate of Eligibility was required to accompany any billfish in the Pacific that was offered for commercial sale in the United States to ensure that it was not from a foreign fishery or from the U.S. Atlantic. The certificate documented the vessel, homeport, port of offloading and date of offloading for the billfish being sold.

Congresswomen Colleen Hanabusa (D-Hawai'i), Madeleine Z. Bordallo (D-Guam) and Aumua Amata Coleman Radewagen (R-American Samoa) said the legislation "will negatively impact the livelihoods of fishermen in Hawai'i, Guam and the Pacific Insular Areas by closing off the only off-island market for U.S. caught billfish." They added, "We support needed conservation efforts in the Atlantic, but do not believe that Pacific fisheries need to be targeted in order to achieve these goals." Unfortunately, their words of reason based on the best scientific information were ignored. Stock assessments showed that, unlike in the Atlantic, billfish in the Pacific (except the western and central North Pacific marlin) were healthy.

"It is disappointing that special interest groups were successful in lobbying Congress to eliminate sustainable U.S. Pacific Island caught billfish sales on the mainland," Council Executive Director Kitty M. Simonds said, in the Council's summer 2018 newsletter, noting that the amendment to the Billfish Conservation Act was inconsistent with the principles and standards of the MSA.

The amendment was supported by a coalition of sportfishing groups including the American Sportfishing Association, the Center for Sportfishing Policy and the International Game Fish Association, among others (Golden 2017). Many of the groups represent East Coast sportfish fishermen who target billfish recreationally in billfish tournaments that provide prize money for the largest billfish landed.

Revisiting the American Samoa Large Vessel Prohibited Area

In 2002, the Council had established the LVPA in American Samoa, which prohibited operations of vessels larger than 50 feet (15.25 meters) from operating within 50 nm of the islands. At the time, about 40 small alia and 24 larger conventional mono-hull longline vessels were targeting albacore tuna using longline gear. In the ensuing decade, the number of alia in the fishery dropped to less than three, and conditions for the larger locally owned longliners worsened, as operation expenses increased and catches and fish prices decreased. Local fisheries throughout the South Pacific experienced similar poor conditions, which many attributed to the increasing number of heavily subsidized Chinese longline vessels in the area. The Chinese vessels catch of South Pacific albacore increased to half of the total catch by all countries, while the catch by American Samoa decreased to 2% of the total catch (WPRFMC 2019, page 15).

In 2014, the Tautai-O-Samoa Fishing Association of U.S. longline vessel owners fishing out of American Samoa asked the Council to change the LVPA regulations to provide relief to their fishery. Temporarily opening the LVPA (with annual reviews) would promote greater fishing efficiency for the Territory, which is boxed in by the EEZs of other countries and thus had a small amount of U.S. waters in which to fish. Opening the LVPA would promote greater fishing efficiency by decreasing the distance and costs associated with fishing. At the time, the alia albacore longline fleet no longer existed, and so the potential for gear conflicts, the impetus for the LVPA, was no longer an issue. After holding several rounds of public meeting, the Council in 2015 took final action to open the LVPA waters 12 to 50 nm from shore to the American Samoa large longline vessels with annual monitoring, and NMFS approved the regulatory amendment in 2016.

The American Samoa government quickly reacted, suing the Council, NMFS and the U.S. Department of Commerce with claims that the action did not take into account the Deeds of Cession, the treaty between the island chiefs and the U.S. government and the importance of local fisheries to the Samoan culture. In 2017 the U.S. District Court for the District of Hawai'i vacated the LVPA exemption for the American Samoa large longline vessels, stating the Deeds of Cession should be treated as other applicable law. NMFS appealed the judgement.

In June 2018, in American Samoa, the Council refined the LVPA regulatory amendment to exclude the large longline vessels permitted under the American Samoa longline limited-entry program from operating within 2 nm around the offshore banks while allowing them to fish within the LVPA seaward of 12 nm around Tutuila, Manu'a and Swains Islands. The exemption would be for four years and would include annual monitoring of the American Samoa longline and troll catch rates, small vessel participation and local fisheries development initiatives.

In September 2020, the Ninth Circuit Court released its decision on NMFS's appeal. Siding with NMFS, the memo stated: "It is of little import that NMFS did not specifically cite the cessions when detailing 'other applicable laws' it consulted, as NMFS considered the consequences of the rule on the alia fishing boats, and rationally determined the effects were not significant."

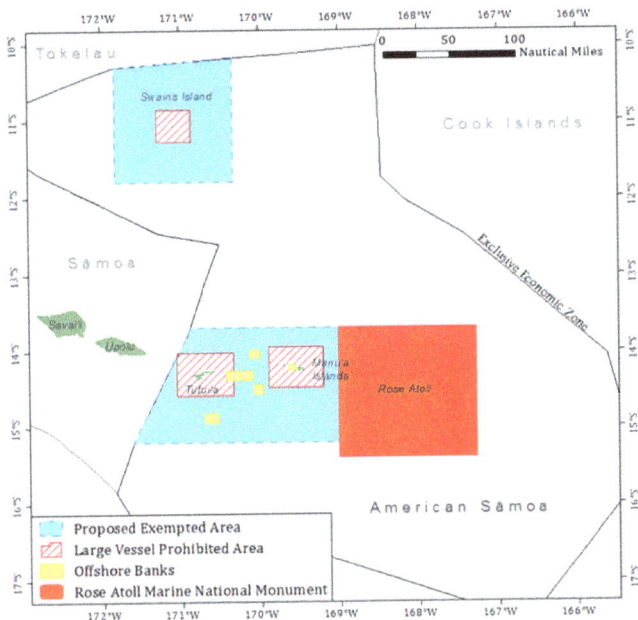

Council recommendations for the American Samoa LVPA include opening of exemption areas 12 to 50 nm from shore (in blue) proposed in 2015 and the retention of closures out to 2 nm around offshore banks (in yellow) proposed in 2018. *WPRFMC illustration.*

Supporting Pelagic Fishing Opportunities for Indigenous Peoples

Passage of the 2006 reauthorized MSA provided the Council with additional sources to fund fishery development projects to support indigenous communities in the region. In 2010, the Council, working with the Americans Samoa DMWR and other partners, completed several fisheries development projects identified in the American Samoa MCP, utilizing funding from the Western Pacific SFF. The projects included the following:

1) Boat ramps at Fagaʻalu and Tafuna on the island of Tutuila;

2) Fishermen facilities on Ofu and Taʻu in the Manuʻa Islands equipped with industrial ice machines, refrigerated storage containers and transportable fuel tanks as well as the establishment of the Tai Samasama (Taʻu) and Faleluaanuʻu (Ofu/Olosega) Fishermen Cooperatives to run the facilities; a fish market in Fagatogo near Pago Pago Harbor;

3) Completion of a small vessel dock on American Samoa government lands fronting Samoa Tuna Processors;

4) Fresh fish handling workshop and training;

5) A demonstration project to help the American Samoa longline fishery for albacore tuna to diversify by targeting other pelagic species, such as yellowfin and bigeye tuna and mahimahi, when the albacore season is low;

6) Evaluation and design of a new multipurpose fishing vessel capable of conducting pelagic longline, troll, handline and bottomfish operations in American Samoa to replace the aging fleet of alia vessels that are limited in their range, fish

hold capacity, onboard fishing gear and safety and navigation equipment;

7) Development of a training program to increase fish catches, introduce simple business management tools and provide the prerequisites to qualify for a fishermen lending scheme;

8) A feasibility study to extend the Malaloa Wharf area for the purpose of accommodating the locally based U.S. longline fleet; and

9) Renovation of the Fagatogo Fish Market.

Council Executive Director Kitty M. Simonds and then Lieutenant Governor H. C. Lemanu Mauga prepare to cut the ribbon during the dedication of the Faga'alu Park Boat Ramp on the island of Tutuila, American Samoa, on March 8, 2013. *WPRFMC photo.*

Dustin Snow, local lead for Island Fisheries Inc., and his family stand ready to reopen the Fagatogo Fish Market on April 5, 2019. The market, which had initially opened in 2010, was closed for two years before Island Fisheries Inc. assumed management of it. *WPRFMC photo.*

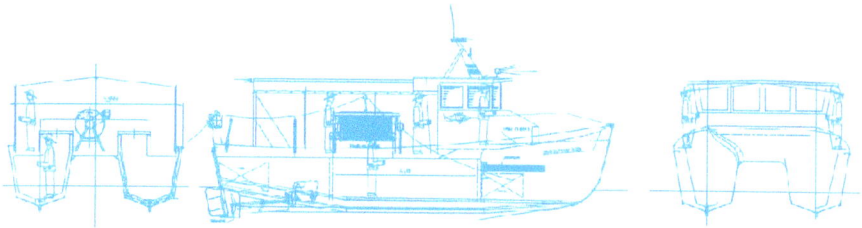

General design specifications are a part of the technical report on a new multi-purpose, small-scale fishing-vessel for American Samoa. The vessel, if constructed, could replace the Territory's aging fleet of alia vessels that are limited in range, holding capacity and safety and navigation equipment.

Continuation of these projects by the American Samoa government is critical to sustaining the fisheries development initiative conducted by the Council.

Another Council effort to help an indigenous Community Development Program project in Hawai'i did not prove successful. The 2006 MSA reauthorization gave the Council regulatory authority to create opportunities for native communities to participate in fisheries managed by the Council through the Western Pacific Community Development Program. The Council attempted to use the program to provide an exemption to deploy a longline within the longline closed area around the main Hawaiian Islands to catch yellowfin tuna as part of a commercial fishing program to train young Native Hawaiians.

Leo Ohai, a Native Hawaiian fisherman who fished commercially for more than 50 years, used a multi-species, multi-gear style of fishing that pre-dates current commercial fishing practices. He had already taught Hawaiians to fish for amaebi (a deep-sea shrimp, *Hetrocarpus* spp.) deploying deepwater traps from a small boat platform as a means to supplement household income. He trained his students to harvest akule and various crab species and to market fish products and value-added products that the students could prepare. For 10 years, the Council unsuccessfully engaged in efforts to obtain a longline permit exemption for this project under the Community Development Program. These efforts included the development and implementation in 2010 of Amendment 1 to the Pelagic and archipelagic FEPs, which established the eligibility requirements and procedures for plans submitted to the Council under the Community Development Program. During this time, Ohai retired from fishing and his children took over his business and created a curriculum for his style of fishing. With recurring delays in attempts to get their plan through the federal bureaucracy, the Ohai family withdrew the Community Development Program proposal in

2016 and planned to work with the University of California at San Diego. Unfortunately, Leo Ohai did not live to see his dream come true.

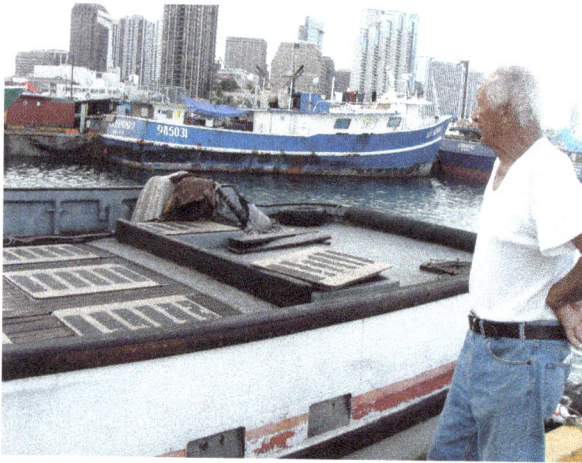

Leo Ohai stands before his vessel docked in Honolulu harbor. Ohai, a Native Hawaiian, unsuccessfully sought a Community Development Program exemption to teach longlining to Native Hawaiian youth in the longline exclusion area that extends seaward 50 nm around the main Hawaiian Islands. *WPRFMC photo.*

Conserving and Managing Sharks

In contrast to the complex issues associated with sea turtle and seabird bycatch, the issues associated with bycatch of sharks were relatively simple. For the past century, only two species of sharks—the thresher and the mako—have been caught and marketed for food in Hawai'i. However, as the longline swordfish fishery expanded, shark bycatch increased as well. This is because the blue sharks (*Prionace glauca*) feed predominantly in the upper water column (about 10 to 30 fathoms). Between 1991 and 1997, the number of blue sharks caught as bycatch by both swordfish and tuna longliners ranged from 71,000 to 154,000 with 86,000 sharks taken in this longline fishery in 1997.

From 1991 through 1995, swordfish fishing contributed to most of the catch of sharks, accounting for as much as 80% of the total catch in some years. However by 1997, the proportion of total shark catch taken by the swordfish fishery dropped to 14%, while those caught by the vessels fishing for tuna increased to 40% (WPRFMC 1998). This was because the efforts made by the swordfish vessels to reduce the bycatch of turtles and seabirds (e.g., changing from squid bait) also reduced the catch of sharks (Walsh et al. 2009). Seabird measures were implemented in 2002 and turtle measures in 2004.

The finning of sharks—cutting off and retaining the fins and discarding the carcass—was not a significant issue in the Hawai'i longline fisheries in the early years. In the period 1991–1992, only 3% of the sharks caught in Hawai'i waters were finned by fishermen. The majority of sharks were cut loose and released. However, as the Chinese economy rapidly developed during the 1990s and NMFS Atlantic shark quotas were reduced, the value of the Hawai'i shark fins increased. By 1997, about 65% of the sharks caught were being routinely finned for sale to Chinese wholesalers (Gillman et al. 2007a). Typically during this period, crewmen on tuna or swordfish vessels earned between $2,375 and $2,850 per year from shark fins, which made up about 10% to 11% of their annual wages. In 1998, approximately 38 tons of shark fins were produced with an ex-vessel value of $1 million (ibid).

In 1999, it was widely publicized that blue sharks were routinely finned in Hawai'i and the carcasses thrown into the sea alive (based on information from NMFS observer coverage). Environmentalists took the position that the practice was unnecessarily cruel, culturally incorrect from the standpoint of Native Hawaiian belief and extremely wasteful. They raised protests with NOAA and demanded that the practice be stopped. The issue received widespread publicity.

In response, the Council defended the fishermen on two points: 1) blue sharks, taken far out at sea, unlike the coastal sharks, were not threatened as the stocks of blue shark were very large and had a higher reproductive rate than other shark species; and 2) the process (which almost always involved killing the shark prior to finning, as finning a shark alive could be very dangerous) was no more wasteful or cruel than many other industrial meat or fish producing processes. However, a number of environmental activists accused the Council of being indifferent to cruelty to marine animals, particularly sharks, despite the fact that Hawai'i fishermen kill the sharks prior to fin removal. Sharks have primitive vertebrae, and post-mortem muscle spasms can be mistaken for them feeling pain. In addition, the Council tried to explain that the local shark fin catch was miniscule relative to the worldwide market for shark fins in Chinese restaurants. Demand for shark fins in Hawai'i's many Chinese restaurants alone resulted in the importation of tens of thousands of dried shark fins, far more than were being brought in by Hawai'i vessels. Moreover, Native Hawaiians regularly hunted sharks, and not all Native Hawaiians thought of sharks as being their 'aumakua (personal spirit god). Only a few specially designated families had relationships with particular sharks. Unfortunately, the more the Council tried to explain its position, the greater grew the sense of indignation about the practice, and opponents perceived the Council's actions to address shark finning and shark fishing as being inadequate.

On April 13, 2000, the House Subcommittee on Fisheries Conservation, Wildlife and Oceans held a hearing on House Resolution 3535 to Amend the MSA to Eliminate the Wasteful and Unsportsmanlike Practice of Shark Finning. Council Chair Jim Cook provided testimony at the hearing via teleconference, one of the first times, if not the first, that testimony has been provided to Congress in this medium. Cook explained that "the Council has adopted measures to restrict the Hawai'i longline fleet to a one-shark-per-trip limit for all non-blue shark species (they are to be landed whole) and a 50,000 annual quota for blue sharks to be adjusted periodically." The Council also banned demersal longline gear, which impacts coastal sharks. Cook noted that "the mortality levels of sharks in the Western Pacific Region where finning is allowed in both federal and state waters is one-tenth the level of the East Coast and the Gulf of Mexico where finning is not allowed in federal waters and most state waters." He further suggested that "an amendment to the [MSA] that would define 'waste' would help and it is preferred to actions that selectively restrict one fishery while allowing other fisheries with similar waste associated with them to continue." He pointed out that observer coverage indicated that 98% of the sharks finned by the Hawai'i longline fleet were done so after they are dead. He said the proposed listing of shark finning as an unlawful act for all U.S. federal waters would lump all shark species and shark fisheries together and distract from the more important shark conservation and management issues such as needed population assessments and international agreements on shark fisheries. He asked the committee to consider maintaining the regional approach to fisheries management.

In December 2000, Congress passed and President Clinton signed into law the Shark Finning Prohibition Act, which outlawed finning in the U.S. EEZ and possession of fins by any U.S.-flagged vessels on international waters. It also prohibited any fishing vessel from landing at a U.S. port with shark fins whose weight exceeded 5% of the total weight of shark carcasses landed or on board.

In the meantime, the Council continued to research fishing techniques that resulted in reducing the bycatch of seabirds and turtles. As these practices were incorporated by fishing boats, the number of shark interactions also fell. In 2006, it was estimated that 99% of sharks caught by longline swordfish vessels in Hawai'i were released alive, as were 97% of sharks caught from longline tuna vessels (Gilman et al. 2007a).

The shark issue was to reappear in a different guise in 2006 when the Council was questioned about the practice of allowing two tour boat operators on the island of O'ahu to gather sharks in federal waters for the amusement of tourists. In response, the Council conducted a public scoping meeting on the issue on the North Shore of O'ahu, where the operations were taking place. As a result of public input, the Council requested that the MSA be amended to ban the aggregation of sharks for tourism.

In May 2010, the State of Hawai'i made it illegal to possess, sell or distribute shark fins. In December of that year, Congress followed suit, passing the Shark Conservation Act, which President Obama immediately signed into law. With respect to the MSA, the 2010 law replaced the 2000 Shark Finning Prohibition Act and made it illegal to remove any shark fin (including the tail) at sea; to have custody, control or possession of any such fin aboard a fishing vessel unless it is naturally attached to the corresponding carcass; to transfer any such fins from one vessel to another at sea or to receive any such fin in such transfer; or to land any such fin that is not naturally attached to the corresponding carcass or to land any shark carcass without such fins naturally attached. The actions by Congress and the State of Hawai'i precluded the need for Council action regarding shark fins. However, because the MSA language did not forbid aggregating sharks, shark feeding tours in Hawai'i continue to this day.

Sharks continued to be an issue for the pelagic fisheries of Hawai'i and American Samoa as well as the U.S. tropical tuna Pacific purse-seine fleet when Defenders of Wildlife petitioned NMFS about the oceanic white tip shark (*Carcharinus lonigmanus*). In response, NMFS in 2018 listed the species as threatened under the ESA. On April 2,

2020, Earthjustice filed a federal lawsuit on behalf of the Conservation Council for Hawai'i and Michael Nakachi, a Native Hawaiian cultural practitioner and owner of a local scuba diving company, to enforce protections for the oceanic white tip shark. The lawsuit noted that, in making the determination to list the oceanic whitetip shark as threatened in 2018, NMFS recognized the overfished status of the species and asked the court to order NMFS to make proper notifications that "would trigger necessary protections as expeditiously as possible." In response, the HLA announced in November 2020 that its vessels would voluntarily switch from steel wire to monofilament leaders, which the sharks can more easily bite through to free themselves. "While the Hawai'i fleet fishes on the margins of oceanic whitetip preferred habitat, and our catch is relatively small compared to foreign fleets, which have much lower levels of monitoring than the Hawai'i fleet, we're taking this significant step to reduce our impact on this species and other sharks," said Eric Kingma, HLA executive director. The Council supported HLA's move and began its own study on ways to reduce the fishery's interactions with the ocean whitetip shark.

New Challenge: False Killer Whales

After the Hawai'i-based longline fishery's success with Hawaiian monk seal, seabird, sea turtle and shark bycatch mitigation efforts, the Council turned its attention to another protected species issue. This time, it was false killer whales in the Hawai'i-based deep-set longline fishery, which primarily targets bigeye tuna.

False killer whales and other toothed whales are known to eat catch and bait off of longline hooks in many parts of the world. These depredation events can be costly to fishermen, as false killer whales will leave only the head of the tuna and other valuable fish. In rare instances, false killer whales may become accidentally hooked or entangled in the process.

Since the observer coverage rate for the deep-set longline fishery increased to more than 20% in 2001, the number of observed false killer whale hookings and entanglements has been less than five in most years. This level of interaction is considered very rare in the fishery, and most interactions resulted in the animal being released alive. Yet, it was estimated that the false killer whale interactions were occurring at an unsustainable levels according to the assessments conducted under the Marine Mammal Protection Act.

When marine mammal interaction levels are found to be unsustainable under the Marine Mammal Protection Act, NMFS is required to convene a Take Reduction Team to develop recommendations for mitigating the impacts. However, in the case of false killer whale interactions in the Hawai'i longline fishery, NMFS delayed the convening of the team due to funding limitations. In the interim, the Council stepped in and formed the Marine Mammal Advisory Committee in 2005.

Three meetings of the committee highlighted the challenges of mitigating false killer whale depredations and interactions, as these rare events appeared to be occurring randomly in time and space. Information on false killer whale stocks around Hawai'i was also limited, and studies aimed at deterring false killer whales from eating bait and catch had not resulted in any effective solutions. As a result, the committee generated recommendations aimed at addressing these challenges and data gaps.

NMFS eventually formed a Take Reduction Team in 2010 that was tasked to develop a draft Take Reduction Plan within six months. Due to the lack of technological solutions to reduce interactions, the team focused its deliberations on ways to minimize the injury if a false killer whale is hooked. The final Take Reduction Plan was implemented in late 2012, requiring the deep-set longline fishery to use "weak" hooks intended to straighten on the weight of a false killer whale but withstand the weight of a large bigeye tuna.

A false killer whale is seen breaching. This oceanic dolphin is known to eat catch and bait off of longline hooks in many parts of the world. *NOAA photo/Southwest Fisheries Science Center.*

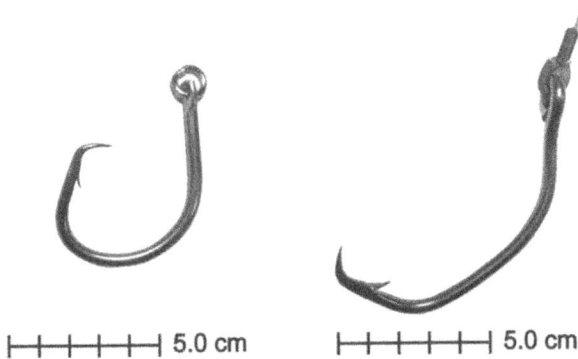

The weak hook, which the Hawai'i deep-set longline fishery has been required to use since 2012, straightens (right image) when taken by a false killer whale but remains intact (left image) when taken by a large bigeye tuna. *NOAA illustration.*

The 2012 Take Reduction Plan by NMFS also created a main Hawaiian Islands Longline Fishing Prohibited Area, commonly referred to as the Southern Exclusion Zone. Deep-set longline fishing is prohibited in this zone when the observed false killer whale mortalities or serious injuries in the fishery operating within the EEZ around Hawai'i reaches two or when the percentage of observer coverage in the fishery for that year exceeds the Hawai'i pelagic false killer whale stock's potential biological removal level. The measure is intended to mitigate impacts on the main Hawaiian Islands insular population of about 170 false killer whales, which was listed as endangered under the ESA in 2012. When the zone is closed, only 17% of the EEZ around Hawai'i remains open to the fishery due to the NWHI marine national monument and the 50-nm longline exclusion zone around the main Hawaiian Islands.

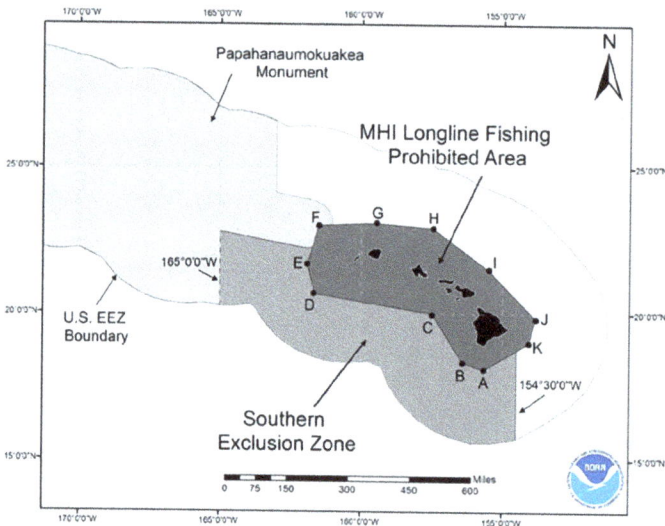

The Southern Exclusion Zone created by NMFS in 2012 is closed to deep-set longline fishing when interaction with false killer whales within the EEZ waters surrounding Hawai'i reaches a certain level. *NOAA illustration.*

Monitoring of the Take Reduction Plan is ongoing, but the effectiveness of the measures intended to minimize injuries had yet to be determined as of 2015. Indeed, the low level of interactions may mean that it may take many years to determine whether measures are working, unlike the significant bycatch reductions seen after the implementation of the Council's seabird and sea turtle measures.

The Council continues to work with the industry and NMFS to search for long-term solutions to the false killer whale depredation and incidental interaction issues. In 2015, the Council supported the industry in their fleet-wide outreach effort to increase awareness of the importance of weak hooks. The Council is also working with the industry to test the commercial viability of a device designed to deter false killer whales from depredating on tuna and other catch.

Addressing the Data Gap in the Hawai'i Noncommercial Pelagic Fishery

From the beginning of its history, the Council has worked to obtain the most complete and timely data on which to scientifically base its management decisions. It has continued these efforts in Hawai'i, the only state in the nation without a requirement for recreational or noncommercial fishermen to have a fishing permit, license or registration. This poses a problem to the Council as it works to manage the Hawai'i non-longline pelagic fishery using the best scientific information available.

Without reliable information about the number of noncommercial trollers and handline fishermen and the effort they expend on fishing, the data used to generate stock assessments is incomplete. Knowing the noncommercial catch of pelagic species is also of increasing concern as quotas for highly migratory species are developed by international regional fishery management organizations. While bigeye tuna has been a concern for these organizations in the past, there are indications that yellowfin tuna quotas may be forthcoming. Yellowfin is an important noncommercial species for Hawai'i fishermen and tournaments, especially during the summer 'ahi run.

Without a source for noncommercial data, the Council has turned to Hawai'i's commercial marine license data as an indicator. According to that data, the commercial troll and handline fishermen exert the majority of their efforts in federal waters and account for about 10% to 20% of the state's total commercial tuna landings. The noncommercial pelagic fishermen are likely to fish in these same federal waters and have catches that are equal to or greater than the commercial troll and handline fishery, based on discussions with fishermen.

In 2016, the Council partnered with the NGO Conservation International to convene a study group of individual from different fishing organizations and interest groups to determine the feasibility of a noncommercial fishing register, permit or license system for Hawai'i. The final report was made public in late 2017 on the Council's website. In 2018, HDAR asked the study group to share its findings more widely with stakeholders. The study group complied, holding a series of meeting in the second half of 2018 throughout the state and soliciting the thoughts, concerns, questions and suggestions from nearly 400 noncommercial fishermen and other interested persons.

The Council held another series of meetings on this topic throughout the state in early 2020 as it continued to address the need for reliable, robust data to optimally manage the Hawai'i small-boat pelagic fishery.

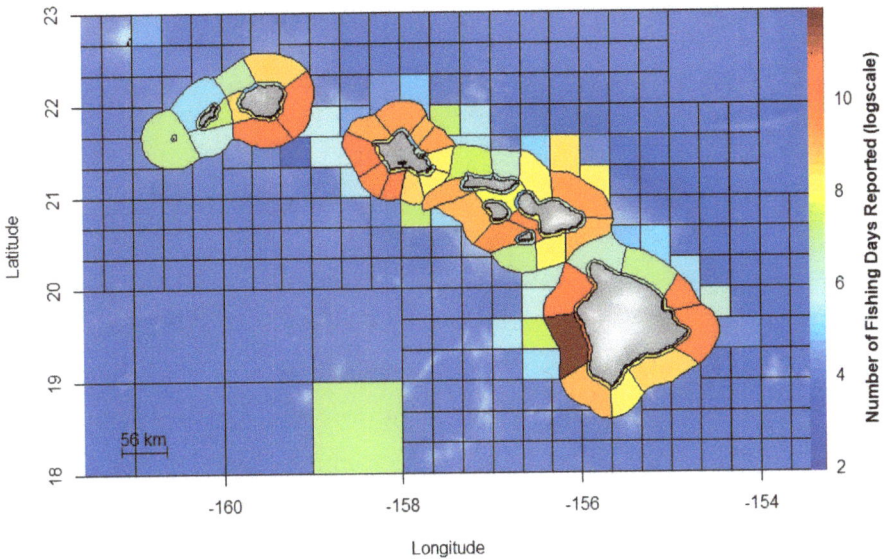

Number of Fishing Days Reported, Troll/handline, 2014-2018

According to the State of Hawai'i's commercial marine license data, the majority of troll and handline fishing effort occurs in federal waters (the large boxes) compared to state waters (thin line around each island). Because noncommercial fishermen do not need to be registered/licensed or to report catches, data from the commercial fisheries are used an indicator for the noncommercial fishery. *Hawai'i Division of Aquatic Resources illustration.*

TABLE 14: PACIFIC PELAGIC FEP: SUMMARY OF COUNCIL ACTIONS

DATE	ACTION	MEASURES
2010 Jan 14	FMP final rule	Restructured all fishing regulations under the previous FMPs in accordance with the new FEP.
2010 Sep 3	Amendment 1	Established eligibility requirements and procedures for reviewing and approving Community Development Program plans.
2011 Jun 27	Amendment 2	Prohibited pelagic longline fishing within approximately 30 nm of the CNMI. NMFS disapproved the Council's recommendation for a purse-seine area closure around the CNMI.
2011 Jul 11	Amendment 3 [disapproved]	NMFS disapproved the proposed establishment of a 75-nm purse-seine closure around American Samoa to prevent localized stock depletion and reduce catch competition and gear conflicts because it was inconsistent with the MSA National Standard 2.
2011 Jun 27	Amendment 4	Established a mechanism for specifying ACLs.

2011 Aug 24	Amendment 5	Established sea turtle mitigation measures for the American Samoa longline fishery to ensure that longline hooks fish deeper than 100 meters (m) to reduce interactions with Pacific green sea turtles. Limited the number of swordfish taken.
2013 Jun 3	Amendment 6	Established management measures for noncommercial and recreational fishing and prohibited commercial fishing within the Marianas Trench, Pacific Remote Islands and Rose Atoll MNM.
2014 Dec 28	Amendment 7	Established a management framework and process for specifying fishing catch and effort limits and accountability measures for pelagic fisheries in the U.S. Pacific territories (American Samoa, Guam and the CNMI). The framework would authorize the government of each territory to allocate a portion of its specified catch or effort limit to a U.S. fishing vessel or vessels through a specified fishing agreement and established criteria that a specified fishing agreement must satisfy.

2020 Sep 17	Amendment 10	Revised the leatherback sea turtle fleet-wide hard cap to 16, removed the loggerhead sea turtle fleet-wide hard cap and established individual trip interaction limits for loggerhead and leatherback turtles for the Hawai'i longline fishery.

Chapter 12

Archipelagic and Pacific Remote Island Fishery Ecosystems

12.1 Hawai'i Archipelago Fishery Ecosystem

New Management Regime for the Deep 7 Bottomfish

With the start of the 2011 fishing year (September 1, 2011–August 31, 2012), the main Hawaiian Islands Deep 7 bottomfish fishery began to be managed under an ACL, as mandated by the MSA, rather than a TAC.

The 2011 ACL was based on a new stock assessment by NMFS, which included input from highline bottomfish fishermen. The assessment incorporated updated bottomfish life history information; adjustments for the introduction of ta'ape (blue-lined snapper) and the loss of kahala in the market due to ciguatera; standardization of the data series that accounted for changes in the fishery from gear modifications, technology developments, shift from multi-day to single-day trips, weather and noncommercial fishing; and other factors that had been raised by fishermen and scientists over the years.

Refinements to the stock assessment model used to help specify the ACL continued to be made with expert input from Hawai'i's high-line bottomfish fishermen. The Pacific Islands Fisheries Science Center developed and implemented the region's only fishery independent survey that has been running for almost a decade. The survey uses fishermen to conduct standardized bottomfish fishing at randomly selected locations throughout the main Hawaiian Islands. The Pacific Islands Fisheries Science Center overlays this fishing information with video taken from baited camera systems used to identify the fish species, their size and numbers. New life history information on 'opakapaka has also become available through the radiocarbon dating of otoliths, which show the fish live longer than previously expected. All this information together with improved catch-per-unit-effort data from fishermen workshops has led to better stock assessments and higher ACLs for fishermen. The main Hawaiian Islands Deep 7 ACL has been nearly 500,000 pounds since 2019.

However, bottomfish fishermen continue to complain to the Council, HDAR and NMFS that the stock assessment does not account for the fish in the BRFAs, and they continue to argue for the removal of these no-take MPAs. In 2019, the State responded by opening four of the 12 BRFAs to fishing.

**Comparison of Main Hawaiian Islands
Deep 7 Bottomfish Landings from 2007 to 9/6/2020**

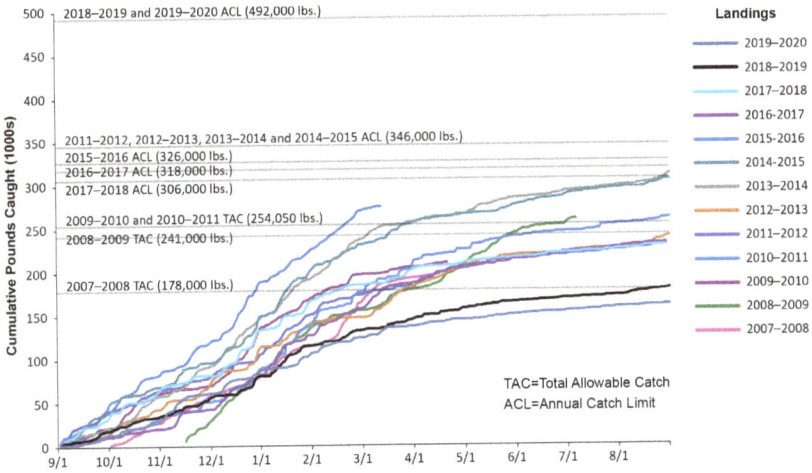

Comparison of main Hawaiian Islands Deep 7 bottomfish landings from 2007 to September 6, 2020. With the start of the 2011 fishing year (September 1, 2011 to August 31, 2012), the fishery began to be managed under an ACL, as mandated by the MSA, rather than a TAC. *Hawai'i Division of Aquatic Resources illustration/WPRMC modification.*

Edwin Ebisui Jr. of Hawai'i chaired the Council from 1994 to 1996 and from 2014 to 2018. He chaired the Council's Advisory Panel from 2007 to 2010. *WPRFMC photo.*

Because the fishery depends on specific knowledge being transferred from one fisherman to the next, documenting its history has been a longtime goal of the Council and its advisory body members. In 2016, the Council began supporting an oral history project of the fishery, funded by a NOAA Preserve America grant, administered by the Pacific Islands Fisheries Group with the guidance from the Council and the Pacific Islands Fisheries Science Center. Those involved in the project hope the information gathered will educate fishery managers and scientists about the nuances of the fishery's operations and thus improve future stock assessments and management efforts.

Paul Bartram, former Council biologist, reflected on the 30-year change in bottomfish policy that had its origins in the Council's Bottomfish FMP. "The lesson here is that Native Hawaiian means of managing bottomfish and other fish was by season and by *koʻa* (specific fishing ground), not by vast areas of the ocean where bottomfish might live or not live. They knew that some *koʻa* might be overfished, some might be under-fished, and some were waiting to be discovered." Bartram said that this is why traditional management followed a system of rotating areas based on the spawning cycle. "Now we have a system that is not based on natural cycles but on number chasing," he said. "The Secretary of Commerce issues a number (the amount by which fishing mortality is expected to be reduced), and everybody tries to meet it."[75]

Continued Closures of the Northwestern Hawaiian Islands Bottomfish and Seamount Groundfish Fisheries

While the main Hawaiian Islands bottomfish fishery has proven to be a sustainable fishery, two other bottomfish and seamount groundfish fisheries

---

[75]   Bartram op. cit.

remain closed. The NWHI bottomfish fishery, which once accounted for half of Hawai'i's Deep 7 landings, remains closed due to the Papahanaumokuakea MNM rules. While President Trump in December 2020 restarted the process to designate the NWHI as a national marine sanctuary, it is to be seen what impact the proclamation of the NWHI as a no-take national marine monument by President Bush in 2006 while have on this new sanctuary designation process and the future of the NWHI bottomfish fishery.

The harvest of seamount groundfish at the Hancock Seamounts remains under a moratorium, which was first established by the Council in 1986 as the stock had been overfished by foreign fleets prior to the establishment of the U.S. EEZ. In 2006, the Council became involved in discussions by Japan, Russia, Korea and the United States to establish mechanisms for the international management of high seas bottom fisheries in the Northwest Pacific Ocean.

In 2010, under the Hawai'i Archipelago FEP, the Council established the Hancock Seamounts Ecosystem Management Area and continued the moratorium on armorhead and other seamount groundfish until the armorhead stock is rebuilt. The area would provide a control site against which to compare fished seamounts and would allow a minimum rebuilding time of 35 years for the U.S. portion of the armorhead stock (NMFS 2010).

In 2012, Japan, Russia, Korea and the United States established a North Pacific Fisheries Commission in response to U.N. Resolution 61/105, which mandates that bottom fisheries operating in international waters be sustainably managed and not cause significant adverse impacts on vulnerable marine ecosystems. The Commission was established by the Convention on the Conservation and Management of High Seas Fisheries Resources in the North Pacific Ocean. It has been most concerned with alfonsin and armorhead, which are seamount ground fish species occurring within the Hawaiian-Emperor Seamount Chain and for which the Council has had a long history of managing.

Continuing Efforts for Improved Data and Research

A well-received, community-oriented project undertaken by the Council was the establishment of a set of community FADs to enhance data collection and research, among other purposes. Akin to the traditional *koʻa*, FADs attract fish to certain locations, making finding and harvesting them easier for fishermen. The State of Hawaiʻi had maintained a FAD program since 1980, and Hawaiʻi fishermen began deploying their own personal FADs without proper authorization from the U.S. Coast Guard and Army Corps of Engineers. The Council recognized the potential for legally permitted community FADs and, in 2006, worked with the fishermen at Hana, Maui, to establish the first legal community FAD in the state.

The Council subsequently entered into a private-public partnership with Mamaʻs Fish House, a premier restaurant specializing in locally sourced seafood, to fund deployment of additional FADs off of Maui. One of the goals of the community FAD project was to support cooperative research. In 2012, the FAD became the site of the first bigeye tuna tagged with a satellite tag in the main Hawaiian Islands. The Council also requested that all fishermen using the FAD submit a voluntary catch log to the Council.

Fishery-independent aerial surveys to estimate the abundance of akule around Oʻahu and an investigation of the survival rate for Kona crab in Hawaiʻi were among the other research efforts funded by the Council in Hawaiʻi during this period.

Concerns about the Future—Marine Planning, Climate Change, Protected Species

The State of Hawaiʻiʻs goal to purchase 100% of its power from renewable resources by 2045 has spurred much development interest, including offshore wind energy. The Bureau of Ocean Energy Management received three unsolicited lease requests to install offshore wind turbines from two potential developers between January and November of 2015. These three projects are scoped for 400 megawatts, or half of the state utility's planned wattage increases. Advisory Panel and Council member concerns include the well-known FAD effects of subsurface structures and the impact this might have on fish migration patterns, particularly of ʻahi, as all proposed wind farm structures appear to be within the summer ʻahi migratory route. Of the three proposed sites, one project is sited west of Kaʻena Point and two are sited south of Barbers Point and Waikiki, approaching the edge of Penguin Bank, a rich bottomfish grounds. The Council's priority is to maintain fishing access to the sites of the potential offshore energy facilities and to monitor the fish aggregations impacts of the wind farms.

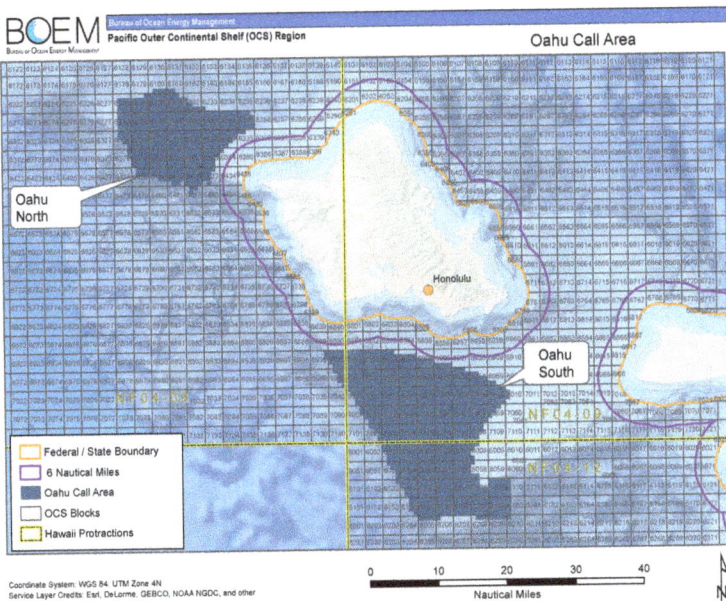

The sites proposed for wind-generated energy facilities in offshore waters around Oʻahu are of concern to fishermen and the Council. One project is sited west of Kaʻena Point, and two are sited south of Barbers Point and Waikiki, approaching the edge of Penguin Bank, a rich bottomfish grounds. All three proposed wind farm structures appear to be within the summer ʻahi migratory route. *Bureau of Ocean Energy Management illustration.*

A longer-term issue relates to environmental variability and global warming. During the 1990s, scientists become aware of the Transition Zone Chlorophyll Front, a large oceanic area that separates subtropical areas from more productive North Pacific waters. When the front shifts or changes in shape, the availability of nutrients in the NWHI rises and falls. Scientists noted that the prevalence of weak undernourished Hawaiian monk seal pups coincided with this phenomenon (Baker et al. 2007). It is quite possible that global warming will result in longer-term changes in the ocean circulation patterns and in the Transition Zone Chlorophyll Front, with adverse implications for monk seals.

The six main populations of monk seals exist on low islands and upon steadily diminishing cays within atolls. As the atolls have eroded, the monk seal mothers and pups have been forced into greater concentrations in the remaining sandy habitat fit for haul out and nesting sites. For example, in 1963, the Whale-Skate Islands were 6.8 hectares in size and were the second largest island complex in French Frigate Shoals. By the late 1990s, the islands had virtually disappeared due to sea level rise and shifting sands, and the animals that were displaced had moved to nearby Trig Island. Scientists believe that the crowding effect of mothers and pups being forced into ever smaller space resulted in so much stress and competition that pups could not grow to a healthy condition. In a weakened state they were more at risk from Galapagos sharks (*Carcharhinus galapagensis*), a species that was not previously even known to eat monk seals (Baker 2006).

Finally, the status of the monk seal population is confounded somewhat by recent shifts in the population distribution. Substantial numbers of healthy monk seals have now migrated from the NWHI habitat to the main Hawaiian Islands, where some pupping has occurred. This has raised potential management problems for the Council as the in-migration has resulted in increased interactions with fishermen, though most if not all such events have been in state waters and thus not in the area under Council jurisdiction. There also have been issues related to public disturbance of monk seals that have hauled out on local beaches in the main Hawaiian Islands. Because of the combination of increasingly stringent monk seal regulation, an ever increasing federal scientific and enforcement presence, and a growing monk seal population in the main islands where there is a significant tourist and fishing population, it is likely that monk seal issues will continue to come before the Council, growing more legally and socially complex over time. The Council continues to monitor this situation and is prepared to respond appropriately if new information demonstrates a need for action.

TABLE 15: HAWAI'I ARCHIPELAGO FEP: SUMMARY OF COUNCIL ACTIONS		
DATE	ACTION	MEASURES
2010 Jan 14	FMP final rule	Restructured all fishing regulations under the previous FMPs in accordance with the new FEP.
2010 Sep 3	Amendment 1	Established eligibility requirements and procedures for reviewing and approving Community Development Program plans.
2010 Nov 10	Amendment 2	Established Hancock Seamounts Ecosystem Management Area and continued the moratorium on armorhead and other seamount groundfish until the armorhead stock is rebuilt.
2011 Jun 27	Amendment 3	Established a mechanism for specifying ACLs.
2016 Feb 12	Amendment 4	Refined the descriptions of EFH and habitat areas of particular concern for bottomfish and seamount groundfish management unit species.
2018 Nov 1	Amendment 5	Reclassified certain management unit species as ecosystem component species.

12.2 American Samoa Archipelago Fishery Ecosystem

In 2005, the Council completed the American Samoa Archipelago FEP. A newly created a special American Samoa Archipelagic FEP Plan Team oversaw its development with input and help from the SSC and the American Samoa REAC as well as various American Samoa community groups. From the outset the FEP Plan Team was tasked to make use of a precautionary approach that links approved fishery activities to adaptive management techniques. This FEP marked a major departure from previous strategies of species-specific management and was one of the first of its kind in the United States.

As an example, it was generally assumed by visiting NOAA biologists from the National Ocean Service and NMFS that the reefs of American Samoa were in poor condition due to the absence of apex predators such as large jacks and the generally low fish abundance. These conditions were assumed to be the result of overfishing. This was an important assumption because it would lead to the conclusion that the FEP should severely limit reef fishing by local residents.

At the 139th Council meeting in 2007, SSC member Marlowe Sabater, chief biologist for the American Samoa DMWR who would later work at the Council, reported that his work did not show a link between the absence of apex predators and relative health of the fishery. Between 2004 and 2007, Sabater analyzed the patterns in species composition in the contemporary fishery surveys, in-water surveys and archaeological midden fish-bone data and found that the historical fish utilization is consistent with what is readily available on the reefs (Sabater and Carroll 2009). Sabater's work led reef biologists to understand that the patterns we see on reefs are driven by not only fishing but also by changes in the ecosystem. The American Samoa FEP takes these ecosystem considerations into account.

Lauvao Stephen Haleck of American Samoa served on the Council
from 2003 to 2012 and as the Council's chair in 2010.

Since 2011, the Council has been specifying ACLs under the
American Samoa Archipelago FEP. This annual specification process
includes ACLs for coral reef fish species, crustaceans, bottomfish and
precious corals.

Addressing Data Needs for Bottomfish Management

In 2019, a stock assessment for the American Samoa bottomfish
fishery was presented by NMFS to the Council indicating that the fishery
was overfished and experiencing overfishing. This was quite a different
picture of the stock than provided in previous assessments. The different
outcomes were attributed largely to the way the data was filtered.

There is no federal permit or reporting requirements and no near-
real-time monitoring for the bottomfish fishery in American Samoa.
The fishery's ACL and accountability measures are specified by NMFS
based on recommendations by the Council using data collected by the

American Samoa DMWR. The local agency relies on creel surveys, in which its staff members interview fishermen about their catch and fishing trip. The data from the Commercial Receipt Book System was used to verify the catch from the creel survey.

The change in stock status required the Council to develop a plan amendment that would rebuild the stock in the shortest time practicable. A federal rebuilding plan will utilize information from the stock assessment and select a level of catch that would allow the bottomfish stock to rebuild within 10 years. This plan and the one for the Guam bottomfish fishery (see below) are the first ever rebuilding plans that the Council has had to develop in its 44-year history.

The Council expressed concerns about the robustness of the data from DMWR used in the 2019 stock assessment. It is working with DMWR to address issues with data collection and the status of the stock. Under a project funded by the Territorial Science Initiative, a contractor worked in DMWR to train major vendors on commercial receipt book process and develop an incentive to promote compliance.

The Council also funded the development and training of an electronic reporting system that allows fishermen to self-report their catch and effort in near real time.

Fisheries Development

In continued efforts to help develop fisheries in American Samoa, the Council helped the community on the island of Aunuʻu with installation of a flake ice machine and storage bin for use by the island's alia fishermen. The Council also hosted open fishing tournaments in Pago Pago on the island of Tutuila to encourage fishermen to fish and to assist the American Samoa DMWR with data collection.

Alia fishing vessels anchor in the harbor of Aunu'u, a small volcanic island located 11 nm from the main island of Tutuila. The Council helped the Aunu'u fishermen with installation of an ice machine and storage bin.

International Management

The South Pacific Regional Fishery Management Organization was formed in 2009 (and came into force in 2012) to fill the gap in international management for non-highly migratory species, such as jack mackerel, occurring in the high seas in the South Pacific Ocean. The Council monitors and cooperates with the organization as necessary to manage shared stocks, such as those that may overlap the waters of American Samoa.

TABLE 16: AMERICAN SAMOA ARCHIPELAGO FEP: SUMMARY OF COUNCIL ACTIONS

DATE	ACTION	MEASURES
2010 Jan 14	FMP final rule	Restructured all fishing regulations under the previous FMPs in accordance with the new FEP.
2010 Sep 3	Amendment 1	Established eligibility requirements and procedures for reviewing and approving Community Development Program plans.
2011 Jun 27	Amendment 2	Established a mechanism for specifying ACLs.
2011 Jun 27	Amendment 3	Established management measures rule for noncommercial and recreational fishing and prohibited commercial fishing within the Rose Atoll MNM.
2018 Nov 1	Amendment 4	Reclassified certain management unit species as ecosystem component species.

12.3 Mariana Archipelago Fishery Ecosystem

Bottomfish Overfished Determination

The NMFS 2019 stock assessment of the bottomfish in the U.S. Pacific territories determined the Guam fishery to be overfished but not subject to overfishing. It is widely believed that the data collection in place did not accurately capture the bottomfish fishery. The creel surveys were designed to capture all fisheries in general rather than specific ones like the bottomfish fishery. The Council is working with NMFS and Guam's government and fishermen to address this issue. As with American Samoa, the status differed from previous assessments due to changes in methodology to filter the data. The Council is working with the parties to improve the collection of accurate catch and effort data from the fishermen, including outreach on the importance of data and how they are used in stock assessments and the setting of ACLs.

Dwindling Access to Safe Fishing Grounds

In the Marianas, a major conflict between fishing and alternative ocean uses relates to military activity. A long-standing issue brought up by the Council's CNMI Advisory Panel has been the naval ships that anchor at 11 prepositioning sites off Saipan and concerns that the anchoring chains are damaging coral and fish habitat.

Naval vessels anchored in prepositioning sites off Saipan (circa 2015) are of concern to CNMI fishermen as the anchoring chains may damage coral and fish habitat. *WPRFMC photo/Sylvia Spalding.*

Issues with the Defense Department escalated in the early 2000s as the United States realigned its strategic focus to the western Pacific. Guam and CNMI—the closest U.S. soil to China—have been the focal points for a massive planning effort to relocate troops, build training ranges and increase the frequency and intensity of military training activities. The incremental loss of fishing access has progressed from the establishment of firing ranges in prime fishing grounds south of Guam to an expansion of the safety zone around Farallon de Medinilla, a prime fishing ground in the CNMI, and the addition of a safety zone at Ritidian Point on Guam. In the spring of 2013, the CNMI Joint Military Training environmental impact statement was published, proposing plans to make a live fire range on the island of Pagan and an amphibious landing ramp on the island of Tinian. The proposal would limit access on Pagan and approximately two-thirds of Tinian, including an area where cultural fishing methods have been traditionally practiced. The Defense Department's planning efforts were challenged twice under the National Environmental Policy Act's requirement for a range of reasonable alternatives. For Guam, the challenge resulted in selection of a training range alternative that did not require additional purchase of federal lands but did cause a portion of the Ritidian Wildlife Refuge to be acquired by the Defense Department. The fates of Tinian and Pagan were unknown at the time of the publication. In 2017, it was announced that preparation of the environmental impact statement continues with an expected release in late 2018 or early 2019. In 2020, no information on its status was available. Continued engagement with the Defense Department on its cultural and natural resource management plans for the Marianas is the most promising public avenue toward improving access and avoiding fish habitat destruction at traditional fishing grounds.

The serious resource use and availability concerns posed by the U.S. Department of Defense's presence and activities in the Mariana Archipelago have been considered by the Council. As early as July 2009, at the Council's 145th meeting in Kailua-Kona, the Council indicated that it did not support a permanent closure for 10 nm around the Farallon de Medinilla for the Navy to use as a bombing target. The Council was responding to a draft environmental impact statement analyzing the proposal's potential impacts and implications, including total closure of the area for local fishermen. The island is 54 nm north of Saipan and an important fishing area for local Saipan and Tinian fishermen for mafute (redgill emperor). The Council supported maintaining the status quo which was a 3 nm closure around the island.

In September 2013, the United States through U.S. Public Law 113-34 conveyed submerged lands extending 3 nm seaward from the 14 islands of the CNMI to the commonwealth. However, after enactment of the law, President Obama through Proclamation 9077 on January 15, 2014, withheld CNMI control of submerged lands adjoining Farallon de Medinilla and Tinian, where military training occurs. Also withheld were the submerged lands adjoining Asuncion, Maug and Uracas (collectively the Islands Unit of the Marianas Trench MNM) until a coordinated management plan for the marine monument was developed.

Arnold Palacios, Council chair and secretary of the CNMI Department of Lands and Natural Resources, took issue with Obama's decision. He told the Council as its March 2014 meeting on Saipan. "[The United States] gave us the submerged lands and then they took it back. … I don't think [the Departments of Commerce and the Interior] are interested in co-management. The Antiquities Act [used to create the monument] doesn't allow the co-management that was promised to us by the White House envoy."

Arnold Palacios of the CNMI served as Council chair in 2013 and 2014. *WPRFMC photo/Edwin Ebisui Jr.*

No-take marine preserves were established by legislative action for Saipan (e.g., Managaha Marine Conservation Area, Forbidden Island Marine Sanctuary and Bird Island Marine Sanctuary), Rota (Sasanhaya Bay Fish Reserve) and Tinian (Tinian Marine Reserve). These conservation areas were in addition to the islands of Guguan, Uracus,

Maug and Asuncion, which were designated as wildlife conservation areas in accordance with CMC §5104(a)(5) and Article XIV(2) of the CNMI Constitution. Additionally, the Lighthouse Reef Trochus Sanctuary and the Laolao Bay Sea Cucumber Sanctuary were established in Saipan.

Guam has five MPAs along its fringing reefs established in 1997: Pati Point Preserve, Tumon Bay Preserve, Piti Bombhole Preserve, Sasa Bay Preserve and Achang Reef Flat Preserve. These MPAs were created by Guam Division of Aquatic and Wildlife Resources with coral reef initiative funds from Washington, D.C. These closures combined with the 30% of the island already closed by the U.S. military have resulted in a large portion of the safest and best shoreline and fringing reefs being off limits to local fishermen. Guamanians say this has polarized their community with environmentalists against indigenous traditional fishermen who face increased fishing hazards at sea.

The expansion of no-take MPAs in Guam has been accompanied by the growth of the ocean-recreation industry. However, few of the fishermen are able to make the transition from fishing to tour boat operations, so these jobs do not necessarily materialize for them. This has created resentment and social discontent. The Council has remained a principal forum in which frustrations are vented and the issues studied and discussed.

Further regionwide pressures have come from environmental advocacy groups such as The Nature Conservancy and the Pew Environmental Group. The Micronesian Challenge has been adopted by several nearby independent island nations as part of a regional effort to effectively conserve 30% of the participants' nearshore coastal waters and 20% of their land by 2020. The Micronesian Challenge was jointly created by The Nature Conservancy and the Ocean Conservancy. It was formally adopted in 2006 by the Republic of Palau, the Federated States of Micronesia, the Republic of the Marshall Islands, and the U.S. territories of Guam and the

CNMI. In 2020, the Micronesian Challenge set a vision to increase the effective management goals from 30% to 50% of the marine resources and from 20% to 30% of terrestrial resources across Micronesia.

MPA CASE STUDY: DROWNING DISPARITIES OF GUAM'S CHAMORRO FISHERMEN

Creation of MPAs in Guam provided an opportunity for the Council to support a study on the effects of MPAs on the safety of the resident population, as promoting safety of human life at sea is one of the 10 national standards of the MSA. In 2010, the Council engaged the National Institute of Occupational Safety and Health (NIOSH) to conduct a study of the impact of MPAs on the safety of fishermen on Guam (Lucas and Lincoln 2010). The objectives of the study were to determine 1) whether the loss of inshore fishing grounds to MPAs had put fishermen at greater risk; 2) whether the drowning rate of fishermen changed after the establishment of the MPAs; and 3) whether the location of drowning changed after the establishment of the MPAs.

The MPA impact on fishermen drowning risk report prepared by NIOSH for the Council was released on December 31, 2010. It was determined that there were no previous studies of drowning rates on Guam, and, in fact, no studies were found for drowning rates for other islands in the Western Pacific Region. Therefore, data were collected on Guam using government documents, death certificates, census data and other sources. The data covered a 24-year period that spanned both pre- and post-MPA designation.

When observed as one population (all residents aggregated), there was no statistically significant difference in drowning rates pre and post the creation of MPAs. However, the differences were significant when the drownings were separated into Chamorro and non-Chamorro residents. The reduction in drowning rates in one population masked the increases in the other. MPAs did not affect the aggregate drowning rate, but the MPAs affected the drowning rates of the two sub-groups differently. For Chamorro fishermen, the drowning rates increased post MPA designation by 125%, while for non-Chamorro fishermen, the drowning rate increased by 50% after MPA designation. Additionally, drowning records were divided into Guam's western coast, considered "safe" by traditional fishermen, and eastern coast, considered "dangerous." Most of the new MPAs were concentrated on the western coast. Analysis showed that 63% of the increased Chamorro drownings occurred on the eastern coast of Guam.

In conclusion, the NIOSH report found that pre-MPA the residents of Guam fished primarily on the protected western Coast. Non-Chamorro residents were primarily recreational users, who scaled back their fishing activities when the MPAs were established. Chamorro residents, on the other hand, were subsistence fishermen. After creation of the MPAs, they were forced to travel further and fish in more hazardous conditions and locations, resulting in a higher risk of drowning. The Council strongly believed that consideration of such social impacts should be part of National Environmental Policy Act analysis for the establishment of MPAs in the U.S. Pacific Islands.

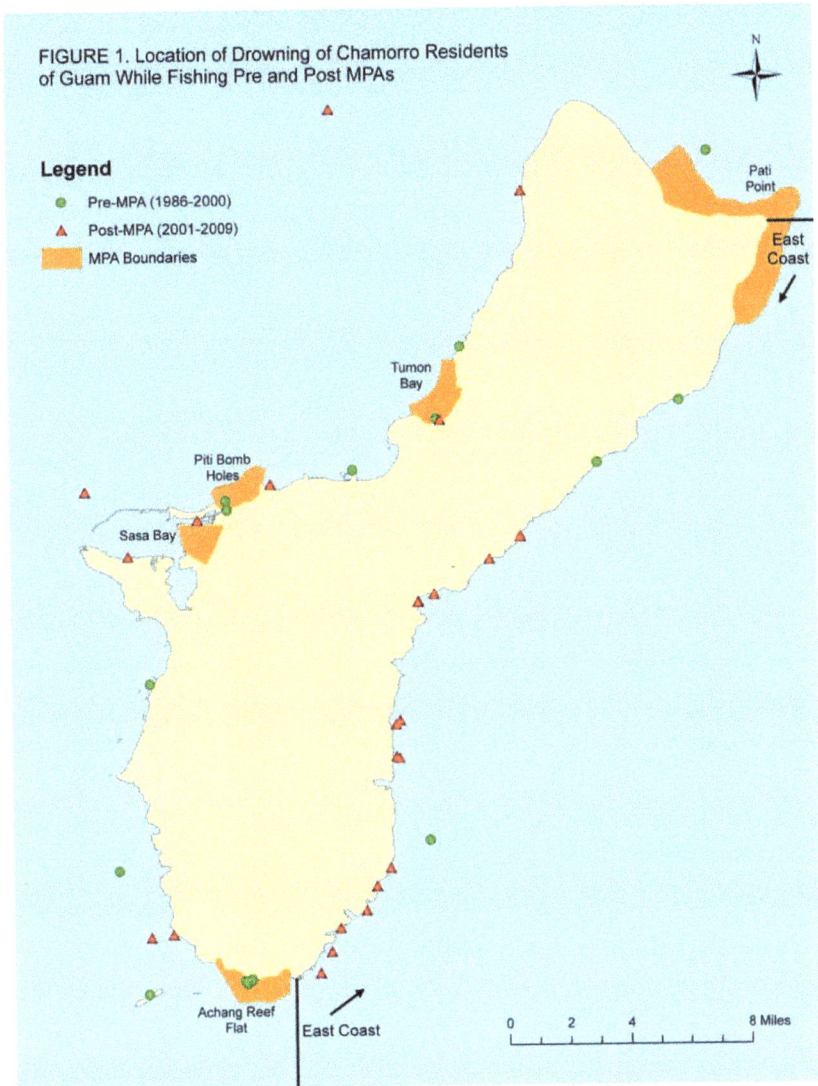

FIGURE 1. Location of Drowning of Chamorro Residents of Guam While Fishing Pre and Post MPAs

Legend
- Pre-MPA (1986-2000)
- Post-MPA (2001-2009)
- MPA Boundaries

Pati Point

East Coast

Tumon Bay

Piti Bomb Holes

Sasa Bay

Achang Reef Flat

East Coast

0   2   4   8 Miles

A National Institute for Occupational Safety and Health study requested by the Council brought to light the impact of no-take MPAs on Guam's Chamorro fishermen. As they subsist on fishing, they traveled further to more hazardous locations, resulting in a 125% drowning rate increase. *NIOSH illustration.*

Marianas Trench Marine National Monument

In January 2009, at the end of his administration, George W. Bush established the Marianas Trench MNM in the CNMI after a controversial campaign with support primarily from mainland environmentalists, including the Pew Environment Group. Ultimately, Bush adopted a plan developed by the indigenous CNMI leaders who negotiated what they could, as it was widely believed that the monument was going to be proclaimed regardless of how the Mariana communities felt about the issue. As in Hawai'i, this led to continued social conflict, principally between those who wished to permanently take away the fishing and extraction rights of island stakeholders and those who wished to continue the sustainable harvest of marine resources.

In the dozen years after the Marianas Trench MNM was proclaimed, it has remained an example of a series of broken promises by the federal government to the people of the Northern Mariana Islands. In July 11, 2017, CNMI Governor Ralph Deleon Guerrero Torres (2015–present) wrote to Interior Department Secretary Ryan Zinke in response to President Trump's Executive Order 13795, Implementing an America-First Offshore Energy Strategy, which directed the Department of Commerce in consultation with the Secretary of the Interior to review the marine national monuments. Torres said:

> In sum, the U.S. Fish and Wildlife Service and NOAA have been working for almost ten years now with the CNMI government, Department of Defense, Department of State, U.S. Coast Guard, and others to develop a monument management plan. Today, the work remains unfinished for a management plan that is supposed to provide for public education programs, traditional access by indigenous persons, scientific

exploration and research, consideration of recreational fishing if it will not detract from the monument, and programs for monitoring and enforcement.

The CNMI's participation and negotiation with our federal counterparts to ensure the inclusion of our preferences and desires have in all honesty been stretch thin throughout the process. Nonetheless, we remain persistent in our efforts because we are greatly concerned about the possibility of losing access to the waters for fishing purposes and to the natural resources contained in the three Units.

For example, at the time of the creation of the Marianas Trench MNM, the CNMI had not yet actually received the formal patent to our submerged lands extending from the islands of Farallon de Pajaros (Uracus), Maug and Asuncion. Even though we have since been granted rights to the submerged lands, access to the natural resources in those submerged lands are now restricted and the Monument's prohibitions on mineral extraction in all three Units takes away potential economic opportunities for the CNMI. Such a result must be reconsidered.

In closing, the federal government has failed to provide the necessary resources to realize the promises made to the CNMI during the negotiations for the creation of the Marianas Trench MNM. This includes the establishment of a co-management plan, an increase in patrols for illegal fishing in the waters of the CNMI, and other incentives such as the building of a visitors' center in the CNMI which could all help mitigate for the loss of access to our natural resources. Although the CNMI does in general support the environmental and protections behind the creation of the [monument] as good stewards, we have since the designation of this monument under the Antiquities Act been required to carry the burden of an over-expansive federal program that should be reassessed.

Likewise, the Senate of the 20th Northern Marianas Commonwealth Legislature on July 11, 2017, passed Senate Resolution No. 20-05 in response to President Trump's Executive Order 13795 and Executive Order 13792, Review of Certain National Monuments Established Since 1996. The resolution noted that the CNMI's opposition to a marine national monument had been conveyed to President George W. Bush by the mayors of Saipan, Tinian and Rota (June/July 2008); by the Senate President Pete P. Reyes and Speaker of the House Arnold I Palacios (August 2008); by Senate Joint Resolution 16-04 (September 2008); by CNMI Governor Felix P. Camacho (October 2008); and by entire membership of the 10th Micronesian Chief Executive's Summit, which is comprised of all of the U.S.-freely associated states and territories in the North Pacific, including the presidents of the Federated States of Micronesia, Republic of Palau and Republic of the Marshall Islands and the governors of Guam, CNMI, Chuuk, Kosrae, Pohnpei and Yap. The resolution's listed the following grievances since Bush's proclaiming the Marianas Trench MNM, in January 2009:

- The responsible federal management authorities failed to meet the January 2011 deadline for development of management plans as required by Presidential Proclamation No. 8335 and these plans are still not complete today.

- The responsible federal management authorities failed to include the CNMI in a meaningful co-management agreement over monument resources, as promised during the assessment period.

- There have been no measurable economic benefits associated with designation of the Marianas Trench MNM, as promised during the assessment period.

- No Marianas Trench MNM visitors' center has been established, as promised during the assessment period.

- There has been no funding increase to the U.S. Coast Guard to step up marine surveillance, protection and enforcement in the Monument waters, as promised during the assessment period.

- President Obama withheld the conveyance of the nearshore marine areas surrounding the Monument islands of Uracus, Maug and Asuncion until such time as the CNMI government agreed to management of these waters under existing Monument guidelines; the federal government inferred unencumbered conveyance of nearshore marine areas surrounding every island in the CNMI during the assessment period.

The Senate resolution called for the restoration of fishing an mineral resource extraction rights to the CNMI that were taken through the proclamation establishing the Marianas Trench MNM and for fishery management with the monument to be returned to the Council as authorized by the MSA while mineral resource extraction be treated as other submerged lands found in the CNMI. The resolution also conveyed disapproval of the establishment of any National Marine Sanctuary in the CNMI.

Traditional Fishing and Community Management

In its efforts to more fully engage indigenous fishing communities in the federal fishery management process under the Mariana Archipelago FEP, the Council beginning in 2010 held regular briefings at the Mayors Council of Guam. Conversations with the mayors and community members revealed that implementation of MPAs had resulted in the loss of access to and gathering of traditional resources and foods. This had created a problem particularly for the people of the village of Malesso (i.e., Merizo), who were angered by the regulations developed by the territorial fishing agency for the MPA at Achang Flats and who thought the NIOSH report (which some employees at the Guam agency viewed as flawed) highlighted additional problems associated with designated and de facto MPAs on Guam.[76] While in MPAs established by Guam the ban on traditional taking of 'i'i (juvenile jack, or trevally), manahak (juvenile rabbitfish) and ti'ao (juvenile goatfish) during their seasonal runs had been exempted since 2006, it was believed that more could be done to support traditional marine resource use and management.

Subsequently, a series of public meetings and workshops in Malesso in 2012 led to an agreement by the Guam Department of Agriculture, the Mayor's Council of Guam, the Mayor of the Village Malesso and the Council to use the village of Malesso as a pilot site for the development of a community-based marine resource plan.Through the coordinated efforts of the partners, a series of workshops were held in the village in 2013, which culminated in the release of the Community-Based Management Plan for Malesso Coastal and Marine Resources in 2014. One of the foremost objectives and actions in the plan was to improve to improve the seasonal run fishing exemption in the Achang Flats MPA. The plan noted the lack of data gathering

---

[76] Charles Ka'ai'ai in discussion with Michael Markrich, March 1, 2022.

and recommended the training of community members to gather this data, thereby improving stock assessments, and also by allowing the mayor to control the seasonal harvest, while the Guam Department of Agriculture would retain authority and oversight. Handing control to the mayor would streamline the process and thus lead to more responsive adaptive management (Malesso Mayor's Office 2014, pages 13, 103-104).

In 2015, the Council met with community members from the CNMI Northern Islands to help them start the process to develop their community-based fishery management plan. Council efforts with the Guam communities continued in June 2016 with a coral reef mapping. Other important cultural work involved Council support for an archaeological analysis of ancient fishery village sites in Guam that point to connections with prehistoric sites in Taiwan.

Fisheries Development

The Council has long held the view that fishery development opportunities exist within the Mariana Archipelago. Between 2010 and 2015, through the SFF, the Council undertook five fishery development projects in the archipelago. It worked with the Guam Port Authority to repair and rehabilitate the Agat Small Boat Marina docks. It partnered with the Secretariat of the Pacific Community to assess the skipjack tuna biomass in the U.S. EEZ around the CNMI. In 2014, it completed a Marianas Seafood Marketing Plan that evaluated the potential fisheries development opportunities in the CNMI and Guam. It worked with the CNMI Department of Lands and Natural Resources to improve the Garapan Fishing Base and completed an evaluation of an engineering plan for a longline dock.

In October 2018 during the 174th Council meeting on Saipan, the Council presented CNMI Governor Torres with a check for $250,000, to assist with the development of the Commonwealth's bottomfish fishery. The Council and CNMI governor's office offered a four-day vessel maintenance and capacity-building training on Saipan in 2019. Funds were also provided for a new ice machine for the Garapan public market.

On Guam, the Council, working with the Guam Department of Agriculture and local fishing organizations, contributed funding from the SFF to the construction of a fishing platform in Hagatna, to provide safe fishing access for Guam anglers, including senior citizens and disabled citizens.

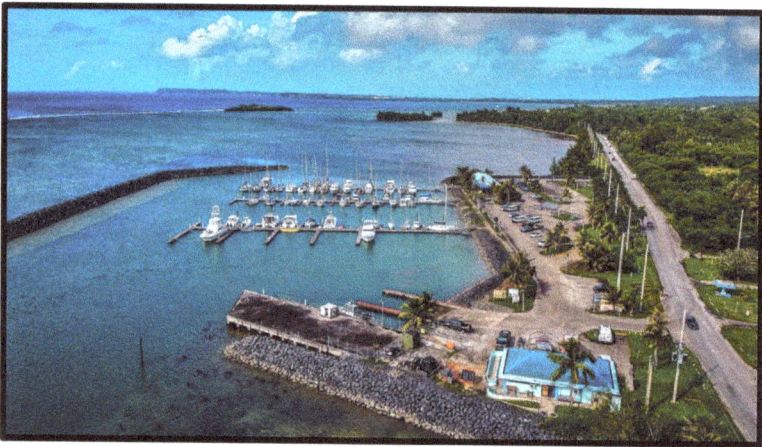

The repair and rehabilitation of Guam's Agat Small Boat Marina dock was one of the projects in the Mariana Archipelago that benefitted from funding from the Western Pacific SFF. *DoubleBlue Images photo/Tim Rock.*

TABLE 17: MARIANA ARCHIPELAGO FEP: SUMMARY OF COUNCIL ACTIONS

DATE	ACTION	MEASURES
2010 Jan 14	FMP final rule	Restructured all fishing regulations under the previous FMPs in accordance with the new FEP.
2010 Sep 3	Amendment 1	Established eligibility requirements and procedures for reviewing and approving Community Development Program plans.
2011 Jun 27	Amendment 2	Established a mechanism for specifying ACLs.
2013 Jun 3	Amendment 3	Established management measures for noncommercial and recreational fishing and prohibited commercial fishing within the Marianas Trench Marine National Monument.
2016 Dec 29	Amendment 4	Removed the CNMI bottomfish prohibited areas for medium and large vessels measuring 40 feet (12 meters) and longer.
2018 Nov 1	Amendment 5	Reclassified certain management unit species as ecosystem component species.

12.4 Pacific Remote Island Areas Fishery Ecosystem

Fisheries targeting species managed under the Pacific Remote Island Areas FEP are limited due to regulations implemented through the Council process, presidential executive orders and the USFWS. The Pacific Remote Island Areas FEP requires fishermen to obtain a special CRE permit, Western Pacific precious coral permit, Western Pacific crustaceans permit or Pacific Remote Island Area bottomfish permit depending on the species targeted. Due to the absence of consistent fisheries data in the Pacific Remote Island Areas, the annual SAFE report for the Pacific Remote Island Area FEP does not at this time include fishery performance data. While information on ecosystem considerations is provided, the data integration chapter is not fully developed.

TABLE 18: PACIFIC REMOTE ISLAND AREAS FEP: SUMMARY OF COUNCIL ACTIONS		
DATE	ACTION	MEASURES
2010 Jan 14	FMP final rule	Restructured all fishing regulations under the previous FMPs in accordance with the new FEP.
2011 Jun 27	Amendment 1	Established a mechanism for specifying ACLs.
2013 Jun 3	Amendment 2	Established management measures for noncommercial and recreational fishing and prohibited commercial fishing within the Pacific Remote Islands MNM.

Chapter 13

Four Decades of Fishery Management—The Legacy of the Council

Fishing conditions in the U.S. Western Pacific were very different when the Council was established in 1976. Foreign fleets could fish up to three miles from shore. Domestic fishing in the islands was primarily for community consumption, and the local fishermen who did fish commercially weren't used to being regulated. There was no federal regulatory process in place to control the communities and the industry from overusing the fishery resources.

Like the nation's other regional fishery management councils, the Western Pacific Council had the authority and the responsibility to work with the fishing communities and industry, environmentalists, varied stakeholders, and the local and federal governments to develop reasonable fishing laws using the process and national standards outlined in the MSA. The Western Pacific Council took on this task when scientific information about the region's fishery resources was only rudimentary. The questions of who was out fishing, how they were fishing and what and how much they caught were unknown. Environmental laws were developing and proliferating. The Western Pacific Council was challenged by transportation, communication, political, social and cultural issues unique to a region comprised of one state, three territories and eight remote unincorporated islands dispersed over wide distances of the Pacific. Each island area had its own indigenous communities with fishing traditions spanning millennia prior to Western conquest; its own legal agreements with its indigenous peoples and with the United States; and its own set of official languages (five in total), with English not being the first language for many.

Faced with these challenges, the Western Pacific Council fulfilled its MSA obligations as a good steward of the region's fisheries for the benefit of the nation and the local communities. Over 44 years, it has promoted science-based management balanced with the rich cultural traditions of the peoples of the region.

The Council's success can be attributed to three factors: First, it made strong efforts to engage the local fishermen and communities in the process of identifying and addressing the region's fishery problems and opportunities. Second, its Council members dedicated themselves to learn about and act on behalf of the region's different island areas to find solutions. And, third, its executive director was experienced in working with local and federal governments.

Consulting with Fishing and Indigenous Communities—The Council Process

When the Council was established, the communities in Hawai'i were arguably the only ones in the region familiar with working with the federal government and within the federal public participatory process. Hawai'i had been a state with Congressional representatives since 1959. By comparison, the first elected delegate to hold office in Congress for Guam was in 1972 and for American Samoa in 1981. The Northern Mariana Islands did not become a commonwealth of the United States until 1978 (the year its constitution came partially in effect; it would become fully effective in 1986) and did not have an elected delegate in Congress until 2009. Like the delegates from Guam and American Samoa, the CNMI delegate is a non-voting member of Congress. Citizens of the U.S. territories cannot vote for the president, and American Samoans are U.S. nationals and not citizens. Even though they have no vote in Congress and cannot vote for the president, the citizens and nationals of the U.S. territories have had their fisheries impacted by the acts and executive orders of these branches of the federal government, as has Hawai'i.

The Council provided the U.S. Pacific Islanders with a system whereby fishermen and other community members could have their voices heard and acted upon in the federal fisheries decision-making process.

Ufagafa Ray Tulafono, a longtime director of the American Samoa DMWR, said what the Council accomplished was giving the local people the opportunity to develop their own recommendations and bring them to the Council to approve, allowing them to come up with what they thought was right for their resources. He recalled the impact of the Council on the islands of Ofu and Olosega in the Manu'a island group:

> One specific meeting that I remember was the one that we held in Ofu and Olosega ... people were talking about it, not only here in Tutuila, but also in Manu'a ... how we [are] able to go to Manu'a and have a meeting with all these people from around the Pacific. ... The people of Manu'a, they were very happy ... they said, "This is the first time that a meeting of this nature has been held in Manu'a." ... The people were feeling really great ... because of the recognition. The Council recognized the people in Manu'a and also the fishermen

in Manuʻa. And that's why they were very happy and very interested. … so many of them asked questions … they contributed to the meeting that we held in Manuʻa. … I did the translation, and the questions that were asked were really great from our native people of the concerns they have in their fishery in the Manuʻa Islands.[77]

Manny Duenas, long-time GFCA president and Council member from Guam, echoed those feelings, noting that one of the MSA national standards is to take into account the importance of fishery resources to fishing communities and to provide for sustained participation of, and to minimize impacts to, such communities consistent with conservation requirements. For him, at the end of the day, it's all about people: "That's the greatest thing about the Council … because it is our communities that make a fishery and it's our communities that depend on the fishery."[78]

Because the Council was regionally based and comprised of members from the region, it was sensitive to the need to especially acknowledge and include the native peoples and, at the same time, to educate the federal government about the islands' indigenous fishing communities. Judith Guthertz, a Council member from Guam, put it this way:

It's probably unique in many ways to [the Western Pacific Council] and also to the [North Pacific Council] that covers Alaska, that we have indigenous people and their traditions and their culture need to be respected. Hawaiʻi, Guam, the Northern Marianas, American Samoa, these are tremendous areas of the United States.

---

[77] Ufagafa Ray Tulafono in discussion with Sylvia Spalding, October 19, 2019.

[78] Manuel Duenas interviewed by Judy Amesbury for the Council, December 15, 2020.

Part of my responsibility, especially when I served as chair for two years, was to educate these federal officials who came to our meetings about the uniqueness of these islands and their traditions and their cultures, and the fact that they historically have always respected the fishery resource, never exploited that. Even today, our fishermen do not exploit the resource. They know what they should and shouldn't do. They know how much fish they should fish for, and they know when they're taking too much. It's others that are fishing now in these island areas who are exploiting their resource, and that's one reason now we have to look at catch limits. We never had to look at catch limits before, especially on Guam.[79]

Judith Gutherz of Guam chaired the Council from 2000 to 2002. *WPFMC photo.*

---

[79]   Judith Guthertz interviewed by Judy Amesbury for the Council, December 15, 2020.

In American Samoa, cultural issues include chiefly titles and land boundaries, which Alo Paul Stevenson, governor of the Eastern District, described as the fabric of the Samoan culture and the reason the Office of Samoan Affairs exists.[80] Disputes about titles go first to this office for settlement. Additionally, Stevenson explained, each village has its own rules which are the law. He couldn't recall a village rule ever being turned down by a court to legislate by adjudication, and he said this is what gives strength to the culture and to the people in the village to manage their own affairs. He added that the Western Pacific Council is important because its management is by people from the islands who understand the local issues and problems. "People always come from outside with new ideas," Stevenson said. "But if it does not embed itself in the culture and the know-how of where we come from, it won't work … you believe your own people. You listen to what they say and you believe it. And I think that sometimes takes a long time."

Alo Paul Stevenson of American Samoa
chaired the Council in 1992.

---

[80]   Alo Paul Stevenson in discussion with Sylvia Spalding, October 18, 2019.

The traditional island way of managing fisheries and that of the federal government clashed in the early years of the MSA, when the latter pushed for the harvest of all fisheries at maximum sustainable yield in the newly established federal waters extending out 200 miles from shore and to allow foreigners to catch what the American fishermen could not catch to reach that target. Paul Callaghan, the Council's SSC chair for three decades, explained that the federal government's early goal, prior to the formation of NMFS when the agency was called the Bureau of Commercial Fisheries, was to expand the American fishing industry. He said that many of the employees of the Bureau of Commercial Fisheries came from the fishing industry and their idea was to take all the possible fish you could, i.e., maximum sustainable yield. "If the American didn't take it," Callaghan recalled, "then we should allow foreigners to take it."[81]

"I asked myself what it would have been like if there had been no Council system," Callaghan continued. "I suspect there would have been more overfishing by the 1980s and 1990s than actually occurred. I think when the Council came along it provided a moderating influence between those who wanted to catch maximum sustainable yields and those who wanted to catch nothing."

Another significant change brought by the Council was inclusion of tuna as a federally managed species. Frank Farm Jr., a Native Hawaiian and early Council member, said when he heard that "tuna is not fish." under the MSA, those words shook him up and he wondered, "What's going on here?"[82]

Tuna was, and clearly continues to be, the most important fish to the Pacific Islands in the sense of economic value. To Callaghan, not

81 Paul Callaghan in discussion with Sylvia Spalding, January 30, 2020.
82 Frank Farm Jr. in discussion with Sylvia Spalding, November 23, 2020.

being able to regulate tuna, as was the case in the early years of the MSA, was tantamount to saying, "Well, you just can't do anything worthwhile in the Pacific." He added that "getting tuna finally under the MSA was a great accomplishment, and one that was fought tooth and nail by the large American tuna purse-seine interests that fished all over the Pacific in other islands' territories."

Open mindedness and sensitivity to not only the indigenous people and fishing communities but also to the industry is part of the Council's legacy for Roy Morioka. In particular, he pointed to the engagement of the Hawai'i commercial bottomfish fishermen in workshops to improve the main Hawaiian Islands bottomfish stock assessment, which he described as "groundbreaking."[83]

Roy Morioka of Hawai'i chaired the Council in 2004 and 2005. *WPRFMC photo.*

The Council's success in working with fishing and indigenous communities was the result of actively inviting them to be part of the Council family. The philosophy was to engage them, to get their input and support and to help them when possible. The Council placed a high

---

[83]   Roy Morioka in discussion with Sylvia Spalding, November 23, 2020.

value on public participation and worked with the media and outreach to invite fishing groups, indigenous communities and the public to participate in Fishers Forums, Council meetings and special events. The Council also educated and engaged the communities through its newsletters, other publications and videos. The Council similarly worked to educate the federal government about the region, its fisheries and culture and what may or may not work in it.

Groundbreaking Problem Solving—The Council Members

The openness of the Council process allowed input from a wide range of sources including commercial, noncommercial and indigenous fishing communities and environmental groups. Instead of allowing various factions to continue to butt heads, the Council's important work and part of its legacy is its ability to blend the ideas and balance the needs of various different communities and mesh them with the recommendations of scientists and the constraints of government in a positive way to find win-win management measures that meet the test of time. Often the solutions were a compromise, which may have been readily available or a long time to accomplish. Regardless, the Council looked at every avenue until it came to a conclusion based on facts.

The term for a Council member is three years, and the MSA allows for a Council member to serve up to three consecutive terms. Those years are needed to groom a seasoned Council member. Jim Cook, a member of the longline industry, recalled that he was just "a guy that caught fish" when he joined the Council.[84] He said it took a long time to learn what goes on. Part of that learning took place when the Council met in its various island areas, which allowed the people from Hawai'i, American Samoa, Guam and the CNMI to acquire a better

---

[84] Jim Cook in discussion with Sylvia Spalding, November 23, 2020.

understanding of each other's local needs and how they are different from those in other areas of the region.

Jim Cook of Hawaiʻi chaired the Council from 1997 to 2000. *WPRFMC photo.*

During Council meetings, the members learned from scientists, Council staff and other guest presenters about fisheries modeling, about competitive foreign fishing fleets and about the law of the sea. Farm explained that, as a Council member, he began to "look at the big picture besides the small picture."

Over the years, Council members participated in meetings and events at the nation's capital, such as the NOAA Fish Fry and Capitol Hill Ocean Week, where they met with NMFS headquarters staff, environmentalists and fishermen from other parts of America, all as a part of their homework to better understand the big picture while at the same time educating those on the U.S. mainland about the local island fisheries and fishing communities thousands of miles away.

Guthertz recalled the working of the Council process in the early years:

> I think people perhaps felt that we were overstepping our authority or they didn't like the sustainability movement that we were starting to move towards, and

they were quite active. They'd come to the Council meetings and protest and were very lively at it, so you had groups that were pro-environment on one end and then you had other groups that wanted more freedom for their commercial fisheries activities at the other end, and we were sort of caught in the middle. Sometimes the environmentalists were a little extreme and sometimes the commercial fishing community was extremely aggressive. In the middle of those groups, you had the indigenous activist groups that were not happy with either group.... Our responsibility was trying to find a happy medium that worked.

For Cook, without the Council and the MSA regulatory process, there would be a "disorganized group of people feeding off the same resource, who are just destined for conflict" and the fishing industry would be involved in "so much crazy conflict with various user groups, including the environmental groups, the government, the recreational fishermen, the small commercial fishermen." He said the Council "did a wonderful job of putting a lid on [the Hawai'i longline] problem, to the satisfaction of all parties. And even today, all this time later, the things that [the Council] did in terms of limited entry, in terms of vessel monitoring, were really issues that needed to be approached and ... are still working ... for almost 30 years now."

Other groundbreaking actions of the Council were its early ban throughout the 1.7 million square nm (5.8 million square kilometers) region of potentially destructive fishing techniques, such as bottom trawling and pelagic gillnetting—methods that are still allowed in other parts of the country—and establishment of the NWHI Protected Species Zone as a result of longline fishermen bringing the issue of bycatch in the area to the attention of the Council. Even the eventual decision to make the NWHI lobster fishery a retain-all fishery was based partially on an indigenous island ethic that avoids wastefulness. Duenas put it this way, "If you say you can keep 25, but you catch number 26, you're going to throw it away, to me it's more wasteful." But the retain-all policy also had its scientific merits, as it provided better data—for example, how many males and how many females there are—which leads to better analysis.

A side benefit to the Council meetings was the improved communication regarding fisheries among the U.S. Pacific Islands and stronger local fisheries management in the territories. "Each island was kind of a separate entity trying to regulate and manage its own fisheries business," Callaghan recalled. "In the early years before the Council, most of the territories virtually had no fisheries management laws that were enforced. With the coming of the Council, it put pressure on the territories to improve their domestic [nearshore] fishing regulations."

Another side benefit to the Council process was the fostering of leadership in each island group. Guthertz noted that many of today's government leaders in Guam and the CNMI had served on the Council, and she attributed their leadership partly to their experiences with the Council. She explained: "Our agricultural leaders, department of agriculture, our governor appointees over the years, they've expanded their storehouse of knowledge about all the subject matter that the Council takes up."

That leadership building for the island areas continues today with the Council's U.S. Pacific Territories Fishery Capacity-Building Scholarship Program, with support from the Pacific Islands Fisheries Science Center and Pacific Islands Regional Office. Graduates from this scholarship program have earned bachelor's and master's degrees in marine science and are required to work for the local fishery agency for a set number of years.

At the 173rd Council meeting, June 2018, on Maui, Council Executive Director Kitty M. Simonds and NOAA Assistant Administrator for Fisheries Chris Oliver (front center) gather with Council members (front from left) Michael Goto, Ryan Okano, Henry Sesepasara, Edwin Ebisui Jr. (chair), Dean Sensui, Matt Sablan; (back row) Lieutenant Commander Adam Disque, Ray Roberto, Brian Peck, Taotasi Archie Soliai, Michael Duenas, John Gourley and Michael Tosatto. *WPRFMC photo.*

Formidable Leadership—The Executive Director

Kitty M. Simonds has served as the executive director of the Council for nearly four decades. A former staff member for U.S. Senator Fong, she understood the politics of Washington, D.C., and the State of Hawai'i, knew how to relate with the various agencies and Congressional groups with which the Council had to deal and was a natural leader. Unlike other Council executive directors who preceded her, she had the advantage of Pacific roots, which enabled her to perhaps relate more

499

personally with regional ocean interests. Also unlike them, she was not a scientist. However, as Callaghan explains, she worked well with scientists because "she was willing to listen and was not intimidated by people who understood things that she didn't understand. She would take people's advice on issues that she thought they were experts in." Her other strength was her strong belief in and defense of the MSA.

These skills helped Simonds see the Council through rough times. Morioka recalled that "as the Council and the nation were experiencing budget cuts and survival challenges … [she] was able to mix and match and patch and keep the Council alive." He said, "Many decisions that were made have hurt her personally, but [they were] for the fishery and for the Council." He found Simonds by nature to be strong-willed, a trait that he said was admirable in a Council executive director "because what the Council feels and the Council believes should be what the executive director carries forward and not what is politically correct." He also found her to be visionary. For example, when ecosystem-based management began, a lot of the Council members "were in denial that needed to be addressed." But, he said, Simonds was steadfast and kept saying, "We've got to address this." As a result, the Council developed the nation's first ecosystem-based fishery management plan.

To kick start the restructuring of the fishery management plans to fishery ecosystem plans, Simonds organized a series of workshops with experts on ecosystem-based management from throughout the United States and the Pacific. She similarly sought the expertise of others in the hosting of the 'Aha Moku *puwalu* on traditional Hawaiian natural resource management, a series of regional data collection workshops and international workshops on tuna management and sea turtle conservation, working with local and national agencies, foreign governments, environmental NGOs and indigenous and fishing communities.

Tulaono appreciated Simonds' responsiveness as an executive director. During his 10 years as director of the American Samoa Department of Marine and Wildlife Resources, he said would call her when he needed help—with the repair of fishing vessel engines, for example—and she would make it happen.

Sincere concern for the fishing and indigenous communities, the willingness to seek and listen to experts and the ability to work with the communities and governance bodies at the local, federal and international levels to find solutions are a legacy of the Council that Duenas attributed to Simonds along with her tenacity and strong will. Though outspoken in her views, Duenas noted, "at the end of the day, she's still willing to compromise ...."

A good friend and colleague Chris Oliver (the long-time executive director of the North Pacific Council and former NOAA Assistant Administrator for Fisheries) expressed similar thoughts. "Kitty has always been the strongest champion for the Council process. As a leading force on the national Council Coordination Committee, she was a fierce advocate for Council autonomy but also knew how to compromise and work hand-in-hand with the National Marine Fisheries Service to achieve the best outcomes for the fisheries management process, both in her region and at the national level. She was a role model for me and for other Council executive directors across the United States."

Addressing Future Problems—The Council Legacy

As the Council approaches its fifth decade and its golden anniversary, the Council is faced with mounting issues.

One is the increasing complexity of federal fisheries management as regulations promulgated though the ESA and other federal laws proliferate, sometimes conflict with each other and make problem solving more difficult. Where in the past problems could be worked out on the Council floor between groups that disagreed, today lawyers are needed to work through the tangled web of competing regulations. Other complications arise from the nature of ecosystem-based management, increased management measures by international regional fishery management organizations and impending MPAs on the high seas from the U.N. Biodiversity beyond National Jurisdiction initiative.

Another concern is what some see as increased pressure from Washington, D.C., to standardize fisheries management so that all regions function alike. Many in the U.S. Pacific Islands see their conditions as different and don't think they should be forced to comply with the rules designed for other areas. These voices call out for regional autonomy they believe the creators of the MSA intended for the councils. They worry that this founding principle may be eroding under the influence of administrative mission creep toward conformity. In their view, requiring ACLs for all federal fisheries, even those in the U.S. Pacific Islands that continue to lack sufficient data to make accurate stock assessments, is an example of imposing one-size-fits-all regulation over varying regions. Such regulations can have a disproportionate impact on small island economies.

Among the unique aspects of the region is the territories' struggle to achieve economic development. The Western and Central Pacific Fisheries Commission recognizes the U.S. Pacific territories as equivalent to small island developing states growing their fisheries. But efforts to nourish sustainable fisheries in the territories face many obstacles. Establishing marine national monuments in 52% of the federal waters closed access to fishing grounds by U.S. domestic commercial fishermen and did away with the Native Hawaiian preferential permits developed for the Northwestern Hawaiian Islands bottomfish fishery. The Billfish

Conservation Act amendment blocked U.S. Pacific Island fisheries from supplying domestic seafood markets.

Concerns also have arisen about the continued need for the Council to assist local fisheries in the face of increased competition from foreign fleets that do not have to comply with the same level of restrictions as the U.S. fisheries. Recognition is widespread that the proper place to regulate tuna is internationally because it is highly migratory, but effective negotiation by the United States for a more equitable share of longline bigeye tuna and purse-seine fishing opportunity has fallen short.

Lastly, concern grows that the federal government has not fully appreciated the cultural aspect of fish and fishing in the Western Pacific Region, and with that has not acknowledged that the U.S. Pacific Islands have limited capacity to feed the people who live there. Unlike the U.S. mainland with vast farmlands and other resources to harvest, U.S. Pacific Islands have depended for millennia on the ocean. Taking that ocean for military use and marine national monuments is seen by residents and allies of the area not only as a lack of respect but also as a threat to their way of life. Western-influenced bans on the use of cultural fishing methods, such as talaya and chenchulu, and denying the taking of culturally important marine species, such as green sea turtle, threaten lasting damage to cultural practices, social cohesion and multigenerational engagement with and knowledge of the ocean and its species.

The Council in its first 44 years has established a legacy of listening empathetically to indigenous and fishing communities, reflecting their concerns, helping solve their problems and providing bold leadership to balance regional needs and national policies.

Epilogue – Addressing Equity and Justice in 2021 and Beyond

The colonization of what would become the U.S. Pacific Islands began with the Spanish takeover of present-day Guam and the Northern Mariana Islands in the 15th century and continued for four centuries to the overthrow of the Hawaiian Kingdom by the United States in the 19th century. Similarly, in the Samoa Archipelago, the Tripartite Convention of 1899 between the United States, the United Kingdom and the German Empire resulted in the Deeds of Cession and the control of the islands of Tutuila and Manu'a being given to the United States. Sadly, uncompensated appropriation of ancestral lands, waters and natural resources of Native Hawaiians, Chamorro, Refaluwasch and American Samoans (collectively, the U.S. Pacific Islanders) has continued to this day.

The unilateral presidential proclamations that designated 52% of the waters and submerged lands in the Western Pacific Region as marine national monuments epitomize this neocolonial policy. The assumption that the value of 1.2 million square miles (3 million square kilometers) of ocean around the U.S. Pacific Islands is limited to the activities of a few fishermen is unjust. These waters have intrinsic value to the indigenous peoples whose ownership over them go back thousands of years.

The only other marine national monument in existence, the Northeast Canyons and Seamounts in New England, comprises less than 5,000 square miles (less than 13,000 square kilometers). The Massachusetts

504

Lobstermen's Association sued the Secretary of Commerce over the monument designation. The case went all the way to the Supreme Court. While the Supreme Court declined to hear the case, Chief Justice John Roberts issued a statement in March 2021 that questioned presidential use of the Antiquities Act to proclaim large marine national monuments. Roberts said, "A statute permitting the President in his sole discretion to designate as monuments 'landmarks,' 'structures,' and 'objects'—along with the smallest area of land compatible with their management—has been transformed into a power without discernible limit to set aside vast and amorphous expanses of terrain above and below the sea."

U.S. Pacific Islanders have questioned why they (who are nonwhite and constitute 0.5% of the nation's population) have been forced by presidential decrees to provide 96% of the U.S. marine monument waters and submerged lands. They have not been compensated for the transgenerational wealth that would normally be associated with access to these resources. This loss without compensation of the economic potential of these resources falls heavily on Native Hawaiians and Pacific Islanders who have among the lowest per capita income of any demographic group in the United States.

This policy of disproportionally appropriating the waters and submerged lands of peoples who have minimal representation in the federal government (e.g., U.S. territories do not have floor-voting members in Congress and are not entitled to electoral votes for Congress) has been supported by privileged environmentalists and their political allies and by mainland big-game sport-fishing associations and companies.

Kitty M. Simonds, the Council's executive director, has noted that, while the United States feeds the world, it has difficulty supporting its own indigenous fishing communities in the Western Pacific.[85] She pointed out that the region, which is the largest in the U.S. fishery management system, lacks adequate representation in federal fisheries policymaking despite the government's rhetoric on the importance of diversity and inclusion. For example, the NOAA Marine Fisheries Advisory Council, which advises the Secretary of Commerce, has not included in recent years any member from the Western Pacific Region.

In 2021, President Joseph R. Biden Jr. announced two executive orders and a proclamation that hold promise of the federal government providing some long overdue relief to indigenous and minority communities. They are Executive Order 13985 Advancing Racial Equity and Support for Underserved Communities through the Federal Government (January 20, 2021); Executive Order 14031 Advancing Equity, Justice and Opportunity for Asian Americans, Native Hawaiian and Pacific Islanders (May 28, 2021); and A Proclamation on Indigenous Peoples Day (October 8, 2021). In his proclamation, Biden said, "The Federal Government has a solemn obligation to lift up and invest in the future of Indigenous people and empower Tribal Nations to govern their own communities and make their own decisions."

Simonds reviewed how these presidential declarations impact U.S. Pacific Island fishing communities in a virtual meeting of the Council Coordination Committee on October 21, 2021. Her presentation "Environmental Justice in Fisheries Management: The Western Pacific" highlighted the problem that the Western Pacific Council has historically experienced melding Western concepts and mandates with those of other cultures in the U.S. Pacific Islands. She described equity barriers such as capacity limits, communication differences, available funding opportunities, regional representation and rigid national policies.

---

[85]   Kitty M. Simonds in discussion with the authors by email, November 1. 2021.

For example, Simonds cited inequities involved in the monitoring and development of stock assessments and how these impact the fishery and economic development in the region. Because there has been an insufficient level of investment to collect appropriate data for stock assessments in the territories, the federal annual catch limit (ACL) requirements don't fit the traditional fisheries management in the territories. Consequently, the American Samoa bottomfish ACL has been slashed from 106,000 to 5,000 pounds and the Guam bottomfish ACL from 66,000 to 31,000 pounds in order to rebuild the stock to the sustainable level.

Another example of a national policy affecting Pacific Island communities has been the proposed federal critical habitat designation for ESA-listed corals in territorial waters around American Samoa, Guam and CNMI, which has created a clash between the national Coastal Zone Management Act federal consistency process and the Territory coastal management programs. The national policy could affect territories' ability to manage their waters, and both the CNMI and Guam have objected to the federal determination that the critical habitat designation would not have an effect on any coastal use or resource.

Speaking to the Council Coordination Committee, Manny Duenas, Western Pacific Council member and president of the Guam Fishermen's Cooperative, stated that the federal government must embrace the diversity of the fishing communities in the region and must engage with them so that policies are not made by the agencies alone.

Two other 2021 White House edicts are being discussed within the Council family. One is President Trump's reopening (with the support of a Democratic Congress) the process to designate the Papahanaumokuakea MNM in the NWHI as a national marine sanctuary (December 27,

WESTERN PACIFIC REGIONAL FISHERY MANAGEMENT COUNCIL

2020). The second is President Biden's Executive Order 14008 Tackling the Climate Crisis at Home and Abroad (January 27, 2021).

The monument proclamation process is top-down, done by presidential fiat. The sanctuary process is a bottoms-up approach to management similar to the MSA and gives the Council "a seat at the table" to propose regulations consistent with the sanctuary goals and objectives and also with monument regulations. The Council is working with NOAA to overlay the NWHI monument with a sanctuary designation. In effect, this would be another layer of protection along with other earlier designations still in existence, such as the State of Hawai'i NWHI Marine Refuge and Kure Atoll Wildlife Sanctuary, the USFWS Hawaiian Islands National Wildlife Refuge, the Council's Protected Species Zone and several no-take refugia promulgated through the MSA, and President Clinton's NWHI Coral Reef Ecosystem Reserve (whose advisory council continues to meet to this day).

The Council has also asked President Biden to allow fishing to be permitted from 50- to 200-nm in the Pacific Remote Island Areas around Johnston Atoll, Jarvis Island and Wake Island subject to Council and NOAA NMFS management through the MSA. Federally managed fishing around these Pacific Remote Island Areas was prohibited when these waters became part of the Pacific Remote Islands MNM.

The Council's June 16, 2021, letter to Biden included details about the effects that the non-science-based proclamations establishing marine national monuments in the Western Pacific Region have had on American fishing industries, seafood consumer and indigenous communities. The waters around the Pacific Remote Island Areas, for example, constituted 12% of the Hawai'i longline fishery's landings and 10% of the U.S. purse-seine fishing effort before being closed by presidential monument proclamations.

Biden's Executive Order 14008 directs his administration to work with "State, local, Tribal and territorial governments, agricultural and forest landowners, fishermen and other key stakeholders, to achieve the goal of conserving at least 30% of [the nation's] lands and waters by 2030." In response to this "30 x 30" provision, the Council on March 29, 2021, wrote to Debra A. Haaland, secretary of the Department of the Interior, noting that 27% of the 30% goal has already been met by 1.19 million square miles (3.07 million square kilometers) of marine national monument waters in the Western Pacific Region.

The Council has created a significant legacy of scientific management of Pacific Island fisheries over the past 40 years. High among its proudest accomplishments has been seeking the views and wisdom of the men *and women* of the U.S. Pacific Islands and providing them with the skills and opportunities to express their views and demand social, cultural and economic justice for all.

A statement by Chief Justice William S. Richardson, Supreme Court Hawai'i, encapsulates the operational philosophy of the Council: "You must make difficult decisions, but, if you make those decisions with the counsel and advice from traditional practitioners and those who are most closely affected by and connected to a particular resource or area, your decisions will be sound" (Simonds et al. 2008, page 17). The Council's proud record of more than 40 years includes respecting the cultural knowledge of the indigenous peoples of Hawai'i, American Samoa, Guam and the Northern Mariana Islands along with Western science in the decision-making process. As we move into the future, neither the importance nor the welfare of the island peoples of the U.S. Western Pacific should be overlooked.

Appendix 1

Council Members, Observers and General Counsel

Council Chairs

1976–1987	Wadsworth Yee (Hawai'i)
1987–1991	William Paty Jr. (Hawai'i)
1992	Alo Paul Stevenson (American Samoa)
1993	Rufo Lujan (Guam)
1994–1996	Edwin Ebisui Jr. (Hawai'i)
1997–2000	Jim Cook (Hawai'i)
2000–2002	Judith Guthertz (Guam)
2002	Frank Farm Jr. (Hawai'i)
2003	Judith Guthertz (Guam)
2003	Manuel Cruz (Guam)
2004–2005	Roy Morioka (Hawai'i)
2005–2006	Frank McCoy (American Samoa)
2007–2009	Sean Martin (Hawai'i)
2010	Lauvao Stephen Haleck (American Samoa)
2011–2012	Manuel Duenas (Guam)
2012	William Aila Jr. (Hawai'i)
2013–2014	Arnold Palacios (CNMI)
2014	William Aila Jr. (Hawai'i)
2014–2018	Edwin Ebisui Jr. (Hawai'i)
2018–present	Taotasi Archie Soliai (American Samoa)

Voting Council Members: Appointed by the Secretary

Territory of American Samoa

1976–1985	Lealaifuaneva Peter Reid Jr. (vice chair 1976–1985)
1983–1986	Alo Paul Stevenson (vice chair 1990–1991,
1989–1998	1993, 1997–1998)
1985–1994	Melvin Makaiwi (vice chair 1989–1990, 1992)
1986–1989	J. Anthony Langkilde
1994–1997	Samuel Puletasi
1997–2003	Aitofele Sunia (vice chair 1999–2003)
1998–2007	Frank McCoy (vice chair 2003–2005, 2007)
2003–2012	Lauvao Stephen Haleck (vice chair 2007–2009, 2011–2012)
2007–2016	William Sword (vice chair 2010, 2013–2016)
2013–2016	Claire Poumele
2016–2019	Christinna Lutu-Sanchez (vice chair 2018–2019)
2016–2020	Taotasi Archie Soliai (vice chair 2016–2018)
2019–present	Howard Dunham (vice chair 2019–present)

Territory of Guam

1976–1978	Isaac Ikehara
1976–1979	Paul Bordallo (vice chair 1976–1979)
1978–1981	Paul Callaghan (vice chair 1979–1981)
1979–1982	John Eads
1980–1983	Steven Amesbury
1981–1984	Betty Guerrero
1982–1985	Robert Smith
1988–1994	Peter Barcinas (vice chair 1993)

1994	Melvin Borja
1994–2003	Judith Guthertz (vice chair 1998–2000, 2002)
2003–2012	Manuel Duenas (vice chair 2004–2010)
2012–present	Michael Duenas (vice chair 2013–present)
2019–present	Monique Genereux Amani

State of Hawai'i

1976–1979	Frank Goto
1976–1979	Peter Fithian
1976–1987	Wadsworth Yee
1976–1988	Louis Agard Jr.
1979–1982	Jay Puffinburger
1980–1989	Gertrude Nishihara (vice chair 1989)
1984–1987	Alika Cooper
1987–1996, 2001–2007, 2012–2018	Edwin Ebisui Jr. (vice chair 1990–1991, 2002, 2006–2007, 2013–2014)
1987–1990, 1996–2005	Roy Morioka
1988–1991, 1996–2005	Frank Farm Jr. (vice chair 1998–2001, 2003–2005)
1989–1992	Clarence Ho'okala
1990–1993	Frank Nibley Jr.
1991–2000	James Cook
1992–1995	Timm Timoney
1995–2001	Thomas Webster
2000–2003	Bryan Ho
2003–2012	Sean Martin (vice chair 2005–2006)
2005–2008	Myrick Rick Gaffney
2005–2011	Frederick Duerr (vice chair 2007–2009)
2007–2010	Peter Young
2008–2011	David Itano (vice chair 2010–2012)

2010–2016	Julie Leialoha
2011–2017	McGrew Rice (vice chair 2014–2017)
2012–2020	Michael Goto
2016–2019	Dean Sensui (vice chair 2018–2019)
2018–present	Edwin Watamura (vice chair 2014–present)

Commonwealth of the Northern Mariana Islands

1996–1999	Arnold Palacios (vice chair 1996–1999, 2012)
1999–2002	Benny Pangelinan (vice chair 1999–2002)
2002–2011	Benigno Sablan (vice chair 2002–2005, 2006–2007)
2011–2014	Richard Seman (vice chair 2013–2014)
2014–present	John Gourley (vice chair 2014–present)
2018–present	McGrew Rice

Voting Council Members: Designated State Officials

Territory of American Samoa

1976–1978	Richard Wass
1978–1984, 1989–1992, 2017–2020	Henry Sesepasara
1985–1988, 1993–1996, 1997–2012	Ufagafa Ray Tulafono (vice chair 1986–1988, 1994–1996, 2006)
1996	Philip Langford

2013–2016	Ruth Matagi–Tofiga
2021–present	Taotasi Archie Soliai

Territory of Guam

1976–1977	Frank Aguon
1978	Rufo Lujan
1979–1984	Harry Kami (vice chair 1981–1984)
1984–1994	Rufo Lujan (vice chair 1984–1992, 1994)
1995–1999	Paul Bordallo (vice chair 1995–1997)
1999	Dot Harris
1999–2002	Isabel Lujan (vice chair 2000–2001)
2003–2004	Manuel Cruz (vice chair 2003)
2004–2007	Adrienne Loerzel
2007–2008	Alberto Lamorena
2009	Paul Bassler
2009–2010	Joseph Torres
2011–2015	Mariquita Taitague (vice chair 2011–2012)
2015–2018	Matthew Sablan
2019–present	Chelsa Muna-Brecht

State of Hawai'i

1976–1977	Michio Takata
1978–1981	Kenji Ego
1982–1987	Henry Sakuda
1987–1999	William Paty Jr. (vice chair 1992–1997)
1999–2000	Timothy Johns
2000–2002	Gilbert Coloma–Agaran
2003–2007	Peter Young
2007	Allen Smith

2007–2010	Laura Thielen
2011–2014	William Aila Jr.
2015	Carty Chang
2015–present	Suzanne Case

Commonwealth of the Northern Mariana Islands

1994–1996	Benigno Sablan
1997	Margarita Wonenberg
1998	Bertha Guerrero
1998–2001	Joaquin Tenorio
2002–2004	Tom Pangelinan
2004–2005	Richard Seman (vice chair 2005)
2006–2011	Ignacio Dela Cruz (vice chair 2006–2007)
2011–2014	Arnold Palacios
2014	Manny Pangelinan
2015–2017	Richard Seman
2017–2019	Raymond Roberto
2019–present	Anthony Benavente

National Marine Fisheries Service

1976–1980	Gerald Howard
1980–1983	Alan Ford
1984–1992	E. Charles Fullerton
1992–1993	Gary Matlock
1994	Rodney McInnis
1994–1996	Hilda Diaz–Soltero
1997–1999	William Hogarth
1999-2000	Rodney McInnis
2000	Rebecca Lent

2001–2002	Rodney McInnis
2003–2004	Samuel Pooley
2004–2010	Bill Robinson
2010–present	Michael Tosatto

Non-voting Council Members

U.S. Coast Guard

1976–1978	Rear Admiral James Moreau
1978–1980	Rear Admiral David Lauth
1980–1983	Admiral Bernie Thompson
1983–1985	Rear Admiral Clyde Robbins
1985–1987	Rear Admiral Alfred Manning
1987–1990	Rear Admiral William Kozlovsky
1990–1993	Rear Admiral William Donnell
1993–1996	Rear Admiral Howard Gehring
1996–1998	Rear Admiral Tom Collins
1998–2001	Rear Admiral Joseph McClelland Jr.
2001–2003	Rear Admiral Ralph Utley
2003–2006	Rear Admiral Charles Wurster
2006–2008	Rear Admiral Sally Brice-O'Hara
2008–2010	Rear Admiral Manson Brown
2010	Rear Admiral Stephen Mehling
2011–2012	Rear Admiral Charles Ray
2013–2015	Rear Admiral Cari Thomas
2015–2018	Rear Admiral Vincent Atkins
2018–present	Rear Admiral Kevin E. Lunday

U.S. Department of State

1976	Lorry Nakatsu
1977–1980	James Carlton Price
1980–1981, 1984–1985	Raymond Arnaudo
1986	Jacob Walles
1986–1987	Stetson Tinkham
1987–1988	Robert Benson
1988–1999	Brian S. Hallman
1999–2015	William Gibbons-Fly
2015–2020	Michael Brakke
2021	David Hogan

U.S. Fish and Wildlife Service

1976–1977	Eugene Kridler
1977–1979	Henry Hansen
1979–1982	Dale Coggeshall
1984–1987	Allan D. Marmelstein
1987	Robert Benson
1988–1990	Allan Marmelstein
1990–2000	Robert Smith
2000–2001	Barbara Maxfield
2001–2002	Paul Henson
2002–2007	Jerry Leinecke
2008–2015	Don Palawaski
2012–2014	Susan White
2014–2018	Matt Brown
2018–present	Brian Peck

Observers

Northern Mariana Islands

1976–1978, 1980, 1982	Joaquin Villagomez
1977	Edward Cabrera
1978	Melinda Sablan
1979–1984	Rufo Lujan
1980–1981	Pedro Dela Cruz
1982	Benigno Sablan
1982–1993	Nicolas Guerrero
1983–1985	Arnold Palacios

NOAA General Counsel

1976–1996	Martin Hochman
1977	William C. Brewer
1979	Eileen Cooney and Elizabeth Mitchell
1996–2005	Judson Feder
2006–2007	Silas DeRoma
2008–present	Fred Tucher

Appendix 2

Scientific and Statistical Committee Members

Chairs

1977–1978	Izadore Barrett (chair)
	Richard Shomura (vice chair)
1979–1980	Doyle Gates (chair)
	Richard Shomura (acting chair)
1981–2011	Paul Callaghan (chair)
2012–2015	Charles Daxboeck (chair)
2016–present	James Lynch (chair)

Members

Stewart Allen (2003–2007)

Judith Amesbury (2006–2015)

Steven Amesbury (1977–1992)

Izadore Barrett (1977–1978)

Anthony Beeching (2002–2004)

Brian Bowen (2007–2010)

Richard Brock (1986–2004)

Karl Brookins (2006–2015)

Raymond Buckley (1986–1988)

Patrick Bryan (1996–1997)

John Byrne (1977)

Debra Cabrera (2011–present)

Paul Callaghan (1977–2017)

Frank Camacho (2011–present)

Chris J. Carr (2001–2002)

Milani Chaloupka (2003–present)

Salvatore Comitini (1979)

Emmanuel Coutures (2003–2004)

Peter Craig (1989–1992)

John Craven (1977)

Flynn Curren (2000–2001)

Jack Davidson (1979)

Gerald Davis (1994–2000)

Charles Daxboeck (1983–2016)

Richard Deriso (1993–2016)

Mary Donohue (2003–2005)

Thomas Dye (1991–1994)

Scott Eckert (2001–2002)

Lucius Eldredge (1978)

Chris Evans (2000–2002)

Douglas Fenner (2004–2005)

Erik Franklin (2011–present)

Doyle Gates (1979–1980)

Richard Grigg (1978)

David Grobecker (2007)

John Hampton (1993–2016)

Shelton Harley (2011–present)

Phillip Helfrich (1979)

Jason Heyler (2020)

Ray Hilborn (2011–present)

Seichi Hirai (1977)

Justin Hospital (2011–present)

Ike Ikehara (1980)

Walter Ikehara (2000–2004)

David Itano (1986–1988, 2012–present)

Harry T. Kami (1978)

Alvin Katekaru (1986–1991)

Tony Kingston (1996–1998)

Pierre Kleiber (2003–2017)

Donald Kobayashi (2010–present)

George Krasnick (1986–1991)

Molly Lutcavage (2007–2016)

James Lynch (2011–present)

Gerald Marten (1979)

Craig MacDonald (1986–1998)

Steve Martell (2011–present)

Alton Miyaski (2013-2017)

Robert Nishimoto (1989–1991, 1996–1998)

Domingo Ochavillo (2011–present)

Francis Oishi (2011–2012)

Ryan Okano (2017–2020)

Minling Pan (2009–2016)

James Parrish (1986–2009)

Graham Pilling (2011–present)

Dan Polhemus (2006–2007)

Samuel Pooley (1986–2002)

Justin Rutka (1979)

Henry Sakuda (1980)

Karl Samples (1986–1991)

Kurt Schaefer (2011–present)

Michael Seki, *ex-officio* (2011–present)

Henry Sesepasara (1986–1988)

Craig Severance (1994–present)

Richard Shomura (1977–1980)

John Sibert (1993–2016)

Robert Skillman (1986–2016)

Alexandra Spoehr (1981–1985)

Lewis Sterry (1979)

Paul Struhsaker (1977)

Tanielu Sua (1996–1999)

Gerald Sumida (1986–1999)

Stanley Swerdloff (1978–1980)

Michael Tenorio (2011–present)

Michael Trianni (2000–2015)

Roy Tsuda (1979–1980)

Ufagafa Ray Tulafono (1989–1991)

Richard Waas (1978–1979)

Jeffrey Walters (2005–2006)

At the 119th SSC meeting, June 9–11, 2015, in Honolulu. Executive Director Kitty M. Simonds (seated, 2nd from right) meets with SSC members (seated from left) Paul Callaghan, PhD (University of Guam retired); John Sibert, PhD (University of Hawai'i at Manoa retired); Craig Severance, PhD (University of Hawai'i at Hilo retired); Pierre Kleiber, PhD (Pacific Islands Fisheries Science Center retired); (standing from left) Frank A. Camacho, PhD (University of Guam); David Itano (fisheries consultant); Minling Pan, PhD (Pacific Islands Fisheries Science Center); Judith Amesbury (Micronesian Archeological Research Services); Molly Lutcavage, PhD (University of Massachusetts Amherst Marine Station); Charles Daxboeck, PhD, SSC chair (BioDax Consulting Tahiti); Alton Miyasaka (HDAR); Robert A. Skillman, PhD (Pacific Islands Fisheries Science Center retired); Donald Kobayashi, PhD (Pacific Islands Fisheries Science Center); and Todd Miller, PhD (CNMI Division of Fish and Wildlife). Not pictured are Milani Chaloupka, PhD (The University of Queensland); Richard Deriso, PhD (Inter-American Tropical Tuna Commission); Erik C. Franklin, PhD (University of Hawai'i, Hawai'i Institute of Marine Biology); John Hampton, PhD (Secretariat of the Pacific Community); James M. Lynch (previously with KL Gates LLP law firm; currently with Sierra Pacific Industries); and Domingo Ochavillo (American Samoa DMWR).

Appendix 3

Plan Team Chairs

Precious Coral Plan Team

1977–2005	Richard "Rick" Grigg
2006–2015	Frank Parrish

Crustacean Plan Team

1986–1989	Craig MacDonald
1989– 1992	Walter Ikehara
1992–2003	Jeffrey Polovina
2003–2015	Gerard Dinardo

Bottomfish and Seamount Groundfish Plan Team

1986–1989	Alvin Katekaru
1989–1992	David Somerton
1992–1993	Gerald Davis
1993	Sam Pooley
1994–2015	Robert Moffitt

Pelagic Plan Team

1987–1992	Robert Skillman
1992–1996	Walter Ikehara
1996–2003	Christofer Boggs
2003–2020	Keith Bigelow

Coral Reef Ecosystem Plan Team

2003–2015	Jeffrey Walters

Archipelagic Plan Team

2012	John Gourley (Territories) and Frank Parrish (Hawai'i)
2013–2014	John Gourley (Territories) and Sam Kahng (Hawai'i)
2015	Sam Kahng
2016–2020	Stefanie Dukes

Appendix 4

Advisory Panel Chairs and Vice Chairs

1977–1978   James W. Sutherland (chair)
                Robert Campbell (vice chair 1977)
                Amituanai Meredith (vice chair 1977)
                Manuel Tenorio (vice chair 1977)

1979–1981   Frank Goto (chair)
                Robert Campbell (vice chair 1980)
                Jed Inouye (vice chair)
                James W. Sutherland (vice chair)
                Charles Yamamoto (vice chair)

1981–1983   Frank Goto (chair)
                Louise K. Agard (vice chair)

1984–1986   Kenji Ego (chair)
                John Eads (vice chair)

1986–1989   Frank Goto (chair)
                Frank Farm Jr. (vice chair)

1989–1990   Gilbert Yanagawa (chair)

1991–1992   Jim Witten (chair)

1993–1999   Frank Farm Jr. (chair 1993–1996)
                Walter Ikehara (chair 1996–1999)
                William Aila Jr. (vice chair 1995–1999)
                Thomas Kraft (vice chair 1995–1996)
                April Romero (vice chair 1996–1999)

1999–2000   Dave Kalthoff (chair)

John Taitano (vice chair)
Pete Tenorio (vice chair)
Wallace Thompson (vice chair)

2001–2003   Jim Cook (chair)
Richard Shiroma (vice chair)

2003–2005   Richard Shiroma (chair)
Sean Martin (vice chair)

2005–2006   Wadsworth Yee (chair)
James Borja (vice chair)
Henry Sesepasara (vice chair 2006)
Stan Taisacan (vice chair)

2007–2008   Edwin Ebisui Jr. (overall chair)
Frank McCoy (overall vice chair)
Nonu Tuisamoa (American Samoa chair)
Ramon Mafnas (CNMI chair)
Jesse Rosario (Guam chair)
Neil Kanemoto (Hawai'i chair)
Frank McCoy (Pelagic chair)
Selaina Vaitautolu (CDPP chair)

2009–2010   Edwin Ebisui Jr. (overall chair)
Edwin Watamura (overall vice chair)
Judith McCoy (American Samoa chair)
Ramon Mafnas (CNMI chair)
Jesse Rosario (Guam chair)
Ray Shirakawa (Hawai'i chair)
William Mossman (CDPP chair)

2011–2014   Edwin Watamura (overall chair)
James Borja (overall vice chair)
Kitara Vaiau (American Samoa chair)
Cecilio Raiukiulipiy (CNMI chair)
Jesse Rosario (Guam chair)
Henry Lau (Hawai'i chair)

	Judy McCoy (Pelagic chair)
2015–2018	Judi Guthertz (overall chair)
	Ed Watamura (overall vice chair 2015–2016)
	McGrew Rice (overall vice chair 2016–2018)
	Christinna Lutu–Sanchez (American Samoa chair 2015–2016)
	Krista Corry (American Samoa chair 2016–2018)
	Richard Farrell (CNMI chair)
	Peter Perez (Guam chair 2015–2017, Guam vice chair 2017–2018)
	Felix Reyes (Guam chair 2017–2018)
	Gary Beals (Hawaiʻi chair)
2019–2022	Clay Tam (overall chair)
	Will Sword (American Samoa vice chair)
	Richard Farrell (CNMI vice chair)
	Ken Borja (Guam vice chair)
	Gil Kualiʻi (Hawaiʻi vice chair)

Appendix 5

Council Executive Directors and Staff*

Executive Director

1976–1978	Wilvan Van Campen
1979	Jack Marr
1980–1982	Svein Fougner
1982–present	Kitty M. Simonds**

The Council staff displays team spirit during a June 2011 retreat held in conjunction with the 52nd Hawaiian International Billfish Tournament in Kailua-Kona, Hawai'i. (Front from left) Sarah Pautzke, Marlowe Sabater, Executive Director Kitty M. Simonds, Jordan Takekawa, Sylvia Spalding, Charles Ka'ai'ai; (back row) Eric Kingma, Elysia Granger, Joshua DeMello, Paul Dalzell, Mark Mitsuyasu and Randy Holmen. *WPRFMC photo.*

Program Staff

Fini Aitaoto	American Samoa on-island coordinator
John Calvo	Guam on-island coordinator
Paul Dalzell	Senior scientist
Carl Dela Cruz	Guam on-island coordinator
Joshua DeMello**	Coral reef ecosystem coordinator
Marcia Hamilton	Economist
Robert Harman	Biologist
Nate Ilaoa	American Samoa on-island coordinator
Asuka Ishizaki**	Protected species coordinator
Charles Ka'ai'ai	Indigenous coordinator
Eric Kingma	International fisheries, enforcement and National Environmental Policy Act coordinator
Jarad Makaiau	Coral reef ecosystem/habitat coordinator
Mark Mitsuyasu**	Fisheries program officer
Jack Ogumoro	CNMI on-island coordinator
Justin Rutka	Economist
Marlowe Sabater**	Marine ecosystem scientist
Robert Schroeder	Coral reef coordinator
Sylvia Spalding	Communications officer
Rebecca Walker	Fisheries analyst/ habitat coordinator

Administrative Staff

Loren Bullard**	Technical assistant
Elysia Granger**	Administrative officer
Bella Hirayama**	Travel and appointments
Randy Holmen**	Fiscal officer
Vera Keala	Clerk typist
Grace Muotka	Fiscal officer

Jane Nakamura   Secretary
Ellen Reformina   Administrative assistant
Kitty M. Simonds   Assistant to the executive director
Jordan Takekawa   Document handler

*Staff lists include members with five or more years of tenure and final position title.
** Indicates current staff member at time of publication.

Appendix 6

Western Pacific Regional Fishery Management Council Meetings

(* indicates a Fishers Forum was held in conjunction with the meeting.)

1st	October 19–21, 1976	Pagoda Hotel, Honolulu, Hawai'i
2nd	December 15–16, 1976	State Capitol, Honolulu, Hawai'i
3rd	February 1–4, 1977	Council office, Honolulu, Hawai'i
4th	April 19–22, 1977	Senate Chamber, Pago Pago, American Samoa
5th	June 27–29, 1977	State Capitol, Honolulu, Hawai'i
6th	August 10–14, 1977	King Kamehameha Hotel, Kailua-Kona, Hawai'i
7th	September 29–30, 1977	Royal Lahaina Hotel, Ka'anapali, Maui, Hawai'i
8th	November 28–29, 1977	State Capitol, Honolulu, Hawai'i
9th	January 10–12, 1978	Continental Hotel, Saipan, Northern Mariana Islands, and Hilton Hotel, Guam
10th	March 15–16, 1978	Kauai Surf Hotel, Kalapaki, Kaua'i, Hawai'i
11th	May 22–23, 1978	Rainmaker Hotel, Pago Pago, American Samoa
12th	August 2–5, 1978	King Kamehameha Hotel, Kailua-Kona, Hawai'i

13th	October 23, 1978	Kona Hilton Hotel, Kailua-Kona, Hawai'i
14th	December 7–8, 1978	State Capitol, Honolulu, Hawai'i
15th	January 10–11, 1979	Sheraton Moloka'i Hotel, Moloka'i. Hawai'i
16th	March 14–16, 1979	Grand Hotel, Saipan, Northern Mariana Islands, and Reef Hotel, Guam
17th	May 3–4, 1979	State Capitol, Honolulu, Hawai'i
18th	June 25–26, 1979	Conference Center, Pago Pago, American Samoa
19th	August 22–25, 1979	King Kamehameha Hotel, Kailua-Kona, Hawai'i
20th	October 15–16, 1979	Pago Pago, American Samoa
21st	November 29–30, 1979	Maui Lu Hotel, Maui, Hawai'i
22nd	February 14–15, 1980	State Capitol, Honolulu, Hawai'i
23rd	April 8–9, 1980	Guam Reef Hotel, Agana, Guam
24th	May 28, 1980	Kaua'i Surf Hotel, Kalapaki, Kaua'i, Hawai'i
25th	July 31–August 1, 1980	Naniloa Surf Hotel, Hilo, Hawai'i
26th	September 15–17, 1980	Full Gospel Church, Ofu, Manu'a Islands, American Samoa
27th	November 12–13, 1980	State Capitol, Honolulu, Hawai'i
28th	February 3–5, 1981	Saipan Continental Hotel, Saipan, CNMI, and Dai-Ichi Hotel, Guam
29th	March 31–April 1, 1981	State Capitol, Honolulu, Hawai'i
30th	June 9–10, 1981	Royal Lahaina Hotel, Lahaina, Maui, Hawai'i
31st	July 29–31, 1981	King Kamehameha Hotel, Kailua-Kona, Hawai'i
32nd	October 8–9, 1981	State Capitol, Honolulu, Hawai'i

33rd	December 1–2, 1981	Pau Hana Inn, Moloka'i, Hawai'i
34th	January 26–28, 1982	Saipan Grand Hotel, Saipan, CNMI, and Pacific Islands Hotel, Tumon Bay, Guam
35th	April 14–15, 1982	Pago Pago, American Samoa
36th	June 24–25, 1982	State Capitol, Honolulu, Hawai'i
37th	August 16–18, 1982	King Kamehameha Hotel, Kailua-Kona, Hawai'i
38th	December 6–7, 1982	Makaha Resort Hotel, Makaha, O'ahu, Hawai'i
39th	February 22–24, 1983	Saipan Grand Hotel, Saipan, CNMI and Cliff Hotel, Agana, Guam
40th	May 23–24, 1983	Ala Moana Americana Hotel, Honolulu, Hawai'i
41st	July 27–28, 1983	Rainmaker Hotel, Pago Pago, American Samoa and Ta'u High School, Manu'a, American Samoa
42nd	September 28–29, 1983	Pacific Islands Hotel, Agana, Guam, and Rota Paupau Hotel, Rota, CNMI
43rd	January 30–31, 1984	Ala Moana Hotel, Honolulu, Hawai'i
44th	April 23–24, 1984	Ala Moana Hotel, Honolulu, Hawai'i
45th	June 13–15, 1984	Intercontinental Hotel, Wailea, Maui, Hawai'i
46th	August 20–21, 1984	Kona Hilton, Kailua-Kona, Hawai'i
47th	December 5–7, 1984	Kauai Surf Hotel, Kalapaki, Kaua'i, Hawai'i
48th	February 21–22, 1985	Naniloa Surf, Hilo, Hawai'i
49th	May 6–9, 1985	Guam Hilton Hotel, Guam
50th	August 7–8, 1985	King Kamehameha Hotel, Kailua-Kona, Hawai'i

51st	November 4–6, 1985	Pago Pago and Ofu Island, American Samoa
52nd	March 3–5, 1986	Ala Moana Americana Hotel, Honolulu, Hawai'i
53rd	May 28–30, 1986	Pagoda Hotel, Honolulu, Hawai'i
54th	August 6–8, 1986	King Kamehameha Hotel, Kailua-Kona, Hawai'i
55th	November 9–11, 1986	Surf Hotel, Saipan, CNMI, and Fujita Hotel, Tumon, Guam
56th	March 9–11, 1986	Ala Moana Hotel, Honolulu, Hawai'i
57th	June 3-5, 1987	Pagoda Hotel, Honolulu, Hawai'i
58th	July 29–30, 1987	Sheraton Kauai-Poipu Beach, Kaua'i, Hawai'i
59th	September 21–22, 1987	Hawai'i Department of Land and Natural Resources office, Honolulu, Hawai'i
60th	November 16–18, 1987	Reef Hotel, Honolulu, Hawai'i
61st	February 24–26, 1988	Ala Moana Hotel, Honolulu, Hawai'i
62nd	August 8–11, 1988	King Kamehameha Hotel, Kailua-Kona, Hawai'i
63rd	November 27–29, 1988	Ala Moana Hotel, Honolulu, Hawai'i
64th	February 15–17, 1989	Ala Moana Hotel, Honolulu, Hawai'i
65th	April 10–12, 1989	Utulei, American Samoa
66th	July 24–26, 1989	Turtle Bay Hilton, Kahuku, O'ahu, Hawai'i
67th*	December 7–8, 1989	Ilikai Hotel, Honolulu, Hawai'i
68th*	April 9–12, 1990	Guam Hilton Hotel/Aqua Resort, Saipan
69th*	June 18–19, 1990	Ala Moana Hotel, Honolulu, Hawai'i
70th	September 27–28, 1990	Ala Moana Hotel, Honolulu, Hawai'i
71st	December 5-7, 1990	Dole Cannery, Honolulu, Hawai'i

72nd	February 27–March 1, 1991	Dole Cannery, Honolulu, Hawai'i
73rd	May 13–16, 1991	Dole Cannery, Honolulu, Hawai'i
74th	August 21–22, 1991	King Kamehameha Hotel, Kailua-Kona, Hawai'i
75th	December 16–18, 1991	Pagoda Hotel, Honolulu, Hawai'i
76th	March 16–17, 1992	Ala Moana Hotel, Honolulu, Hawai'i
77th	July 1–2, 1992	Ala Moana Hotel, Honolulu, Hawai'i
78th	September 21–23, 1992	Ala Moana Hotel, Honolulu, Hawai'i
79th	November 30–December 2, 1992	Ala Moana Hotel, Honolulu, Hawai'i
80th	April 26–29, 1993	Ala Moana Iotcl, Honolulu, Hawai'i
81st	September 14–16, 1993	Hale Koa Hotel, Honolulu, Hawai'i
82nd	December 13–15, 1993	Ala Moana Hotel, Honolulu, Hawai'i
83rd*	April 25–27 1994	Pacific Star Hotel, Tumon Bay, Guam, and Saipan Diamond Hotel, Saipan, CNMI
84th	November 7–9, 1994	Hawaiian Regent Hotel, Honolulu, Hawai'i
85th	August 3–5, 1994	Royal Kona Resort, Kailua-Kona, Hawai'i
86th*	April 18–21, 1995	Rainmaker Hotel, Pago Pago, American Samoa
87th*	August 8–10, 1995	Sheraton Makaha, Wai'anae, O'ahu, Hawai'i
88th*	December 6–8, 1995	Kauai Coconut Beach Resort, Kapa'a, Kaua'i, Hawai'i
89th*	April 24–26, 1996	Ala Moana Hotel, Honolulu, Hawai'i
90th*	August 7–9, 1996	Kaluakoi Hotel, Moloka'i, Hawai'i
91st	November 18–21, 1996	Ala Moana Hotel, Honolulu, Hawai'i

92nd	April 23–25, 1997	Ala Moana Hotel, Honolulu, Hawai'i
93rd*	August 19–21, 1997	Ala Moana Hotel, Honolulu, Hawai'i
94th*	November 12–14, 1997	Ala Moana Hotel, Honolulu, Hawai'i
95th*	April 14–16, 1998	American Samoa Legislature, Fagatogo, American Samoa
96th	May 8, 1998	Council office, Honolulu, Hawai'i
97th*	July 27–28, 1998	King Kamehameha Kona Beach Hotel, Kailua-Kona, Hawai'i
98th	December 1–3, 1998	Hawaii Prince Hotel, Honolulu, Hawai'i
99th*	March 15–18, 1999	Hilton Guam Hotel, Guam, and Saipan Diamond Hotel, Saipan
100th*	June 15–18, 1999	Ala Moana Hotel, Honolulu, Hawai'i
101st*	September 18–21, 1999	Sheraton Waikiki Hotel, Honolulu, Hawai'i
102nd	October 18–21, 1999	Sheraton Waikiki Hotel, Honolulu, Hawai'i
103rd	May 1, 2000	Council office, Honolulu, Hawai'i
104th	June 14–15, 2000	Maui Prince Hotel, Makena, Hawai'i
105th	July 10–11, 2000	Midway Atoll, NWHI, Hawai'i
106th	July 13, 2000	Dole Cannery Ballroom, Honolulu, Hawai'i
107th*	November 28–December 1, 2000	Ala Moana Hotel, Honolulu, Hawai'i
108th	February 12–15, 2001	Ala Moana Hotel, Honolulu, Hawai'i
109th	March 13, 2001	Ala Moana Hotel, Honolulu, Hawai'i
110th	June 19–21, 2001	Ala Moana Hotel, Honolulu, Hawai'i
111th	October 24–26, 2001	Hawai'i Convention Center, Honolulu, Hawai'i
112th	March 18–21, 2002	Ala Moana Hotel, Honolulu, Hawai'i

113th	June 24–27, 2002	American Samoa Convention Center, Pago Pago, American Samoa
114th	August 29, 2002	Virtual meeting via teleconference
115th	October 15–17, 2002	Aloha Tower Pier 11, Honolulu, Hawai'i
116th	December 16, 2002	Virtual meeting via teleconference
117th	February 11–13, 2003	Multipurpose Center, Saipan, CNMI
118th	June 10–13, 2003	Ala Moana Hotel, Honolulu, Hawai'i
119th	September 23, 2003	Virtual meeting via teleconference
120th*	October 20–23, 2003	Pagoda Hotel, Honolulu, Hawai'i
121st	November 25, 2003	Virtual meeting via teleconference
122nd*	March 23–25, 2004	Hawai'i Convention Center, Honolulu, Hawai'i
123rd*	June 21–24, 2004	Ala Moana Hotel, Honolulu, Hawai'i
124th	October 13–15, 2004	Pagoda Hotel, Honolulu, Hawai'i
125th	January 26, 2005	Virtual meeting via teleconference
126th*	March 14–17, 2005	Ala Moana Hotel, Honolulu, Hawai'i
127th*	May 31–June 2, 2005	Ala Moana Hotel, Honolulu, Hawai'i
128th	November1, 2005	Virtual meeting via teleconference
129th*	November 9–11, 2005	Hilton Guam, Tumon Bay, Guam
130th	December 20, 2005	Virtual meeting via teleconference
131st*	March 14–16, 2006	Ala Moana Hotel, Honolulu, Hawai'i
132nd	April 20, 2006	Virtual meeting via teleconference
133rd*	June 13–15, 2006	Utulei Convention Center, Pago Pago, American Samoa
134th	August 30, 2006	Virtual meeting via teleconference
135th*	October 16–19, 2006	Ala Moana Hotel, Honolulu, Hawai'i
136th	December 21, 2006	Virtual meeting via teleconference
137th*	March 13–16, 2007	Ala Moana Hotel, Honolulu, Hawai'i
138th	June 20–22, 2007	Ala Moana Hotel, Honolulu, Hawai'i
139th*	October 9–12, 2007	Pagoda Hotel, Honolulu, Hawai'i

140th*   March 17–21, 2008   Guam Hilton Hotel, Tumon Bay, Guam and Fiesta Resort, Saipan, CNMI

141st   April 14, 2008   Virtual meeting via teleconference

142nd*   June 16–19, 2008   Ala Moana Hotel, Honolulu, Hawai'i

143rd*   October 14–17, 2008   Pagoda Hotel, Honolulu, Hawai'i

144th*   March 24--26, 2009   Governor H. Rex Lee Auditorium, Pago Pago, American Samoa

145th*   July 22–25, 2009   King Kamehameha Hotel, Kailua-Kona, Hawai'i

146th*   October 20–23, 2009   Laniakea YWCA-Fuller Hall, Honolulu, Hawai'i

147th*   March 22–23, 25–26, 2010   Fiesta Resort, Saipan, CNMI and Guam Hilton Hotel, Tumon, Guam

148th*   June 28–July 1, 2010   Laniakea YWCA-Fuller Hall, Honolulu, Hawai'i

149th*   October 11–14, 2010   Laniakea YWCA-Fuller Hall, Honolulu, Hawai'i

150th*   March 8–10, 2011   Governor H. Rex Lee Auditorium, Pago Pago, American Samoa

151st*   June 16–18, 2011   Waikiki Beach Marriot Resort & Spa, Honolulu, Hawai'i

152nd*   October 19–22, 2011   Laniakea YWCA-Fuller Hall, Honolulu, Hawai'i

153rd*   March 5–6, 8–9, 2012   Fiesta Resort, Saipan, CNMI and Guam Hilton Hotel, Tumon, Guam

154th*   June 26–28, 2012   Laniakea YWCA-Fuller Hall, Honolulu, Hawai'i

155th*   October 29–November 1, 2012   Laniakea YWCA-Fuller Hall, Honolulu, Hawai'i

156th*	March 12–14, 2013	Governor H. Rex Lee Auditorium, Pago Pago, American Samoa
157th*	June 25–28, 2013	Laniakea YWCA-Fuller Hall, Honolulu, Hawai'i
158th*	October15–18, 2013	Laniakea YWCA-Fuller Hall, Honolulu, Hawai'i
159th*	March 17–21, 2014	Fiesta Resort, Saipan, CNMI and Guam Hilton Hotel, Tumon, Guam
160th*	June 24–27, 2014	Laniakea YWCA-Fuller Hall, Honolulu, Hawai'i
161st*	October 20–23, 2014	Laniakea YWCA-Fuller Hall, Honolulu, Hawai'i
162nd*	March 16–18, 2015	Laniakea YWCA-Fuller Hall, Honolulu, Hawai'i
163rd*	June 16–18, 2015	Harbor View Center, Honolulu, Hawai'i
164th*	October 17–22, 2015	Governor H. Rex Lee Auditorium, Pago Pago, American Samoa
165th*	March 14–17, 2016	Laniakea YWCA-Fuller Hall, Honolulu, Hawai'i
166th*	June 6–10, 2016	Fiesta Resort, Saipan, CNMI and Guam Hilton Hotel, Tumon, Guam
167th	August 3, 2016	Virtual meeting via teleconference
168th*	October 12–14, 2016	Laniakea YWCA-Fuller Hall, Honolulu, Hawai'i
169th*	March 21–23, 2017	Ala Moana Hotel, Honolulu, Hawai'i
170th*	June 20–22, 2017	Laniakea YWCA-Fuller Hall, Honolulu, Hawai'i
171st*	October 17–19, 2017	Sadie's by the Sea, Utulei, American Samoa

172nd*	March 14–16, 2018	Laniakea YWCA-Fuller Hall, Honolulu,
173rd*	June 11–13, 2018	Wailea Beach Resort, Wailea, Maui, Hawai'i
174th*	October 23–24, 26–27, 2018	Fiesta Resort, Saipan, CNMI/Guam Hilton Hotel, Guam
175th	December 17, 2018	Virtual meeting via teleconference
176th*	March 18–21, 2019	Laniakea YWCA-Fuller Hall, Honolulu, Hawai'i
177th	April 12, 2019	Virtual meeting via teleconference
178th*	June 25–27, 2019	Laniakea YWCA-Fuller Hall, Honolulu, Hawai'i
179th	August 8, 2019	Virtual meeting via teleconference
180th*	October 22–24, 2019	Governor Tauese P.F. Sunia Ocean Center, Pago Pago, American Samoa
181st*	March 10–12, 2020	Laniakea YWCA-Fuller Hall, Honolulu, Hawai'i
182nd	June 22–25, 2020	Virtual meeting via teleconference
183rd	September 15–17, 2020	Virtual meeting via teleconference
184th	December 2–4, 2020	Virtual meeting via teleconference

Appendix 7

Special Events Organized or Hosted by the Council

October 27–28, 1978	4th Council Chairmen's Meeting, Kailua-Kona, Hawai'i
October 30–31, 1984	Pacific Fishery Development Foundation Fisheries Officers' Workshop, Honolulu, Hawai'i
February 25–28, 1985	13th Council Chairmen's Meeting, Hilo, Hawai'i
August 6, 1985	U.S. Tuna Association Meeting, Kailua-Kona, Hawai'i
January 27–February 1, 1986	Western Pacific Regional Tuna Negotiations, Sixth Round, Kona, Hawai'i
October 13–16, 1987	North Pacific Rim Fishermen's Conference on Marine Debris, Kailua-Kona, Hawai'i
July 31–August 5, 1988	International Billfish Symposium, Kailua-Kona, Hawai'i
February 6, 1991	Inter-Agency Task Force Meeting, Hawai'i

August 14–15, 1997	A Workshop on the Magnuson-Stevens Community Development Program: Perspectives from Alaska and the Western Pacific, Honolulu, Hawai'i
July 14–15, 1998	26th Council Chairmen's Meeting, Maui, Hawai'i
October 8-10, 1998	Black-footed Albatross Population Biology Workshop, Honolulu, Hawai'i
February 10–19, 1999	Fourth Session of the Multilateral High-Level Conference for the Conservation and Management of Highly Migratory Fish Stocks in the Central and Western Pacific Ocean (MHLC), Honolulu, Hawai'i
September 6–15, 1999	Fifth MHLC Session, Honolulu, Hawai'i
November 13 1999	Fisheries Forum 2000, Honolulu, Hawai'i
April 12–19, 2000	Sixth MHLC Session, Honolulu, Hawai'i
August 6–11, 2000	1st International Conference on the Sources, Impacts, Mitigation and Prevention of Marine Debris, Honolulu, Hawai'i
August 30–September 5, 2000	Seventh MHLC Session finalizing the convention for signatures, Honolulu, Hawai'i

February 5–8, 2002	1st Western Pacific Sea Turtle Cooperative Research and Management Workshop, Honolulu, Hawai'i
November 19–22, 2002	Second International Fishers Forum, Honolulu, Hawai'i
January 13–15, 2004	Asia-Pacific Economic Cooperation Derelict Fishing Gear and Related Marine Debris Seminar, Honolulu, Hawai'i
January 13–16, 2004	Workshop on the Development of Bottomfish Resource Assessment Methodologies for the U. S. Central and Western Pacific Fisheries, Honolulu, Hawai'i
February 10–13, 2004	Coral Reef Fish Stock Assessment Workshop, Honolulu, Hawai'i
April 13–15, 2004	32nd Council Chairs and Executive Director's Meeting, Lihue, Hawai'i
May 17–21, 2004	2nd Western Pacific Sea Turtle Cooperative Research and Management Workshop on West Pacific Leatherback and Southwest Pacific Hawksbill Sea Turtles, Honolulu, Hawai'i
December 9–10, 2004	Inaugural Session of the Western and Central Pacific Fisheries Commission (WCPFC), Pohnpei, Federated States of Micronesia

March 2–3, 2005	2nd Western Pacific Sea Turtle Cooperative Research and Management Workshop on North Pacific Loggerhead Sea Turtles, Honolulu, Hawai'i
April 11–14, 2005	Technical Assistance Workshop on Sea Turtle Bycatch Reduction Experiments in Longline Fisheries, Honolulu, Hawai'i
July 25–29, 2005	International Tuna Fishers Conference on Responsible Fisheries & Third International Fishers Forum, Yokohama, Japan
April 18–19, 2006	Black Coral Science and Management Workshop, Honolulu, Hawai'i
August 15–17, 2006	Ho'ohanohano I Na Kupuna, Puwalu No Na Lae'ula, Honolulu, Hawai'i
September19–21, 2006	South Pacific Albacore Longline Fisheries Workshop, Honolulu, Hawai'i
October 24–25, 2006	Data Workshop for Developing Fishery Ecosystem Plans, Honolulu, Hawai'i
November 8–9, 2006	Ho'ohanohano I Na Kupuna, Ke Kumu Ike Hawaii, Waikiki, Hawai'i
December 19–20, 2006	Ho'ohanohano I Na Kupuna, Puwalu 'Ekolu: Lawena Aupuni, Honolulu, Hawai'i
February 15–16, 2007	Asia and Pacific Islands Bycatch Consortium, Honolulu, Hawai'i

July 17–20, 2007	Bellagio Sea Turtle Conservation Initiative: Strategic Planning for Long-Term Financing of Pacific Leatherback Conservation and Recovery, Terengganu, Malaysia
October 22–23, 2007	Third Inter-Governmental Meeting on Establishment of New Mechanism for Management of High Seas Bottom Fisheries in the North Western Pacific Ocean, Honolulu, Hawai'i
October 24–26, 2007	Third Inter Governmental Meeting on Management of High Seas Bottom Fisheries in the Northwestern Pacific Ocean, Honolulu, Hawai'i
October 30–31, 2007	Asia and Pacific Islands Bycatch Consortium, Honolulu, Hawai'i
October 31–November 1, 2007	Ho'ohanohano I Na Kupuna, Puwalu 'Elima, Honolulu, Hawai'i
November 12–15, 2007	Fourth International Fishers Forum, Puntarenas, Costa Rica
December 19–20, 2007	North Pacific Loggerhead Sea Turtle Expert Workshop, Honolulu, Hawai'i
November 12–14, 2008	National Scientific and Statistical Committee (SSC) Workshop, Honolulu, Hawai'i
January 20–22, 2009	Technical Workshop on Minimizing Sea Turtle Interactions in Coastal Net Fisheries, Honolulu, Hawai'i
August 10–12, 2009	Pacific Islands Bio-sampling Workshop, Agana, Guam

October 2009	Technical Workshop Effects of Pelagic Fisheries on Seamount Ecosystems, Noumea, New Caledonia
November 17–19, 2009	Western Pacific Region Fisheries Data Workshop, Honolulu, Hawai'i
July 20, 2010	Community-Based Monitoring Workshop on Reef Fish Tagging, Bio-sampling, Ecosystem Indicator Sampling, Community FADs and Online Data Reporting, Merizo, Guam
July 24, 2010	Community-Based Monitoring Workshop on Reef Fish Tagging, Bio-sampling, Ecosystem Indicator Sampling, Community FADs and Online Data Reporting, Santa Rita, Guam
August 3–5, 2010	Fifth International Fishers Forum on Marine Spatial Planning and Bycatch Mitigation, Taipei
November 19–20, 2010	Ho'o Lei Ia Pae 'Aina Puwalu, Honolulu, Hawai'i
December 6–10, 2010	Seventh Regular Session of the WCPFC, Ko'olina, Hawai'i
January 25–27, 2011	Mariana Archipelago Green Turtle Workshop, Saipan, CNMI
January 26–27, 2011	Open Ocean Cage Culture Symposium, Saipan, CNMI

February 1–4, 2011	Workshop on Establishing Annual Catch Limits for Coral Reef Fisheries, Honolulu, Hawai'i
December 7, 2011	Hawai'i Non-Commercial Fishery Data Collection Workshop, Honolulu, Hawai'i
February 29, 2012	Village of Malesso Community-Based Monitoring Meeting, Merizo, Guam
April 30–May 3, 2012	Council Coordination Committee, Kohala Coast, Hawai'i
March 26-27, 2013	Expert Workshop to Develop Alternative Model to Evaluate Impacts of Fisheries on Hawai'i False Killer Whale Populations, Honolulu, Hawai'i
August 24, 2013	Community-Based Management Plan for Marine Resources of Malesso – Workshop 1, Merizo, Guam
November 20, 2013	Community-Based Management Plan for Marine Resources of Malesso – Workshop 2, Merizo, Guam
March 5, 2014	Community-Based Management Plan for Marine Resources of Malesso— Final Report, Merizo, Guam
April 22-24, 2014	Bigeye Tuna Movement Workshop, Honolulu, Hawai'i
September 18–20, 2014	Disproportionate Burden Workshop, Honolulu, Hawai'i

December 17–19, 2014   Social Scientist in Regional Fisheries Management Workshop, Honolulu, Hawai'i

February 23–25, 2015   National SSC Workshop V, Honolulu, Hawai'i

April 7, 2015   Western and Central Pacific Ocean (WCPO) Longline Management Information Meeting, Honolulu, Hawai'i

April 8–10, 2015   WCPO Purse-Seine Bigeye Management Workshop, Honolulu, Hawai'i

April 28, 2015   Hawai'i Longline Catch Shares Informational Meeting, Honolulu, Hawai'i

August 19–21, 2015   WCPO Purse-Seine Bigeye Management Workshop II, Majuro, Republic of Marshall Islands

June 11, 2016   Guam Coral Reef Participatory Mapping Workshop, Tumon, Guam

October 18–20, 2016   Rare Events Bycatch Workshop Series, Honolulu

December 6–17, 2016   Public Scoping Meetings on Proposed Fishing Regulations for the Monument Expansion Area in the Northwestern Hawaiian Islands, various Hawai'i locations

August 22–24, 2017   WCPFC Intersessional Meeting on Tropical Tuna, Special Session, Honolulu

November 7–9, 2017	Albatross Workshop, Honolulu
February 6, 2018	Western Pacific Stock Assessment Review of the Draft 2017 Benchmark Stock Assessments for Guam Coral Reef Fish, Honolulu
April 13, 2018	Western Pacific Stock Assessment Review (WPSAR) Steering Committee Meeting, Honolulu
December 9–14, 2018	15th Regular Session of the WCPFC, Honolulu
April 15–19, 2019	WPSAR Territory Bottomfish Meeting, Honolulu
September 10–14, 2018	WPSAR of the 2018 Benchmark Stock Assessment for the Main Hawaiian Islands Kona Crab, Honolulu
April 3, 2019	WPSAR Steering Committee Meeting, Honolulu
September 15–20, 2019	OceanObs'19 (Indigenous Delegation), Honolulu
February 4–13, 2020	Hawai'i Pelagic Small-boat Fisheries Public Scoping Meetings, conducted statewide at different venues
February 24–28, 2020	WPSAR of a 2020 Benchmark Stock Assessment for Hawai'i Gray Jobfish (Uku), Honolulu
April 30, 2020	WPSAR Steering Committee, virtual meeting via teleconference
May 27–28, 2020	Council Coordination Committee, virtual meeting via teleconference

June 15–17, 2020	International Workshop on Area-Based Management of Blue Water Fisheries, virtual meeting via teleconference
August 27, 2020	Hawai'i Small-Boat Fisheries Management Scoping, virtual meeting via teleconference
September 23–24, 2020	Council Coordination Committee, virtual meeting via teleconference
December 16–17, 2020	WPSAR of a 2020 Stock Assessment Update for Seven Deepwater Bottomfish Species in the Main Hawaiian Islands, virtual meeting via teleconference

Appendix 8

Council Publications and Reports

*funded/co-funded by the Council and/or authored/co-authored
or edited/co-edited by Council staff.*

Books, Proceedings, Journal Series

Dutton PH, Squires D, Ahmed M, editors. 2011. Conservation of Pacific sea turtles. Honolulu: Univ Hawaii Pr. 481 p.

Gilman E, editor. 2006. Proceedings of the Symposium on Mangrove Responses to Relative Sea Level Rise and Other Climate Change Effects; 2006 July 13. Catchments to Coast. The Society of Wetland Scientists 27th International Conference; 2006 Jul 9–14; Cairns, Australia. Honolulu: WPRFMC. 81 p.

———. 2009. Proceedings of the Technical Workshop on Mitigating Sea Turtle Bycatch in Coastal Net Fisheries; 2009 Jan. 20–22; Honolulu. Honolulu: WPRFMC. 76 p.

Gilman E, Clarke S, Nigel B, Alfaro-Shigueto J, Mandelman J, Mangel J, Petersen S, Piovano S, Thomson N, Dalzell P, Donoso M, Goren M, Werner T. 2007. Shark depredation and unwanted bycatch: industry practices and attitudes, and shark avoidance strategies. Honolulu: WPRMC. 148 p.

Gilman E, Ishizaki A, Chang D, Liu WY, Dalzell P, editors. 2011. Proceedings of the Fifth International Fishers Forum on Marine Spatial Planning and Bycatch Mitigation; 2010 Aug 3–5; Taipei. Honolulu: WPRFMC. 300 p.

Glazier E. 2019. Tradition-based natural resource management: practice and application in the Hawaiian Islands. Cham, Switzerland: Palgrave Macmillan. 281 p.

Glazier E, editor. 2011. Ecosystem-based fisheries management in the Western Pacific. Chichester, UK: Wiley-Blackwell. 280 p.

Impact Assessment Inc. 2006. Proceedings of the Ecosystem Social Science Workshop; 2006 Jan 17–20; Honolulu. Final report for the Western Pacific Fishery Management Council. 138 p.

Karam A, compiler. n.d. Resources assessment investigation of the Mariana archipelago, 1980–1985: Compilation of published manuscripts, reports and journals. Honolulu: WPRFMC.

Kinan I, editor. 2002. Proceedings of the Western Pacific Sea Turtle Cooperative Research and Management Workshop; 2002 Feb 5–8; Honolulu. Honolulu: WPRFMC. 300 p.

———. 2005. Proceedings of the Second Western Pacific Sea Turtle Cooperative Research and Management Workshop; 2004 May 17–21;

Honolulu. Volume I: West Pacific leatherback and Southwest Pacific hawksbill sea turtles. Honolulu: WPRFMC. 118 p.

———. 2006. Proceedings of the Second Western Pacific Sea Turtle Cooperative Research and Management Workshop; 2005 Mar 2–3; Honolulu. Volume II: North Pacific loggerhead sea turtles. Honolulu: WPRFMC. 96 p.

Miller ML, Daxboeck C, Dahl C, Kelly K, Dalzell P, editors. 2001. Proceedings of the 1998 Pacific Island Gamefish Tournament Symposium; 1999 July 29–Aug 1; Kailua-Kona, HI. Honolulu: WPRFMC. 301 p.

Parks NM. 2003. Proceedings of the Second International Fishers Forum; 2002 Nov 19–22; Honolulu. Honolulu: WPRFMC. 210 p.

Simonds KM, Spalding S, Kaʻaiʻai C, editors. 2008. Hoʻohanohano I Na Kapuna Puwalu and the International Pacific Marine Educators Conference. Current: J Mar Educ 24(2). 57 p.

Spalding S, Witherell D, Gilden J, editors. 2009. U.S. regional fishery management councils: Providing sound stewardship of our nation's fishery resources. Current: J Mar Educ 25(3). 64 p.

Steering Committee, Bellagio Conference on Sea Turtles. 2004. What can be done to restore Pacific turtle populations? The Bellagio blueprint for action on Pacific sea turtles. 24 p.

Steering Committee, Bellagio Sea Turtle Conservation Initiative. 2008. Strategic planning for long-term financing of Pacific leatherback conservation and recovery: Proceedings of the Bellagio Sea Turtle Conservation Initiative; 2007 Jul 17–20; Terengganu, Malaysia. Worldfish Center Conference Proceedings 1805. Penang, Malaysia: Worldfish Center. 79 p.

Western Pacific Regional Fishery Management Council. 2003. Executive Summary of the Second International Fishers Forum; 2002 Nov 19–22; Honolulu. Honolulu: WPRMC. 32 p.

———. 2006. Proceedings of the International Tuna Fishers Conference on Responsible Fisheries and Third International Fishers Forum; 2005 Jul 25–29; Yokohama, Japan. Honolulu: WPRFMC. 296 p.

———. 2008. Proceedings of the Fourth International Fishers Forum; 2007 Nov 12–14; Puntarenas, Costa Rica. Honolulu: WPRFMC. 234 p.

WPRFMC [Western Pacific Regional Fishery Management Council]. 2006. Proceedings of the International Tuna Fishers Conference on Responsible Fisheries & Third International Fishers Forum; 2005 July 25–29; Yokohama, Japan. Honolulu: WPRMC. 156 p.

———. 2009. Proceedings of the Fourth International Fishers Forum; 2007 Nov. 12–14; Puntarenas, Costa Rica. Honolulu: WPFRMC. 234 p.

Monographs

Dalzell P. December 2020. University of Hawai'i Pelagic Fisheries Research Program. Pacific Islands Fishery Monographs 11. Honolulu: WPRFMC. 35 p.

DeMello J, Maciaz M, Kurokawa J, Sabater M, editors. 2016. Western Pacific coral reef fisheries: a decade of research in the U.S. Pacific Islands. Pacific Islands Fishery Monographs 8. Honolulu: WPRFMC. 22 p.

Fougner S. 2010. Ten years and counting: the first 10 years of the Western and Central Pacific Fisheries Commission. Pacific Islands Fishery Monographs 2 Honolulu: WPRFMC. 16 p.

Fougner S, Fitchett M. Western and Central Pacific Fisheries Commission, the second decade: the evolution of modern management. Pacific Islands Fishery Monographs 15. Honolulu: WPRFMC. Forthcoming.

Grigg R. 2010. The precious corals fishery management plan of the Western Pacific Regional Fishery Management Council. Pacific Islands Fishery Monographs 1. Honolulu: WPRFMC. 9 p.

Ishizaki A. 2015. Protected species conservation by the Western Pacific Regional Fishery Management Council. Pacific Islands Fishery Monographs 4. Honolulu: WPRFMC. "21 p.

Ka'ai'ai C. 2016. Western Pacific community development program and Western Pacific community demonstration project program. Pacific Islands Fishery Monographs 7. Honolulu: WPRFMC. 21 p.

Kingma E. 2016. Fisheries development projects in American Samoa, Guam, and the Northern Mariana Islands, 2010–2015. Pacific Islands Fishery Monographs 6. Honolulu: WPRFMC. 11 p.

Markrich MI, Hawkins C. 2016. Western Pacific Region fishing fleets and fishery profiles. Pacific Islands Fishery Monographs 5. Honolulu: WPRFMC. 34 p.

Markrich ML. 2020. Northwestern Hawaiian Islands Lobster Fishery. Pacific Islands Fishery Monographs 9. Honolulu: WPRFMC. 31 p.

———. 2020. History of the billfish fisheries and their Management in the Western Pacific Region. Pacific Islands Fishery Monographs 10. Honolulu: WPRFMC. 36 p.

Martell L, Spalding S. 2020. Fishery ecosystem management in the Western Pacific Region. Pacific Islands Fishery Monographs 12. Honolulu: WPRFMC. 34 p.

Sabater M. 2021. Fishery data collection systems: evasive as an elusive fish. Pacific Islands Fishery Monographs 13. Honolulu: WPRFMC. 19 p.

Sabater M, Tulafono R. 2011. American Samoa archipelagic fishery ecosystem report. Pacific Islands Fishery Monographs 3. Honolulu: WPRFMC. 31 p.

Spalding S, Vandehey A. Public involvement and outreach to sustain fisheries in the Western Pacific Region. Pacific Islands Fishery Monographs 14. Honolulu: WPRFMC. Forthcoming.

Journal Articles and Book Chapters

Adams T, Dalzell P, Ledua E. 1999. Ocean resources. In: Rapaport M, editor. The Pacific Islands. Honolulu: Bess Pr. 366–81.

Adams TJH, Dalzell P, Farman R. 1997. Status of Pacific Island reef fisheries. In: Lessios HA, Macintyre IG, editors. Proceedings of the 8th International Coral Reef Symposium; 1996 Jun 24–29; Panama City, Panama. Vol. II. Balboa, Panama: Smithsonian Trop Res Inst. p 1977–80.

Bailey TP. 2012. The Aha Moku: an ancient Native Hawaiian resource management system. In: Courtney C et al., editors. Fishing people of the North: cultures, economies and management responding to change. 27th Lowell Wakefield Fisheries Symposium; 2011 Sept. 14–17; Anchorage, Alaska. Fairbanks, Alaska: Alaska Sea Grant. P 171–6.

Chaloupka M, Balazs G. 2007. Using Bayesian state-space modeling to assess the recovery and harvest potential of the Hawaiian green sea turtle. Ecol Modelling 205: 93–109.

Cousins KL, Dalzell P, Gilman E. 2000. Managing pelagic longline-albatross interactions in the North Pacific Ocean. In: Cooper J, editor. Albatross and Petrel Mortality from Longline Fishing International Workshop; 2000 May 11–12; Honolulu. Report and presented papers. Mar Ornith 28:159–74.

Dalzell P. 1997. Current status of pelagic fisheries of the Western Pacific Region. In: D Zachary, C. Sterling and L Karolot, editors. Towards a prosperous Pacific: building a sustainable tuna industry in the Pacific Islands. Maui Pacific Center 7th Annual Conference; 1997 Nov 5–8; Makena, Maui, HI. p 7.

———. 1998. The role of archaeological and cultural-historical records in long-range coastal fisheries resources management strategies and policies in the Pacific Islands. Ocean Coast Manage 40:237–52.

——. 2001. Marlin management in Hawaii: Are there interactions between longline vessels and charter vessels targeting blue and striped marlin? In: Miller ML, Daxboeck C, Dahl C, Kelly K, Dalzell P, editors. Proceedings of the 1998 Pacific Island Gamefish Symposium: facing the challenges of resource conservation, sustainable development and the sportfishing ethic; 1998 Jul 29–Aug 1; Kailua-Kona, HI. Honolulu: WPRFMC. p 247–55.

Dalzell P, Boggs C. 2003. Pelagic fisheries catching blue and striped marlins the U.S. Western Pacific islands. Mar Fresh Res 54:419–24.

Dalzell P, Laurs M, Haight W. 2008. Case study: catch and management of pelagic sharks in Hawai'i and the Western Pacific region. In: Camhi M, Thomas J, editors. Sharks of the open ocean. Oxford, U.K.: Blackwell Sci Pub. p 268–74.

Dalzell P, Pautzke C. 2004. Directions in marine research. In: Witherell D, editor. Managing our nation's fisheries: past, present and future. Proceedings of a Conference on Fisheries Management in the United States; 2003 Nov 13–15; Washington, DC. p 190–5.

Dalzell P, Schug D. 1997. Coping with instability: Pelagic fisheries issues and the Western Pacific Fishery Management Council. In: D Zachary, C. Sterling and L Karolot, editors. Towards a prosperous Pacific: building a sustainable tuna industry in the Pacific Islands. Maui Pacific Center 7th Annual Conference; 1997 Nov 5–8; Makena, Maui, HI. p 82–6.

Dalzell P, Smith A. 1998. Rapid data collection and assessment of the biology of Acanthurus nigrofuscus at Woleai Atoll, Micronesia. J Fish Biol 52: 386–97.

DeMello JK. 2004. Commercial marine landings from fisheries on the coral reef ecosystem of the Hawaiian archipelago. In: Friedlander AM, editor. Status of Hawai'i's coastal fisheries in the new millennium. Proceedings of the 2001 Fisheries Symposium. American Fisheries Society, Hawai'i Chapter; 2001 Nov 1; Honolulu. Honolulu: Hawai'i Audubon Society. p 157–70.

Friedlander A, Dalzell P. 2003. A review of the biology and fisheries of two large jacks, ulua (Caranx ignobilis) and omilu (Caranx melampygus), in the Hawaiian archipelago. In: Friedlander A, editor. Status of Hawaii's coastal fisheries in the new millennium. Proceedings of the 2001 Fisheries Symposium. American Fisheries Society, Hawai'i Chapter; 2001 Nov 1; Honolulu. Honolulu: Hawai'i Audubon Society. p 174–89.

Gilman E, Brothers N, McPherson G, Dalzell P. 2006. A review of cetacean interactions with longline gear. J Cetacean Res Manage 8(2):215–23.

Gilman E, Chaloupka M, Ishizaki A et al. 2021. Tori lines mitigate seabird bycatch in a pelagic longline fishery. Rev Fish Biol and Fisheries 31:653–61.

Gilman E, Chaloupka M, Fitchett MD et al. 2029. Ecological responses to blue water MPAs. PLoS One 15. e0235129.

Gilman E, Chaloupka M, Read A, Dalzell P, Holetschek J, Curtice C. 2012. Hawai'i longline tuna fishery temporal trends in standardized catch rates and length distributions and effects on pelagic and seamount ecosystems. Aqua Conserv: Mar Fresh Ecosys 22(4):446–88.

Gilman E, Clarke S, Brothers N, Alfaro-Shigueto J, Mandelman J, Mangel J, Petersen S, Piovano S, Thomson N, Dalzell P, Donoso M, Goren M, Werner T. 2008. Shark interactions in pelagic longline fisheries. Mar Pol 32:1–18.

Gilman E, Dalzell P, Martin S. 2006. Fleet communication to abate fisheries bycatch. Mar Pol 30(4):360–6.

Gilman E, Gearhart J, Price B, Eckert S, Milliken H, Wang J, Swimmer Y, Shiode D, Abe O, Peckham SH, Chaloupka M, Hall M, Mangel J, Alfaro-Shigueto J, Dalzell P, Ishizaki A. 2009. Mitigating sea turtle bycatch in coastal passive net fisheries. Fish and Fisheries 11(1):57–88.

Gilman E, Kobayashi D, Swenarton T, Dalzell P, Kinan I, Brothers N. 2007. Reducing sea turtle interactions in the Hawai'i-based longline swordfish fishery. Biol Conserv 139:19–28.

Hilborn R, Agostini V, Chaloupka M et al. 2021. Area-based management of blue water fisheries: Current knowledge and research needs. Fish and Fisheries 00:1–27. Available at: https://doi.org/10.1111/faf.12629.

Ishihara T, Kamezaki N, Matsuzawa Y, Ishizaki A. 2014. Assessing the status of Japanese coastal fisheries and sea turtle bycatch [in Japanese]. Wildlife and Hum Soc 2(1):23–35.

Ishizaki A. 2009. Protected species conservation and fishery management. Current: J Mar Ed 25(3):32–5.

Ishizaki A, Simonds KM. 2010. Sea turtle conservation and management in the United States: The case of Hawai'i longline fishery [in Japanese]. Aquabiol 32(5):463–6.

Ka'ai'ai C, Spalding S. 2009. Ho'ohanohano I Na Kupuna Puwalu. Current: J Mar Ed 24(2):2–6.

Kinan I, Dalzell P. 2005. Sea turtles: flagship species for conservation and fishery management. Presented at the People and the Sea II Conference; 2003 Sep 4–6; Amsterdam. MAST 3(2) and 4(1):195–212.

Matsuzawa Y et al. 2016. Fine-scale genetic population structure of loggerhead turtles in the northwest Pacific. Endang Spec Res 30:83–93.

Miller ML, Daxboeck C, Dahl C, Kelly K, Dalzell P, editors. Proceedings of the 1998 Pacific Island Gamefish Symposium: facing the challenges of resource conservation, sustainable development and the sportfishing ethic; 1998 Jul 29–Aug 1; Kailua-Kona, HI. Honolulu: WPRFMC. 301 p.

Moore BR, Allain V et al. 2020. Defining the stock structures of key commercial tunas in the Pacific ocean II: Sampling considerations and future directions. Fish Res 230 (2020) 105524. 21 p.

Nicol SJ, Allain V, Pilling GM, Polovina J, Coll M, Bell J, Dalzell P et al. 2012. An ocean observation system for monitoring the effects of climate change on the ecology and sustainability of pelagic fisheries in the Pacific ocean. Clim Change 119(1):131–45.

Pons M, Watson JT, Ovando D et al. 2022. Trade-offs between bycatch and target catches in static versus dynamic fishery closures. Proc Nat Acad Sci 119(4).

Sabater M, Kleiber P. 2014. Augmented catch-MSY approach to fishery management in coral-associated fisheries. In: Bortone SA, editor. Interrelationships between corals and fisheries. Boca Raton, FL: CRC Press. p 199–218.

Simonds KM. 2008. Call for a community and cultural consultation process. Current: J Mar Ed 24(2):13–4.

Spalding S. 2008. International Pacific marine educators conference. Current: J Mar Ed 24(2):23–5.

———. 2009. Working cooperatively: International fisheries management in the 21st century. Current: J Mar Ed 25(3):36–8.

Spalding S, Dalzell P. 2009. Unique entities: US regional fishery management councils. In: Current, J Mar Educ 25(3):4–7.

Wiley J, Sabater M, Langseth B. 2021. Aerial survey as a tool for understanding bigeye scad (Selar crumenophthalmus) dynamics around the island of Oʻahu, Hawaiʻi. Fish Res 236:105866.

Reports and Presentations

Amesbury JR. 2005. Information on indicators for the Mariana archipelago fishery ecosystem plan. A report prepared for the WPRFMC. 102 p.

Amesbury JR, Hunter-Anderson RL. 1989. Native fishing rights and limited entry in Guam. Prepared for the WPRFMC. 88 p.

———. 2003. Review of archaeological and historical data concerning reef fishing in the US flag islands of Micronesia: Guam and the Northern Mariana Islands: Final report for the WPRFMC. 166 p.

———. 2008 May. Analysis of archaeological and historical data on fisheries for pelagic species in Guam and the Northern Mariana Islands. Prepared for the Pelagic Fisheries Research Program. Guam: Micronesian Archaeological Research Services.

Amesbury JR, Hunter-Anderson RL, Wells EF. 1989. Native fishing rights and limited entry in the CNMI. Prepared for the WPRFMC. 129 p.

Ault J, Begg G, Gribble N, Kulbicki M, Mapstone B, Medley P. 2004. Coral Reef Fish Stock Assessment Workshop; 2004 Feb 10–13; Honolulu. Interim final panel report for the WPRFMC. 30 p.

Bartram P. 2003. Findings and recommendations on pilot studies supporting the proposed Mariana Archipelago fishery ecosystem plan. A report prepared for the WPRFMC. 10 p.

———. 2012. Assessment of freshwater impacts on coral reef fisheries of Hawai'i. Honolulu: WPRFMC.

Boggs C, Dalzell P, Essington T, Labelle M, Mason D, Skillman R, Wetherall J. 2000. Recommended overfishing definitions and control rules for the Western Pacific Regional Fishery Management Council's pelagic fishery management. NOAA NMFS SWFSC Admin rep H-00-05. 18 p.

Chaloupka M, Gilman E, Carne M, Ishizaki A et al. 2021. Could tori lines replace blue-dyed bait to reduce seabird bycatch risk in the Hawai'i deep-set longline fishery? Honolulu: WPRFMC.

Dalzell P. 1997. The influence of incidental catch and protected species interactions on the management of the Hawaii-based longline fishery. 48th Annual Tuna Conference; 1997 May 9–22; Lake Arrowhead, CA. 13 p.

Dalzell P, Schug DM. 2002. Synopsis of information relating to sustainable coastal fisheries. Strategic action programme for the international waters of the Pacific small island developing states. Tech rep 2002/04. Apia, Samoa: South Pacific Regional Environmental Programme. 38 p.

Del Raye G, Weng K. 2012. Habitat usage and movement behavior of five ubiquitous mesopredators in Palmyra Atoll. Honolulu: WPRFMC.

DeMello J. 2002. Time series of commercial marine landings from fisheries on coral reef ecosystems of the Hawaiian archipelago. A report prepared for the WPRFMC.

DeMello JK, editor. 2006. A report on the 2006 Black Coral Science and Management Workshop; 2006 Apr 18–19. Honolulu: WPRFMC. 48 p.

Dye TS, Graham TR. 2004. Review of archeological and historical data concerning reef fishing in Hawai'i and American Samoa. Final report for the WPRFMC. 160 p.

Fugro Pelagos Inc. 2004. Bathymetry and habitat mapping: Farallon de Medinilla, Northern Mariana Islands survey report. Honolulu: WPRFMC.

Gilman E, Clarke S, Brothers N, Alfaro-Shigueto J, Mandelman J, Mangel J, Piovano S, Peterson S, Watling D, Dalzell P. 2007. Strategies to reduce shark depredation and unwanted bycatch in pelagic longline fisheries: industry practices and attitudes, and shark avoidance strategies. Honolulu: WPRFMC. 148 p.

Gilman E, Ishizaki A, editors. 2018. Report of the workshop to review seabird bycatch mitigation measures for Hawai'i's pelagic longline fisheries; 2018 Sept. 18–19; Honolulu. Honolulu: WPRFMC.

Gilman E., Kobayashi D, Swenarton T, Dalzell P, Kinan I, Brothers N. 2006. Efficacy and commercial viability of regulations designed to reduce sea turtle interactions in the Hawai'i-based longline swordfish fishery. Honolulu: WPRFMC. 45 p.

Gilman E, Naholowaa HA, Ishizaki A, Chaloupka M et al. 2021. Practicality and efficacy of tori lines to mitigate albatross interactions in the Hawai'i deep-set longline fishery. Honolulu: WPRFMC. 48 p.

Gilman E, Zollett E, Beverly S, Nakano H, Davis K, Shiode D, Dalzell P, Kinan I. 2006. Reducing sea turtle by-catch in pelagic longline fisheries. Fish and Fisheries 7:2–23.

Green A. 1997. An assessment of the status of the coral reef resources and their patterns of use in the US Pacific Islands. Final report prepared for the WPRFMC. 278 p.

Haight WR, Dalzell P. 2000. Catch and management of sharks in pelagic fisheries in Hawai'i and the Western Pacific region. Presented at the International Pelagic Shark Workshop; 2000 Feb 14–17; Pacific Grove, CA.

Hawaii Cooperative Research Unit. 2008. Report on biology of parrotfish of Hawai'i. Honolulu: WPRFMC.

Hawhee J. 2007. Western Pacific coral reef ecosystem: a report prepared for the WPRFMC. 185 p.

H.R. 4576 Ensuring Access to Pacific Fisheries Act: Hearing before the Subcommitee on Water, Power and Oceans, of the House Committee on natural Resources. 114th Cong., 2d sess. (2016) (testimony of Kitty M. Simonds). https://naturalresources.house.gov/hearings/hearing-on-hr-4576.

Hunter CL. 1995. Review of coral reefs around American flag Pacific Islands and assessment of need, value and feasibility of establishing a coral reef fishery management plan for the Western Pacific Region. Final report prepared for the WPRFMC. 30 p.

Hunter-Anderson RL. 2005. Ecological notes in support of designing ecosystem-based fishery management plans for the Mariana Islands, Micronesia. A report prepared for the WPRFMC. 51 p.

Iversen RTB, Dye T, Palu ML. 1990. Native Hawaiian fishing rights phase 1: the Northwestern Hawaiian Islands. Honolulu, HI: WPRFMC.

——. 1990. Native Hawaiian fishing rights phase 2: Main Hawaiian Islands and the Northwestern Hawaiian Islands. Honolulu, HI: WPRFMC.

Kahng SE. 2007. Ecological impacts of Carijoa riisei on black coral habitat. Final report for the WPRFMC. 5 p.

Kapur MR, Fitchett MD, Carvalho FC, Yau AJ. 2019. 2018 Benchmark stock assessment for the main Hawaiian Islands Kona crab fishery. NOAA Tech Memo. NMFS-PIFSC. 104 p.

Kelly K, Messer A. 2005. Main Hawaiian Island lobsters: commercial catch and dealer data analysis (1984–2004). Final report for the WPRFMC. 42 p.

Kinch J. 2006. A Socio-economic assessment of the Huon Coast leatherback turtle nesting beach project, Morobe Province, Papua New Guinea. A technical final report to the WPRFMC.

Lammers MO, Au WWL. 2011. Spatial and temporal patterns of deep-water acoustic chorus observed off O'ahu, Hawai'i. Honolulu: WPRFMC.

Lammers MO, Kelley C. 2007. In situ acoustic records of Hawaiian deep-reef slopes and seamounts. Honolulu: WPRFMC.

Luck D. 2012. An approach for assessing essential fish habitat for coral reef species in Hawai'i. Honolulu: WPRFMC.

Luck D, Dalzell P. 2010. Western Pacific region reef fish trends: a compendium of ecological and fishery statistics for reef fishes in American Samoa, Hawaiʻi and the Mariana archipelago, in support of annual catch limit implementation. Honolulu: WPRFMC.

Maciasz M. 2012. Analysis of commercial fish trap data in Hawaiʻi. Honolulu: WPRFMC.

Mason D, Dalzell P, Boggs C. 1999. The collection of information necessary to create a control rule for albacore, bigeye, skipjack and yellowfin tunas in the Pacific Ocean: materials and estimation methods. Honolulu: WPRFMC. 16 p.

McNamara B, Torre L, Kaaialii G. 1999. Hawaiʻi longline mortality mitigation project. Honolulu: WPRFMC.

Micronesian Environmental Services. 2005. Guam bottomfish archipelagic fishery ecosystem plan: a preliminary plan based on the adaptive management approach. A report prepared for the WPRFMC. 10 p.

Milne N. 2012. Hawaiʻi coral reef dealer study. Honolulu: WPRFMC.

Munro JL. 2003. The assessment of exploited stock of coral reef fishes. Final report for the WPRFMC. 55 p.

Ochavillo D. 2012. Coral reef fishery assessment in American Samoa. Honolulu: WPRFMC.

Parrish JD, Schumach BD. 2005. Feeding interactions of the introduced blue-line snapper with important native fishery species in Hawaiian benthic habitats. Final report for the WPRFMC. 31 p.

Paty W, Dalzell P. 1996. The importance and uniqueness of fisheries in Western Pacific Region. Western Pacific Regional Fishery Management Council 91st Meeting; 1996 Nov 18–21; Honolulu. 10 p.

Polovina J et al. 2016. Pacific Islands regional action plan: NOAA Fisheries climate science strategy. U.S. Dep Commer, NOAA Tech Memo. NOAA-TM-NMFS-PIFSC-49. 33 p.

Polunin N, Graham N. 2003. Review of the impacts of fishing on coral reef populations. Final report for WPRFMC. 45 p.

Pylman K. 2012. Fish traps as an indicator of fish abundance. Honolulu: WPRFMC.

Ralston S, Cox S, Labelle M, Mees C. 2004. Bottomfish Stock Assessment Workshop; 2004 Jan 13–16; Honolulu. Final panel report for the WPRFMC.

Rice J, Carvalho F, Fitchett MD, Harley S, Ishizaki A. 2021. Future stock projections of oceanic whitetip sharks in the western and central Pacific ocean. 17th Regular session of the Science Committee to the Western and Central Pacific Fisheries Commission. WCPFC-SC17-2021/SA-IP-21.

Sabater MS. 2010. Mapping and assessing critical habitats for the wrasse (Cheilnus undulates). Honolulu: WPRFMC.

Sabater M, Dalzell P, editors. 2016. Fifth National Meeting of the Regional Fishery Management Councils' Scientific and Statistical Committees. Report of a national SSC workshop on providing scientific advice in the face of uncertainty: from data to climate and ecosystems. Honolulu: WPRFMC. 70 p.

Schug D, Clarke R, Dalzell P. 1998. Use of area closures to mitigate interactions between large-scale and small-scale tuna fisheries. Presented at the Second Pacific Community Fishery Management Workshop; 1998 Oct 12–16; Noumea, New Caledonia.

Severance CJ, Franco R. 1989 Dec. Justification and design of limited-entry alternatives for the offshore fisheries of American Samoa and an examination of preferential fishing rights for Native people of American Samoa within a limited-entry context. Honolulu: WPRFMC.

Simonds KS. 2021. Environmental Justice in Fisheries Management: The Western Pacific. Presented at the Council Coordination Committee meeting; 2021 Oct 21, via webinar. Available at: https://www.wpcouncil.org/wp-council-environmental-justice_ppt-f/

Spalding S, Ka'ai'ai C. 2008. Getting traditional practitioner informants to cooperate with reserchers. International Pacific Marine Educators Network conference; 2008 Oct. 17–19; Townsville, Australia [delivered remotely].

———. 2010. Developing a literacy guide to perpetuate traditional knowledge. AGU fall meeting; 2010 Dec. 13–17; San Francisco, Calif.

State of Fisheries: Hearing before the Subcommittee on Water, Oceans, and Wildlife, of the House Committee on Natural Resources. 116th Cong., 1st sess. (2019) (testimony of Kitty M. Simonds). https://naturalresources.house.gov/hearings/the-state-of-fisheries.

Thomas L. 2012. Assessment of non-fishery impacts on the catch and effort of the Kona crab fishery in Hawai'i. Honolulu: WPRFMC.

———. 2015. Characterizing the Kona crab (Ranina ranina) fishery in the main Hawaiian Islands. Coral reef report to the WPRFMC.

Veron JEN. 2014. Results of an update of the Coral of the World Information Base for the listing determination of 66 coral species under the Endangered Species Act. Honolulu: WPRFMC.

Wagner D, Toonen RJ. 2010. Reproductive characteristics of the Hawaiian black coral species Antipathes griggi with implications for future management. Honolulu: WPRFMC.

Walker R, Ballou L, Wolfford B. 2012. Non-commercial coral reef fishery assessments for the western Pacific region. Honolulu: WPRFMC.

Weng K, Dalzell P. 1999. The akule fishery in Hawai'i. Hawai'i Fisheries Forum 2000; 1999 Nov 13; Honolulu. 10 p.

Western Pacific Regional Fishery Management Council. 2003. Our voyage continues … managing marine fisheries of Hawai'i and the U.S. Pacific Islands—past, present and future. Honolulu: WPRFMC. 16 p.

——. 2004. Strategic plan for the conservation and management of marine resources in the Pacific Islands region: summary. Honolulu: WPRFMC. 20 p.

——. 2006. Ecosystem Science and Management Planning Workshop: development of ecosystem-based approaches to marine resource management in the Western Pacific Region; 2005 Apr 18–22; Honolulu. 158 p.

——. 2006. Hawaiʻi, a center for Pacific sea turtle research and conservation. Honolulu: WPRFMC. 48 p.

Western Pacific Regional Fishery Management Council, National Marine Fisheries Service. 2008. Report on the North Pacific loggerhead sea turtle expert workshop; 2007 Dec. 19–20; Honolulu. Honolulu: WPRFMC.

Wiles P, Sabater MS, Jacob L. 2010. Current surveys between potential marine managed areas in American Samoa. Coral reef report to the WPRFMC.

Witherell D, Dalzell P, editors. 2009. First National Meeting of the Regional Fishery Management Councils' Scientific and Statistical Committees. Report of a workshop on developing best practices for SSCs; 2008 Nov 12–14; Honolulu. Honolulu: WPRFMC. 62 p.

Woodworth-Jefcoats PA, Ellgen S, Jacobs A, Lumsden B, Spalding S. 2019. Report from the 2nd annual collaborative climate science workshop; 2018 Sept. 4–6; Honolulu. NOAA Admin Rep H-19-02. 16 p.

Woodsworth-Jefcoats PA, Ellgen S, Garrison M, Jacobs A, Lumsden B, Marra J, Sabater M. 2020. Summary report from the 3rd annual collaborative climate science workshop; 2019 Sept. 11–12; Honolulu. NOAA Admin Rep H-20-03. 35 p.

Woodsworth-Jefcoats PA, Ellgen S, Jacobs A, Lumsden B, Marra J, Mooney A, Sabater M. 2021. Summary report from the 4th annual collaborative climate science workshop; 2020 Oct. 13–15; Honolulu. NOAA Admin Rep H-21-03. 35 p.

Zeller D, Booth S, Pauly D. 2005. Reconstruction of coral reef- and bottom-fisheries catches for US flag island areas in the Western Pacific, 1950–2002. Coral reef report to the WPRFMC. 113 p.

Glossary

ACCOUNTABILTY MEASURES are controls to prevent annual catch limits from being exceeded and to correct or mitigate overages of the catch limits that may occur. There are two categories of accountability measures. In-season measures include quota closures, trip or bag limit changes, gear restrictions, individual fishing quotas and catch shares. Post-season measures take effect the following season and include seasonal closures, reduced trip or bag limits and shortened fishing seasons implemented in the subsequent year.

BIOLOGICAL OPINION is the document that states the conclusion of NMFS or the USFWS whether a federal action is likely to jeopardize the continued existence of a listed species or result in the destruction or adverse modification of critical habitat.

CIGUATERA is an illness caused by eating fish containing toxins produced by the marine microalgae *Gambierdiscus toxicus*.

CONTROL DATES are set by regional fishery management councils or NMFS for determining future access to fisheries. Persons entering the fishery after that date may not necessarily continue participation if a limited-entry program is subsequently implemented for the fishery.

CUSTOMARY EXCHANGE means the non-market exchange of marine resources between fishermen and community residents, including family and friends of community residents, for goods, and/ or services for cultural, social or religious reasons. Customary exchange may include cost recovery through monetary reimbursements and other means for actual trip expenses, including but not limited to ice, bait, fuel or food, that may be necessary to participate in fisheries in the western Pacific. Actual trip expenses do not include expenses that a fisherman would incur without making a fishing trip, including expenses related to dock space, vessel mortgage payments, routine vessel maintenance, vessel registration fees, safety equipment required by the U.S. Coast Guard and other incidental costs and expenses normally associated with ownership of a vessel (50 CFR § 665.12).

EX VESSEL is the price that the fishermen receive directly for their catch, meaning the price at which the catch is sold when it first enters the supply chain.

EXCLUSIVE ECONOMIC ZONE is a nation's waters extending 200 nautical miles from its shores, within which countries may limit and control both domestic and foreign fishing but do not have the right to prohibit or limit freedom of navigation or overflight, subject to very limited exceptions.

FRAMEWORK MEASURES are fishing controls examined in a fishery management plan for future implementation. This allows an administratively simpler process than that for a full fishery management plan amendment.

HAWAIIAN–EMPEROR SEAMOUNT CHAIN is a mostly undersea mountain range in the Pacific Ocean.

HIGH GRADING is selectively harvesting fish so that only the best quality fish are landed. The practice allows fishermen to get higher prices for their catch but can be environmentally destructive if the discarded fish returned to the ocean die.

MAXIMUM SUSTAINABLE YIELD is the largest long-term average catch that can be harvested from a stock under existing environmental conditions.

NATIONAL ENVIRONMENTAL POLICY ACT requires federal agencies to assess the environmental effects of federal projects or actions There are three levels of analysis: a categorical exclusion, an environmental assessment and an environmental impact statement.

NAUTICAL MILE is equal to 1.151 land-based (or statute) miles (1.852 meters).

OPTIMUM YIELD is the amount of fish harvested that will provide the greatest benefit to the nation with respect to food production and recreational opportunities, while protecting marine ecosystems.

OVERFISHED is when the biomass of a stock or stock complex declines to a level that jeopardizes its capacity to produce maximum sustainable yield on a continuing basis.

OVERFISHING occurs when the fishing mortality or total annual catch of a stock or stock complex jeopardizes its capacity to produce maximum sustainable yield on a continuing basis.

POTENTIAL BIOLOGICAL REMOVAL means the maximum number of animals (not including natural mortalities) that may be removed from a marine mammal stock while allowing that stock to reach or maintain its optimum sustainable population. The potential biological removal level is the product of the following factors: (1) the minimum population estimate of the stock; (2) one-half the maximum theoretical or estimated net productivity rate of the stock at a small population size; and (3) a recovery factor of between 0.1 and 1.0.

PURSE SEINE is a method of fishing that deploys a large wall of netting around a school of fish.

RECRUITMENT is the number of fish which transition from young, small fish to fishable stock (those fish above the minimum legal size) or spawning stock (those fish which are sexually mature).

REGULATORY AMENDMENT changes the regulations that implement a fishery management plan but doesn't change the underlying plan.

SOUTH PACIFIC TUNA TREATY is an ongoing agreement between the United States and 16 Pacific island countries. The treaty allows for U.S. purse-seine vessels to fish in the EEZs of the Pacific island countries party to the treaty. Entering into force in 1988, the Treaty has lasted more than 30 years.

SPAWNING POTENTIAL RATIO is the proportion of the unfished reproductive potential left at any given level of fishing pressure, and it is commonly used to set fishing targets and reference points for fisheries.

SPAWNING STOCK BIOMASS PER RECRUIT is the estimated lifetime reproductive potential of an average recruit.

TAKE is defined by the Endangered Species Act as to harass, harm, pursue, hunt, shoot, wound, kill, trap, capture or collect any species listed under the Act as threatened or endangered species.

TERRITORIAL SEA is the waters adjacent to a coastal state over which the state has full sovereignty. The sovereignty (generally the area 0 to 3 nautical miles from shore) extends to the air space over the territorial sea as well as to its bed and subsoil.

References

Aila W, Chandler B, Paul L, Winterner R, Ziegler M. 2007 Jun 20. Environmental & cultural organization call for Wespac congressional hearing and the resignation of Wespac executive director, Kitty Simonds. Press release.

Alverson DL, June JA (editors). 1988. Proceedings of the North Pacific Rim Fisherman's Conference on Marine Debris, Kailua-Kona, HI, 13–16 Oct 1987. Seattle, WA: Natural Resources Consultants. 460 p.

Amesbury JR, Hunter-Anderson RL. 1989a Sep. Native fishing rights and limited entry in the CNMI. Honolulu, HI: WPRFMC.

———. 1989b Sep. Native fishing rights and limited entry in Guam. Honolulu, HI: WPRFMC.

———. 2003. Review of archaeological and historical data concerning reef fishing in the US flag islands of Micronesia: Guam and the Northern Mariana Islands: Final report for the WPRFMC. 166 p.

———. 2008 May. An analysis of archaeological and historical data on fisheries for pelagic species in Guam and the Northern Mariana Islands. Report for the Pelagic Fisheries Research Program. Guam: Micronesian Archaeological Research Services.

Antonelis G, Baker JD, Johanos TC, Braun RC, Harting AL. 2006. Hawaiian monk seal (Monachus schauenslandi): status and conservation issues. Atoll Res Bull 543:75–101.

[AS DOC] American Samoa Government Dept. of Commerce. N.d. Comprehensive Economic Development Startegy 2018–2022. Available at: https://pacificbasindevelopment.org/.

Babbitt B. 2006 Jun 15. Plans for Hawaiian monument laid in 1999. All Things Considered. Hawai'i Public Radio. Available at: https://www.npr.org/templates/story/story.php?storyId=5488812.

Baird SJ (compiler and editor). 2001 Mar. Report on the International Fishers' Forum on solving incidental capture of seabirds in longline fisheries, Auckland, NZ, 6–9 Nov 2000. Wellington, NZ: Dept. of Conservation. 63 p.

Baker JD. 2006 Jun. The Hawaiian monk seal: Abundance, estimation, patterns in survival and habitat issues [unpublished PhD theses]. Univ of Aberdeen, Aberdeen, U.K. 182 p. Available at: https://origin-apps-pifsc.fisheries.noaa.gov/library/pubs/Baker_thesis_PhD.pdf.

Baker JD, Harting AL, Wurth TA. Johanos TC. 2010. Dramatic shifts in Hawaiian monk seal distribution predicted from divergent regional trends. Mar Mam Sci 27(1):78–93. doi: 10.3354/meps09987.

Baker JD, Howell EA, Polovina JJ. 2012. Relative influence of climate variability and direct anthropogenic impact on a sub-tropical Pacific top predator, the Hawaiian monk seal. Mar Ecol Prog Series 469:175–89.

Baker JD, Polovina JJ, Howell EA. 2007. Effect of variable oceanic productivity on the survival of an upper trophic predator, the Hawaiian monk seal *Monachus schauinslandi*. Mar Ecol Prog Series 346:277–83.

Bandarin F. 2020 May 27. [Letter to H.E. Mr. David Killion, Permanent delegation of the USA to UNESCO.]

Barayuga D. 2000 Nov 18. Judge extends closure of northwest isles lobster fishery. Honolulu Star-Bulletin. Available at: archives. starbulletin.com/2000/11/18/news/story2.html.

Bartram PK, Kaneko JJ. 2004. Catch to bycatch ratios, comparing Hawai'i's longline fishery with others. SOEST Publication 04-05, JIMAR Contribution 04-352. 40 p.

Bellagio Conference on Sea Turtles Steering Committee. 2004. What can be done to restore Pacific turtle populations? The Bellagio blueprint for action on Pacific sea turtles. Penang, Malaysia: Worldfish Center. 24 p.

Bellagio Sea Turtle Conservation Initiative Steering Committee. 2008. Strategic planning for long-term financing of Pacific leatherback conservation and recovery: proceedings of the sea turtle conservation initiative, Terengganu, Malaysia; July 2007. Worldfish Conference Proceedings 1805. Penang, Malaysia: WorldFish Center. 79 p.

Bienfang P. n.d. Chapter 6. Living resources of the sea. Oceanography 331. Available at: http://www.soest.hawaii.edu/oceanography/courses_html/OCN331/

Boggs CH, Ito RY. 1993. Hawaiʻi's pelagic fisheries. Mar Fish Rev 55(2):69–82.

Bonk K, Owens T, Kaaumoana M, Paul L. 2007 Jun 20. Hawaiʻi environmental leaders file complaints against the Western Pacific Regional Fishery Management Council. Press release.

Brown W. 2006 Jun 18. Sanctuary, a victory long in the making. Honolulu Advertiser. B1.

Campagnoni BA. 2006 July 6. [Letter from commander and chief, Enforcement Branch, 14th US Coast Guard District. to Simonds K]. Located at: WPRMFC office, Honolulu.

Carr CJ. 2004. The legal status of highly migratory species, 1970–2000: a case study of debate and innovation in international fisheries law. [PhD dissertation]. Univ of Calif, Berkeley. 203 p. Available at: https://escholarship.org/uc/item/0789w222.

Carretta JV, Forney KA, Lowry MS, Barlow J, Baker J, Johnston D, Hanson B, Muto MM, Lynch D, Carswell L. 2009 January. U.S. Pacific marine mammal stock assessments: 2008. NOAA-TM-NMFS-SWFSC-434. U.S. Dept. Comm., NOAA, NMFS, Southwest Fisheries Science Center.

Carretta JV, Oleson EM, Forney KA, Muto MM, Weller DW, Lang AR, Baker J, Hanson B, Orr AJ, Barlow J, Moore JE and Brownell RL Jr. 2021. Draft U.S. Pacific marine mammal stock assessments: 2021. NOAA-TM-NMFS-SWFSC-XXX. U.S. Dept Comm, NOAA, NMFS Southwest Fisheries Science Center.

Case E. 2005 May 16. Speech upon the introduction of the H.R. 2376 NWHI National Marine Refuge Act of 2005 to the 109th Congress.

Cousins K, Cooper J. 2000 April. The population biology of the black-footed albatross in relation to the mortality caused by longline fishing. Honolulu, HI: WPRFMC. 120 p.

[CRER Advisory Council] NWHI Coral Reef Ecosystem Reserve Advisory Council. 2001. Transcript of the 2001 Apr 5 NWHI CRE-RAC meeting.

Crustacean Plan Team. 1991 Jan 31. Recommendations regarding management of the lobster fishery. Located at: WPRFMC office, Honolulu, HI.

DiNardo GT, DeMartini EE. 2002. Estimates of lobster-handling mortality associated with the Northwestern Hawaiian Islands lobster-trap fishery. Fish Bull 100:128–33.

Dowie M. 2009. Conservation refuges: The hundred-year conflict between global conservation and native peoples. Cambridge, MA: MIT Press.

Drazen JC, Moriwake V, Sackett D, Demarke C. 2014. Evaluating the effectiveness of restricted fishing areas for improving the bottomfish fishery in the main Hawaiian Islands. [Final report to HDAR].

Dutton P, Squire D, Ahmed M. 2011. Conservation of Pacific sea turtles. Honolulu: Univ of Hawai‘i Pr. 481 p.

Eberhardt LL, Garrott RA, Becker BL. 1999. Using trend indices for endangered species. Mar Mam Sci 15:766–85.

[EPAP] Ecosystem Principles Advisory Committee. 1999. Ecosystem-based fishery management: a report to Congress by the Ecosystem Principles Advisory Panel. Silver Spring (MD): U.S. Dept. Comm., NOAA, NMFS.

Evans W, Douglas J Jr., Powell B. 1987. NMFS program development plan for ecosystem monitoring and fisheries management. U.S. Dept. Comm., NOAA, NMFS.

Everson A, Skillman R, Polovina J. 1992. Evaluation of rectangular and circular escape vents in the Northwestern Hawaiian Islands lobster fishery. North Am J Fish Management 12:161–71.

Fougner S. 2010 Dec. Ten years and counting: the first 10 years of the Western and Central Pacific Highly Migratory Fish Stocks Convention. Pacific Islands Fishery Monographs 2. Honolulu: WPRFMC.

Gates PD, Samples KC. 1986. Dynamics of fleet composition and vessel fishing patterns in the Northwestern Hawaiian Islands commercial lobster fishery: 1983–6. Admin Rep H-86-17C. Honolulu: NOAA, NMFS, Southwest Fisheries Science Center, Honolulu Lab.

Gilden J (editor). 2013. Managing our nation's fisheries 3: advancing sustainability. Proceedings of a conference of fisheries management in the United States. 2013 May 6–9. Washington, DC. Portland (OR): Pacific Fishery Management Council.

Gilman E, Clarke S, Brothers N, Alfaro-Shigueto J, Mandelman J, Mangel J, Petersen S, Piovano S, Thomson N, Dalzell P, Donoso M, Goren M, Werner T. 2007a. Shark depredation and unwanted bycatch in pelagic longline fisheries: industry practices and attitudes, and shark avoidance strategies. Honolulu, HI: WPRFMC.

Gilman E, Kobayashi D, Swenarton T, Brothers N, Dalzell P, Kinan-Kelly I. 2007b. Reducing sea turtle interactions in the Hawaii-based longline swordfish fishery. Biol Conserv 139(1):19–28.

Gilman E, Kobayashi D, Chaloupka M. 2008. Reducing seabird bycatch in the Hawaii longline tuna fishery. Endanger Spec Res 5(2-3):309–23.

Gilmartin WG. 1983 March. Recovery plan for the Hawaiian monk seal (Monachus schauislandi). [In cooperation with the Hawaiian Monk Seal Recovery Team.] NOAA, NMFS, Southwest Fisheries Science Center, Honolulu Lab. 29 p.

Glazier EW. 2011. Ecosystem-based fisheries management in the western Pacific. Chichester, UK: John Wiley & Sons. 280 p.

———. 2019. Tradition-based natural resource management: practice and applications in the Hawaiian Islands. Palgrave studies in natural resource management. Cham, Switzerland: Palgrave Macmillian. 281 p.

Golden D. 2017 Oct 4. Billfish conservation act's Hawai'i amendment passes in Senate. Sport Fishing Magazine.

Gooding RM. 1985. Predation on released spiny lobster, Panulirus marginatus, during tests in the Northwestern Hawaiian Islands. Mar Fish Rev 47(1):27–35.

Govan H et al. 2008. What marine education does the South Pacific really need? In: Simonds K, Spalding S, Kaʻaiʻai C (editors). 2008. Current, J Mar Educ 24(2):34–39.

Green A. 1997. An Assessment of the status of the coral reef resources and their patterns of use in the U.S. Pacific Islands. Honolulu, HI: WPRFMC.

Grigg R. 2006. The history of marine research in the Northwestern Hawaiian Islands: lessons from the past and hopes for the future. In: Atoll Res Bull 543:13–22.

———. 2010. The precious coral fishery management plan of the Western Pacific Regional Fishery Management Council. Pacific Islands Fishery Monographs 1. Honolulu, HI: WPRFMC.

Grigg R and Pfund RT. 1980. Proceedings of the Symposium on Status of Resource Investigations in the Northwestern Hawaiian Islands; 1980 Apr 24–25; Honolulu. Sea Grant misc rep UNIHI-SEAGRANT-MR-80-04. Honolulu: Hawaii Sea Grant Coll Prog.

Grover J. 2016 Feb 3. Oral history of Robert P. Smith. USFWS Retirees Assoc. Available at: https://digitalmedia.fws.gov/digital/collection/document/id/2147/.

Hawaii Seafood Council. n.d. Longline Fishing. Available at: https://www.hawaii-seafood.org/hawaii-fishing-industry/longline-fishing/.

Hawai'i State. 1979. Hawai'i Fisheries Development Plan. Honolulu: Hawai'i Dept. Land and Natural Resources. 297 p.

Henderson J. 2001 Jul. A pre and post Marpol Annex V, summary of Hawaiian monk seal entanglements and marine debris accumulation in the Northwestern Hawaiian Islands 1982–98. Mar Pollu Bull 42(7):584–9.

Hogarth W. 1999 Mar 11. [Letter to Jim Cook, WPRFMC chair]. Located at: WPRFMC office, Honolulu, HI.

Honolulu Advertiser. 2000 Nov 14. President Clinton to delay preserve for Northwestern Hawaiian Islands.

Hospital J, Beaver C. 2011. Management of the main Hawaiian Islands bottomfish fishery: fishers' attitudes, perception and comments. Pac Is Fish Sci Cent Admin rep H-11-06. Honolulu: NMFS, NOAA. 46 p + appendices.

Howell EA, Kobayashi DR, Parker DM, et al. 2008. TurtleWatch: a tool to aid in the bycatch reduction of loggerhead turtles Caretta caretta in the Hawai'i-based pelagic longline fishery. Endang Spec Res 5:267–78. doi: 10.3354/esr00096

Hunter CL. 1995. Review of status of coral reefs around American flag Pacific islands and assessment of need, value and feasibility of establishing a coral reef fishery management plan for the Western

Pacific region. [Final report prepared for the WPRFMC]. Honolulu, HI: WPRFMC. 39 p.

Ishizaki A. 2015. Protected species conservation by the Western Pacific Regional Fishery Management Council. Pacific islands fishery monographs 4. Honolulu: WPRFMC.

Iversen RTB. 2020. Swimming with Fishes. British Columbia, Canada: Tellwell Talent.

Iversen RTB, Dye T, Palu ML. 1990a July. Native Hawaiian fishing rights phase 1: the Northwestern Hawaiian Islands. Honolulu, HI: WPRFMC.

———. 1990b July. Native Hawaiian fishing rights phase 2: Main Hawaiian Islands and the Northwestern Hawaiian Islands. Honolulu, HI: WPRFMC.

Iverson S, Piche J, Blanchard W. 2011. Hawaiian monk seals and their prey: assessing characteristics of prey species fatty acid signatures and consequences for estimating monk seal diets using quantitative fatty acid signature analysis. Tech Memo NOAA-TM-NMFS-PIFSC 23. U.S. Dept. Comm., NOAA. 114 p. + Appendices.

Ka'ai'ai C. 2016. Western Pacific indigenous fishing communities. Pacific islands fishery monographs 7. Honolulu: WPRFMC.

Kaneko JJ, Bartram PK. 2005. Operational profile of a highliner in the American Samoa small-scale (alia) longliner albacore fishery. SOEST Pub 05-03, JIMAR Contribution 05-357. 34 p.

Kawamoto KE, Pooley SG. 2000. Annual report of the 1998 Western Pacific lobster fishery (with preliminary 1999 data). Admin Rep H-00-02. NOAA, NMFS, Southwest Fisheries Science Center. 38 p.

Keeney TRE. 2002 Jun 27. Statement presented at the oversight hearing before the Subcommittee on Fisheries, Conservation, Wildlife and Oceans of the Committee on Resources, US House Rep, 107th Congress, 2nd session.

Kikiloi KST. 2012. Kukulu manamana: ritual power and religious expansion in Hawai'i: the ethnohistorical and archaeological study of Mokumanamana and Nihoa islands. [PhD dissertation]. Univ of Hawai'i at Manoa. 365 p. Available at: https://scholarspace.manoa.hawaii.edu/bitstream/10125/101211/Kikiloi_Scott_r.pdf.

Kirch PV. 2000. On the road of the winds: an archaeological history of the Pacific Islands. Berkeley (CA): Univ of Calif Pr.

Koberstein P. 2003 Fall. Plundering the Pacific. Cascadia Times 54.

———. 2006 Spring. Rogues of the Pacific. Special Report Northwestern Hawaiian Islands. Cascadia Times 58.

Landgraf K, Pooley S. May 1990. Annual report of the 1989 Western Pacific lobster fishery. Admin Rep H-90-06. NOAA, NMFS, Southwest Fisheries Center, Honolulu Lab.

Leung PS, Loke M. 2008 Dec. Economic impacts of increasing Hawai'i's food self-sufficiency. Economic Issues 6. Univ of Hawai'i Coll Trop Ag Human Resources. In: Hawai'i Dept. Business Economic

Development & Tourism. 2012 Oct. Increased food security and food self-sufficiency strategy. Available at: https://files.hawaii.gov/dbedt/op/spb/INCREASED_FOOD_SECURITY_AND_FOOD_SELF_SUFFICIENCY_STRATEGY.pdf

Levine A, Allen S. 2009. American Samoa as a fishing community. Tech Memo NOAA-TM-NMFS-PIFSC-19. U.S. Dept. Comm., NOAA, NMFS. 74 p.

Lewis S. 2005 Sep 30. Hawaii safeguards Northwestern Hawaiian Islands. Environment News Service.

Lovejoy WS. 1977. BFISH: A population dynamics analysis of the impact of several alternative fishery management policies in the Hawaiian Fishery Conservation Zone on the Pacific stocks and Hawaiian sport fishing yields of blue and striped marlins. WPRFMC contract report WP-77-107. Southwest Fisheries Science Center Admin Rep 12H.

———. 1981. BFISH revisited. [Contract report to the WPRFMC]. Located at: WPRFMC office, Honolulu, HI.

Lucas DL, Lincoln JM. 2010. The impact of marine preserve areas on the safety of fishermen on Guam. [A report prepared for the WPRFMC]. Available at: http://www.wpcouncil.org/wp-content/uploads/2019/05/Lucas-and-Lincoln.-2010.-Impact-of-MPAs-on-Guam-Fishermen-Safety-Adobe-Indesign-version.pdf.

Maeda T. 1992 Feb. A study of tuna transshipment operation by longline vessels in Guam. Honiara, Solomon Islands: Forum Fisheries

Agency. Available at: http://www.spc.int/DigitalLibrary/Doc/FAME/FFA/Reports/FFA_1992_025.pdf.

Marianas Conservation.org. 2008 Oct 20. From James Connaughton's visit. Available at: http://www.marianasconservation.org/connaughtons-visit.htm.

Markrich MI, Hawkins C. 2016. Fishing fleets and fishery profiles. Pacific islands fishery monographs 5. Honolulu: WPRFMC.

Markrich ML. 2020. Northwestern Hawaiian Islands lobster fishery. Pacific islands fishery monographs 9. Honolulu: WPRFMC.

Martell L, Spalding S. 2020. Fishery ecosystem management in the western Pacific region. Pacific islands fishery monographs 12. Honolulu: WPRFMC. 34 p.

Mawae JK. 2000 Nov 2. [Letter to George T. Frampton, chairman of the Council of Environmental Quality chairman]. Located at: WPRFMC office, Honolulu, HI.

Mendoza M, Mason M. 2016 Sep 8. Foreign fishermen confined to boats catch Hawaiian seafood. Associated Press.

Miller SL, Crosby MP. 1998. The extent and condition of U.S. coral reefs. In: NOAA's state of the coast report. Silver Spring, MD. pp. 1–34.

Monaco ME, Anderson SM, Battista TA, Kendall MS, Rohmann SO, Wedding LM, Clarke AM. 2012. National summary of NOAA's shallow-water benthic habitat mapping of U.S. coral reef ecosystems.

NOAA Tech Memo NOS NCCOS 122. Prepared by the NCCOS Center for Coastal Monitoring and Assessment Biogeography Branch. Silver Spring, MD. 83 p.

MPA News. 2006 Jul. U.S. designates "world's largest" MPA in Northwestern Hawaiian Islands. MPA News 8(1):1–2. Available at: https://mpanews.openchannels.org/sites/default/files/mpanews/archive/MPA76.pdf.

[MREDA] Marine Resources and Engineering Development Act of 1966, Pub. L. 89-454, title I, June 17, 1966, 80 Stat. 203 (33 U.S.C. 1101 et seq.).

Myers N. 2003. Biodiversity hotspots revisited. BioSci 53(10):796–7.

Nelson J. n.d. [Letter to Scott Foster]. Available at: http://www.scottfoster.org/doc/PewLOR.jpg. Accessed 26 Aug 2016.

Nickerson R. 1976 Nov 12. Hawaii situation unique with 200-mile limit. Maui News.

[NMFS] National Marine Fisheries Service. 1977 Jan. Final environmental impact statement/preliminary fishery management plan. Seamount groundfish fishery resources (pelagic armorheads and alfonsins). 27 p. Available at: https://www.govinfo.gov/content/pkg/CZIC-ql377-c5-u55-1977/html/CZIC-ql377-c5-u55-1977.htm.

———. 2000 June 26. Final rule; emergency closure. Fed Reg 65(123):39314–8.

——. 2002a June 12. Final rule. Fed Reg 67(113):40232–8.

——. 2002b. Endangered Species Act section 7 consultation biological opinion. NMFS Southwestern region, Pacific Islands area office.

——. 2007. Recovery plan for the Hawaiian monk seal (Monachus schauinslandi): Revision. Available at: https://repository.library.noaa.gov/view/noaa/3521.

——. 2010 Nov 10. Final rule. Amendment 2 to the Hawaii Archipelago Fishery Ecosystem Plan. Fed Reg 75(217): 69015. Available at: https://www.gpo.gov/fdsys/pkg/FR-2010-11-10/pdf/2010-28413.pdf.

——. 2018 Jun 14. Final rule. Pacific Islands fisheries; 5-year extension of moratorium on harvest of gold corals. Fed Reg 83(115):27716-7.

——. 2020 April 15. Hawaiian monk seal (Neomonachus schauinslandi) stock assessment report. Available at: https://media.fisheries.noaa.gov/dam-migration/2019_sars_monkseal.pdf.

[NOAA] National Oceanic and Atmospheric Administration. 1978 Sept 14. Notice of withdrawal of PMP for Pacific Billfish and Oceanic Sharks. Fed Reg 43(179):41062. Available at: https://tile.loc.gov/storage-services/service/ll/fedreg/fr043/fr043179/fr043179.pdf.

——. FishWatch. https://www.fishwatch.gov/sustainble-seafood/the-global-picture.

[NOAA et al.] National Oceanic and Atmospheric Administration, Hawai'i Department of Land and Natural Resources, U.S. Fish and

Wildlife Service–Pacific Islands. 2006 May 19. State, federal agencies enhance joint efforts to protect and manage Northwestern Hawaiian Islands. [News Release].

[NOAA NMSP] National Oceanic and Atmospheric Administration, National Marine Sanctuary Program. 2004 Sep 20. Proposed NWHI national marine sanctuary advice and recommendations on development of draft fishing regulations under the National Marine Sanctuaries Act section 304(a)(5).

[NWS] National Weather Service. 2003 April. Service Assessment Super Typhoon Pongsona December 8, 2002. Available at: https://www.weather.gov/media/publications/assessments/Pongsona.pdf.

Oakley D. 2000 Apr 19. Aloha sets scheduled service to Midway Islands. Travel Weekly. Available at: https://www.travelweekly.com/Destinations2001-2007/Aloha-sets-scheduled-service-to-Midway-Islands.

O'Connor J. 2021 Jan 18. Last longline fishing agent leavers. The Guam Daily Post.

Opresko D. 2009. A new name for the Hawaiian antipatharian coral formerly known as Antipathes dichotoma (Cnidaria: Anthozoa: Antipatharia). Pac Sci 63(2):277–91.

Orlowski A. 2020 Apr 15. Hawai'i marine monument expansion's impact on fishing debated 5 years later. SeaFood Source News.

Parke M. 2007. Linking Hawaii fisherman reported commercial bottomfish catch data to potential bottomfish habitat and proposed restricted fishing areas using GIS and spatial analysis. U.S. Dept. Comm., NOAA Tech Memo NOAA-TM-PIFSC-11. 37 p.

[Pew] Pew Charitable Trusts. 2006 May 22. The Northwestern Hawaiian Islands story. [Fact sheet]. Available at: https://www.pewtrusts.org/en/research-and-analysis/fact-sheets/2006/05/22/fact-sheet-the-northwestern-hawaiian-islands-story.

———. 2007. Pew Prospectus 2007.

Polovina JJ. 1981. Planning document for the assessment of marine resources around Guam and the Commonwealth of the Northern Mariana Islands. NOAA admin report H-81-10. NMFS, Southwest Fisheries Science Center Honolulu Lab.

———. 2005. Climate variation, regime shifts and implications for sustainable fisheries. Bull Mar Sci 76(2):233–44.

Polovina J, Harman R. 1989 Nov 20. Crustacean Plan Team report on overfishing. Presented to the WPRFMC.

Pooley S. 1993. Hawai'i's marine fisheries: some history, long-term trends and recent developments. Mar Fish Rev 55(2):7–19. Available at: https://spo.nmfs.noaa.gov/sites/default/files/pdf-content/MFR/mfr552/mfr5523.pdf.

———. 2000 Dec 15. Preliminary Estimates of the Effects of the NWHI Executive Order on the Bottomfish Fishery. [Email attachment]. Located at: WPRFMC office, Honolulu.

———. 2005 Oct 27. NOAA Pacific Islands Fisheries Science Center response to question concerning Hawai'i's bottomfish populations. 3 p.

Rausser G, Hamilton S, Kovach M, Stifter R. 2009. Unintended consequences: the spillover effects of common property regulations. Mar Pol 33(1): 24-39.

Rizzuto J. 1977 Apr 7. Fishery management council has growing pains. West Hawaii Today.

Rohmann SO, Hayes JJ, Newhall RC, Monaco ME, Grigg RW. 2005. Area of potential shallow-water tropical and subtropical coral ecosystems in the United States. Coral Reefs 24:370–83.

Sabater MG. 2021. Fishery data collection systems: evasive as an elusive fish. Pacific islands fishery monographs 13. Honolulu: WPRFMC.

Sabater MG, Carroll BP. 2009. Trends in reef fish population and associated fishery after three millennia of resource utilization and a century of socio-economic changes in American Samoa. Rev Fish Sci 17(3):318–35.

Schmitten RA. 1996 Jul 31. [Letter to Council Chair Edwin Ebisui Jr.]. Located at: WPRFMC office, Honolulu, HI.

Schug DM. 2001 Jan. Hawai'i's commercial fishing industry: 1820–1945. The Hawaiian J Hist 35:15–34.

Seki MP, Polovina JJ, Kobayashi DR, Bidigare RR, Mitchum GT. 2002. An oceanographic characterization of swordfish (Xiphias gladius) longline fishing grounds in the springtime subtropical North Pacific. Fish Oceanogr 11(5):251–66.

Sette OE, POFI staff. 1954. Progress in Pacific Oceanic Fishery Investigations, 1950–53. Washington, DC: Interior Department, Fish and Wildlife Service.

Severance CJ, Franco R. 1989 Dec. Justification and design of Limited entry alternatives for the offshore fisheries of American Samoa and an examination of preferential fishing rights for Native people of American Samoa within a limited entry context. Honolulu: WPRFMC.

Shinsato H. 1973 May 9. [Letter to Mr. Findlay, regional director of sport fisheries and wildlife, Portland, OR]. Available at: WPRFMC office, Honolulu, Hawai'i, HI.

Shomura RS. 1987. Hawai'i's marine fishery resources: yesterday (1900) and today (1986). Admin Rep H-87-2 I. NOAA, NMFS, Southwest Fisheries Science Center, Honolulu Lab. 14 p.

Shomura RS (editor). 1978 Jul. Summary report of the Billfish Stock Assessment Workshop. Admin Rep 5H. NOAA, NMFS, Southwest Fisheries Science Center, Honolulu Lab. 62 p.

Shomura RS. 1987. Hawai'i's marine fishery resources: yesterday (1900) and today (1986). Admin Rep H-87-2 I. NOAA, NMFS, Southwest Fisheries Science Center, Honolulu Lab. 14 p.

Shomura RS, Tagami DT. 1984. Groundfish fisheries in the vicinity of seamounts in the North Pacific Ocean. Mar Fish Rev 46(2):117.

Simonds KM. 1990a May 14. Memo to Maile Luuwai, re: critique of East Coast Tuna Association proposed Senate MFCMA language. In: Carr CJ. 2004. The legal status of highly migratory species, 1970–2000: a case study of debate and innovation in international fisheries law. [PhD dissertation]. Univ of Calif, Berkeley.

——. 1990b May 20. Memo to Maile Luuwai, re: last shot at tuna inclusion—Pacific. In: Carr CJ. 2004. The legal status of highly migratory species, 1970–2000: a case study of debate and innovation in international fisheries law. [PhD dissertation]. Univ of Calif, Berkeley.

——. 2011. Managing marine turtles and pelagic fisheries on the high seas. In: Dutton P, Squires D, Ahmed M, editors. Conservation of Pacific sea turtles. Honolulu: Univ Hawai'i Pr. pp. 226–47.

——. 2021. Environmental Justice in Fisheries Management: The Western Pacific. Presented at the Council Coordination Committee meeting; 2021 Oct 21, via webinar. Available at: https://www.wpcouncil.org/wp-council-environmental-justice_ppt-f/

Simonds KM, Spalding S, Ka'ai'ai C (editors). 2008. Ho'ohanohano I Na Kapuna Puwalu and the International Pacific Marine Educators Conference. Current: J Mar Educ 24(2). 57 p.

Skillman R, Christofer B, Pooley S. 1993. Fishery Interaction between the tuna longline and other pelagic fisheries in Hawaii. Tech memo NOAA-TM-NMFS-SWFSC-189. NOAA, NMFS, Southwest Fisheries Science Center.

Smith J. 2016 Oct 14. Tri Marine: tuna prices, supply concerns influenced cannery closure. Undercurrent News.

Spalding S, Dalzell PJ. 2009. Unique entities—U.S. regional fishery management councils. Current: J Mar Educ 25(3):4–7.

Spalding S, Witherell D, Gilden J (editors). 2009. U.S. regional fishery management councils: providing sound stewardship of our nation's fishery resources. Current: J Mar Educ 25(3). 64 p.

[SPC] Pacific Community. 2020. Mariana Islands (CNMI). Pacific Community, Statistics for Development Division. Available at: https://sdd.spc.int/mp.

Tenbruggencate J. 1988 Jul 10. Precious corals built up over ages disappear in a commercial instant, environmental update. Honolulu Advertiser.

Tillman M. 2006 Feb 3. Area specific estimates of the exploitable population of lobster in the Northwestern Hawaiian Islands. NOAA, NMFS, Southwest Fisheries Science Center.

U.S. Code. 2001. Title 46 (shipping), subtitle II (vessels and seamen), part H (identification of vessels), section 12108 (fishery endorsements). Available at: https://law.justia.com/codes/us/2001/title46/subtitleii/parth/chap121/sec12108

[US CMSER] U.S. Commission on Marine Science, Engineering and Resources. 2016 Jan. Our nation and the sea: a plan for national action: report of the commission on marine science, engineering and resources. Washington, DC: US Gov Print Office.

[US DOI] U.S. Department of the Interior. 2006 Aug 29. Agencies publish rules on Northwestern Hawaiian Islands monument. Press release. Available at: https://www.doi.gov/sites/doi.gov/files/archive/news/archive/06_News_Releases/060829b.html.

[U.S. DOJ] U.S. Department of Justice. 2021. Task force on human trafficking in fishing in international waters. Report to Congress. Available at: https://www.justice.gov/crt/page/file/1360366/download.

[USFWS] U.S. Fish and Wildlife Service. 2008 Dec. Midway Atoll National Wildlife Refuge, Battle of Midway National Memorial and Midway Atoll special management area. Appendix B visitor services plan.

U.S. House of Representatives. 113th Congress 2nd session. 2014 July 17. Rept. 113–540 Part 1: Endangered Species Litigation Reasonableness Act report together with dissenting views to accompany H.R. 4318.

Van Fossen L. 2007. Annual report on seabird interactions and mitigation efforts in the Hawaii longline fishery for 2006. Honolulu: NMFS, Pacific Islands Regional Office.

Walsh WA, Bigelow BA, Sender KL. 2009. Decreases in shark catches and mortality in the Hawaii-based longline fishery as documented by fishery observers. Mar Coast Fish1(1):270–82.

Ward HEM. 2010. Creating the Papahanaumokuakea Marine National Monument: Discourse, Media, Place-making and Policy Entrepreneurs. [PhD dissertation]. East Carolina Univ. 144 p.

Watson JW, Epperly SP, Shah AK, Foster DG. 2005. Fishing methods to reduce sea turtle mortality associated with pelagic longlines. Can J Fish Aquat Sci 62:965–81.

Watson TK, Kittinger JN, Walters JS, Schofield DT. 2011. Culture, conservation and conflict: assessing the human dimensions of Hawaiian monk seal recovery. Aqua Mam 37(3):386–396.

[WPRFMC] Western Pacific Regional Fishery Management Council. 1976a Oct 19–21. Minutes of the first Council meeting. Located at: WPRFMC office, Honolulu, HI.

——. 1976b Dec 15–16. Minutes of the second Council meeting. Located at: WPRFMC office, Honolulu, HI.

——. 1977a Feb 1–4. Minutes of the third Council meeting. Located at: WPRFMC office, Honolulu, HI.

——. 1977b Jun 27–29. Minutes of the fifth Council meeting. Located at: WPRFMC office, Honolulu, HI.

——. 1977c Sep 29–30. Minutes of the seventh Council meeting. Located at: WPRFMC office, Honolulu, HI.

——. 1978a Jan 10–12. Minutes of the ninth Council meeting. Located at: WPRFMC office, Honolulu, HI.

——. 1978b Mar 15–16. Minutes of the 10th Council meeting. Located at: WPRFMC office, Honolulu, HI.

——. 1978c May 22–23. Minutes of the 11th Council meeting. Located at: WPRFMC office, Honolulu, HI.

——. 1978d Aug 2–5. Minutes of the 12th Council meeting. Located at: WPRFMC office, Honolulu, HI.

——. 1979 Mar 14–16. Minutes of the 16th Council meeting. Located at: WPRFMC office, Honolulu, HI.

——. 1981 Mar 31–Apr 1. Minutes of the 29th Council meeting. Located at: WPRFMC office, Honolulu, HI.

——. 1984a. Council has two jobs: fishery development AND conservation. Pacific Islands Fishery News.

——. 1984b Apr 23–24. Minutes of the 44th Council meeting. Located at: WPRFMC office, Honolulu, HI.

——. 1984c Jun 13–15. Minutes of the 54th Council meeting. Located at: WPRFMC office, Honolulu, HI.

——. 1985a May 6–8. Summary of the 49th Council meeting. Located at: WPRFMC office, Honolulu, HI.

——. 1985b Nov 4–6. Summary of the 51st Council meeting. Located at: WPRFMC office, Honolulu, HI.

——. 1986a Mar 3–5. Minutes of the 52nd Council meeting. Located at: WPRFMC office, Honolulu, HI.

———. 1986b Aug 6–8. Minutes of the 54th Council meeting. Located at: WPRFMC office, Honolulu, HI.

———. 1986c Oct 31. Amendment 4 and environmental assessment for the fishery management plan for lobster fisheries of the Western Pacific region. Honolulu, HI: WPRFMC. 18 p. Available at: http://www.wpcouncil.org/wp-content/uploads/2013/10/Crustaceans-FMP-Amendment-4.pdf.

———. 1987a Jun 3–5. Minutes of the 57th Council meeting. Located at: WPRFMC office, Honolulu, HI.

———. 1987b Sep 21–22. Minutes of the 59th Council meeting. Located at: WPRFMC office, Honolulu, HI.

———. 1988a Aug 8–11. Minutes of the 62nd Council meeting. Located at: WPRFMC office, Honolulu, HI.

———. 1988b Nov 27–29. Minutes of the 63rd Council meeting. Located at: WPRFMC office, Honolulu, HI.

———. 1990 Sep 27–28. Minutes of the 70th Council meeting. Located at: WPRFMC office, Honolulu, HI.

———. 1991 Feb. Amendment 2 and environmental assessment for the fishery management plan for pelagic fisheries of the Western Pacific region. Available at: http://www.wpcouncil.org/pelagic/Documents/FMP/Amendment2.pdf.

——. 1992a Aug. Pelagic fisheries of the Western Pacific region, 1991 annual report. Honolulu, HI: WPRFMC. Located at: WPRFMC office, Honolulu, HI.

——. 1992b Sep 21–23. Minutes of the 78th Council meeting. Located at: WPRFMC office, Honolulu, HI.

——. 1994 Aug. Bottomfish and seamount groundfish fisheries of the Western Pacific region, 1993 annual report. Honolulu, HI: WPRFMC. Located at: WPRFMC office, Honolulu, HI.

——. 1998 Apr 14–16. Minutes of the 95th Council meeting. Located at: WPRFMC office, Honolulu, HI.

——. 1999a July. The Value of the Fisheries in the Western Pacific Fishery Management Council s Area. Available at: http://www.wpcouncil.org/documents/value.pdf.

——. 1999b Oct 18–21. Minutes of the 112nd Council meeting. Located at: WPRFMC office, Honolulu, HI.

——. 2000. Living the legacy: the Northwestern Hawaiian Islands. [video 19 min.]

——. 2004a Jun 30. Bottomfish and seamount groundfish fisheries of the Western Pacific region, 2003 annual report. Honolulu: WPRFMC.

——. 2004b Jun 30. Pelagic fisheries of the Western Pacific region, 2003 annual report. Honolulu: WPRFMC.

———. 2005a Feb 25. Summary of written comments received on draft fishing regulations for the proposed NWHI national marine sanctuary. Located at: WPRFMC office, Honolulu, HI.

———. 2005b. Annual stock assessment and fishery evaluation report for the U.S. Pacific island pelagic fisheries ecosystem plan 2004. Honolulu: WPRFMC.

———. 2006. Summary report of the first workshop on South Pacific albacore longline fisheries, 19–21 Sep 2006, Honolulu, HI. Available at: http://www.wpcouncil.org/wp-content/uploads/2019/05/Council-staff.-2006.-SP-albacore-wkshop.pdf.

———. 2007a Aug 21. International program that unites fishermen to protect sea turtles convenes in Honolulu, hopes to save hundreds of loggerheads. [Press release]. Honolulu, HI: WPRFMC.

———. 2007b Dec 19. Amendment 14 to the fishery management plan for the bottomfish and seamount groundfish fisheries of the Western Pacific region. Honolulu, HI: WPRFMC.

———. 2008a Apr 15. Federal fishery managers reaffirm purse seine closures for U.S. Pacific Islands. [Press release]. Honolulu, HI: WPRFMC.

———. 2008b Apr 24. International roundtable calls for improved regional management and use of market-driven incentives for sustainable tuna production. [Press release]. Honolulu, HI: WPRFMC.

———. 2010. Bottomfish fisheries in Hawai'i archipelago. [Fact sheet]. Honolulu: WPRFMC.

———. 2015 Oct. Shutdown, the effects of arbitrary quotas on the U.S. longline fleet based in Hawai'i. [video]. Honolulu: HI.

———. 2016 Spring. Can past glories of the Mariana fishing industry be revived? Pacific Islands Fishery News. Honolulu, HI: WPRFMC.

———. 2019 April. Impacts of marine national monument fishing prohibitions on U.S. Pacific Island fisheries. Honolulu: WPRFMC.

———. 2020a. Annual stock assessment and fishery evaluation report for the American Samoa archipelago fishery ecosystem plan 2019. Remington et al. (editors). Honolulu: WPRMFC. 161 p + appendices.

———. 2020b. Annual stock assessment and fishery evaluation report for the Pacific pelagic fishery ecosystem plan 2019. Remington et al. (editors). Honolulu: WPRMFC. 382 p + appendices.

Wilkinson EB, Abrams K. 2015 Nov. Benchmarking the 1999 EPAP recommendations with existing fishery ecosystem plans. NOAA Tech Memo NMFS-OSF-5. U.S. Dept. Comm., NOAA, NMFS.

Witherell D (editor). 2004. Managing our nation's fisheries: past, present and future. Proceedings of a conference on fisheries management in the United States. 2003 Nov 13–15. Washington, DC. USA: North Pacific Fishery Management Council.

———. 2005. Managing our nation's fisheries: past, present and future. Proceedings of a conference on fisheries management in the United States. 2005 Mar 24–26. Washington, DC. USA: North Pacific Fishery Management Council.

Worm B et al. 2009. Rebuilding global fisheries. Science 325(5940): 578–85.

Wright W. 2000 Dec 12. Future of bottom-fish market uncertain; coral reef reserve may limit fishing; regulation: tightening supply could raise market prices. Honolulu Advertiser.

Wu N. 2020 Jul 8. $2.6M in federal aid committed to new program with hopes to help keep Hawai'i's fishing industry afloat. Star Advertiser.

Yee W. 1977 Feb 7. [Letter to Juanita Kreps, Commerce secretary]. Available at: WPRFMC office, Honolulu, HI.

Young P. 2020 Jan 6. Hawaiian Tuna Packers. Images of old Hawai'i. Available at: http://imagesofoldhawaii.com/hawaiian-tuna-packers/.

Biographical Notes

Michael L. Markrich is the former public information officer for the State of Hawai'i Department of Land and Natural Resources; communications officer for State of Hawai'i Department of Business, Economic Development and Tourism; columnist for the *Honolulu Advertiser*; socioeconomic analyst with John M. Knox and Associates; and consultant/owner of Markrich Research. He holds a bachelor of arts degree in history from the University of Washington and a master of science degree in agricultural and resource economics from the University of Hawai'i. Among his previous publications are *Northwestern Hawaiian Islands Lobster Fishery* and *History of the Billfish Fisheries and Their Management in the Western Pacific Region*, monographs 9 and 10 of the Pacific Islands Fishery Monographs series. He has received an award for outstanding research from the University of Hawai'i.

Michael L. Markrich

Sylvia M. Spalding retired from the Western Pacific Regional Fishery Management Council after serving as its communications director for nearly two decades. She held similar positions at the Marine Aquarium Council and the University of Hawai'i, Pacific Business Center Program. She also worked as a fishery development assistant at the South Pacific Commission, as a reporter/editor for *Hawaii Fishing News* and the Pelagic Fisheries Research Program and as a U.S. Coast Guard certified captain/charterboat owner. She holds a bachelor's degree in journalism from the University of Hawai'i and a master's degree in nonfiction writing from the University of Iowa. Among her previous works are the *Fishing Hawaii Style* book series (editor), *Living the Legacy: The Northwestern Hawaiian Islands* video (project manager/writer) and The Aquarium Industry and Reef Conservation chapter in *Marine Ornamental Species: Collection, Culture and Conservation* (co-author). Her recognitions include the Carol Burnett Ethics in Journalism Award from the University of Hawai'i and the James Centorino Award for informal marine education (2009) and Honorary Membership Award (2022) from the National Marine Educators Association.

Sylvia M. Spalding

Additional books about the Western Pacific Regional Fishery Management Council and U.S. Pacific Islander fishery issues.

Dutton PH, Squires D, Ahmed M, editors. 2011. Conservation of Pacific sea turtles. Honolulu: University of Hawaii Press. 481 p. ISBN 978-0-8248-3407-4.

Glazier E, editor. 2011. Ecosystem-based fisheries management in the Western Pacific. Chichester, UK: Wiley-Blackwell. 312 p. ISBN 978-0-8138-2154-2.

Glazier E. 2019. Tradition-based natural resource management: practice and application in the Hawaiian Islands. Cham, Switzerland: Palgrave Macmillan. 281 p. ISBN 978-3-030-14841-6; ISBN 978-3-030-14842-3.

www.ingramcontent.com/pod-product-compliance
Lightning Source LLC
Chambersburg PA
CBHW040751220326
41597CB00029BA/4722